METHODS IN MOLECULAR BIOLOGY

Series Editor
John M. Walker
School of Life and Medical Sciences
University of Hertfordshire
Hatfield, Hertfordshire, AL10 9AB, UK

For further volumes:
http://www.springer.com/series/7651

Breast Cancer

Methods and Protocols

Edited by

Jian Cao

School of Medicine, Stony Brook University, Stony Brook, NY, USA

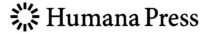 Humana Press

Editor
Jian Cao
School of Medicine
Stony Brook University
Stony Brook, NY, USA

ISSN 1064-3745   ISSN 1940-6029 (electronic)
Methods in Molecular Biology
ISBN 978-1-4939-3442-3   ISBN 978-1-4939-3444-7 (eBook)
DOI 10.1007/978-1-4939-3444-7

Library of Congress Control Number: 2015954358

Springer New York Heidelberg Dordrecht London

Humana Press is a brand of Springer
Springer Science+Business Media LLC New York is part of Springer Science+Business Media (www.springer.com)

Preface

The use of an appropriate method for studying or diagnosing breast cancer is imperative for basic research and clinical practice. Although there are many books tackling breast cancer on the market already, none are written systematically for the modern molecular biological technologies spanning from basic research to clinical practice. The focus of this book is to provide resources, ideas, and bench manuals for the study of breast cancer.

This book contains five parts including methods used in clinical laboratory for diagnosis (Detection of Molecular Markers of Breast Cancer), methods used in both clinical and research laboratories for testing genetic alterations (Genetic Detection for Breast Cancer), methods used to isolate breast cancer cells including circulating cancer cells and breast cancer stem cells (Isolation of Breast Cancer Cells), methods used to study the behavior of breast cancer cells (In Vitro Experimental models for Breast Cancer), and methods used for mimicking human breast cancer in a living organism (In Vivo Experimental Models for Breast Cancer). Each part includes 3–11 different assays for readers to use based on their preferences. This book also includes several recently developed techniques for the study of breast cancer progression.

Each chapter of this book was written by professionals who have extensive experience on the corresponding techniques. Although most of these techniques can be found in the published literature, the chapters described here are more comprehensive with detailed step-by-step procedures so readers can successfully carry out the experiments without difficulty. In addition, extensive explanations for critical steps are described in the Notes section of each chapter to ensure the successful completion of the experiments described. These notes are usually essential for dependable results.

This book will be a valuable handbook for both graduate and advanced undergraduate students of biological sciences as well as scientists, technicians, and physicians working in the academic, hospital, or pharmaceutical industry aimed at studying or diagnosing breast cancer. Our goal is to provide this book as a handbook for researchers who routinely work on breast cancer research. I ask the readers to send any corrections or missing information that can be revised in future editions.

Finally, I would like to thank the many authors for their thoughtful and timely contributions. I also like to extend my appreciation to the Series Editor, Professor John Walker, and the Springer staff, especially Dr. David Casey, for their contributions to the development of this book.

I hope you find this book to be valuable for your research.

Stony Brook, NY, USA   *Jian Cao, M.D.*

Contents

Contributors

VINCENT M. ALFORD • *Division of Cancer Prevention, Department of Medicine, Stony Brook University, Stony Brook, NY, USA*

JACQUELYN J. AMES • *Center for Molecular Medicine, Maine Medical Center Research Institute, Scarborough, ME, USA; Graduate School of Biomedical Science and Engineering, University of Maine, Orono, ME, USA*

JOEL ANDREWS • *Mitchell Cancer Institute, Mobile, AL, USA*

DALE B. BOSCO • *Institute of Molecular Biophysics, Florida State University, Tallahassee, FL, USA*

PETER C. BROOKS • *Center for Molecular Medicine, Maine Medical Center Research Institute, Scarborough, ME, USA*

MARTIN BROWN • *Departments of Pathology and Laboratory Medicine, University of Tennessee Health Science Center, Memphis, TN, USA*

MING CAI • *Department of Integrative Biology and Pharmacology, The University of Texas Health Science Center at Houston, Houston, TX, USA; Department of General Surgery, Union Hospital, Tongji Medical College, Huazhong University of Science and Technology, Wuhan, Hubei Province, China*

JIAN CAO • *Division of Cancer Prevention, Department of Medicine, Stony Brook University, Stony Brook, NY, USA*

JILLIAN CATHCART • *Department of Cellular and Molecular Pharmacology, Stony Brook University, Stony Brook, NY, USA*

LONGHUA CHEN • *Department of Radiation Oncology, Nanfang Hospital, Southern Medical University, Guangzhou, China*

WEN-TIEN CHEN • *Vitatex Inc., Stony Brook, NY, USA; Division of Gynecologic Oncology, Stony Brook Medicine, Stony Brook, NY, USA*

MASSIMO CRISTOFANILLI • *Department of Cancer Biology, Sidney Kimmel Cancer Center, Thomas Jefferson University, Philadelphia, PA, USA*

HUAN DONG • *Vitatex Inc., Stony Brook, NY, USA*

GUANGWEI DU • *Department of Integrative Biology and Pharmacology, The University of Texas Health Science Center at Houston, Houston, TX, USA*

NIKKI A. EVENSEN • *Department of Pediatrics, NYU Medical School, New York, NY, USA*

MEIYUN FAN • *Departments of Pathology and Laboratory Medicine, University of Tennessee Health Science Center, Memphis, TN, USA*

GREGG B. FIELDS • *Department of Chemistry & Biochemistry, Florida Atlantic University, Jupiter, FL, USA; Department of Chemistry, The Scripps Research Institute/Scripps Florida, Jupiter, FL, USA; Departments of Chemistry and Biology, Torrey Pines Institute for Molecular Studies, Port St. Lucie, FL, USA*

BIANA GODIN • *Department of Nanomedicine, Houston Methodist Research Institute, Houston, TX, USA*

ALLEN M. GOWN • *Department of Molecular Pathology, PhenoPath Laboratories, Seattle, WA, USA*

RAJARSI GUPTA • *Department of Pathology, Stony Brook University Hospital, Stony Brook, NY, USA*

KESTER HAYE • *Department of Pathology, Stony Brook University Hospital, Stony Brook, NY, USA*

JINGQUAN HE • *Department of Integrative Biology and Pharmacology, The University of Texas Health Science Center at Houston, Houston, TX, USA*

TERRY HENDERSON • *Center for Molecular Medicine, Maine Medical Center Research Institute, Scarborough, ME, USA*

HARRY C. HWANG • *Department of Molecular Pathology, PhenoPath Laboratories, Seattle, WA, USA*

XUANMAO JIAO • *Department of Cancer Biology, Sidney Kimmel Cancer Center, Thomas Jefferson University, Philadelphia, PA, USA*

ZAHRAA I. KHAMIS • *Department of Chemistry and Biochemistry, Florida State University, Tallahassee, FL, USA*

CANAN KUSCU • *Department of Biochemistry and Molecular Genetics, School of Medicine, University of Virginia, Charlottesville, VA, USA*

CEM KUSCU • *Department of Biochemistry and Molecular Genetics, School of Medicine, University of Virginia, Charlottesville, VA, USA*

FRANSISCA LEONARD • *Department of Nanomedicine, Houston Methodist Research Institute, Houston, TX, USA*

WEIYI LI • *Department of Materials Science and Engineering, Stony Brook University, Stony Brook, NY, USA*

YIYI LI • *Division of Cancer Prevention, Department of Medicine, Stony Brook University, Stony Brook, NY, USA; Department of Radiation Oncology, Nanfang Hospital, Southern Medical University, Guangzhou, China*

XIAO LIANG • *Department of Integrative Biology and Pharmacology, The University of Texas Health Science Center at Houston, Houston, TX, USA; Shanghai Institute of Digestive Disease, Shanghai Renji Hospital, Shanghai Jiao Tong University School of Medicine, Shanghai, China*

LUCY LIAW • *Center for Molecular Medicine, Maine Medical Center Research Institute, Scarborough, ME, USA*

LAURIE E. LITTLEPAGE • *Department of Chemistry and Biochemistry, Harper Cancer Research Institute, University of Notre Dame, Notre Dame, IN, USA*

JINGXUAN LIU • *Department of Pathology, Stony Brook University Hospital, Stony Brook, NY, USA*

TY J. LIVELY • *Department of Chemistry and Biochemistry, Florida State University, Tallahassee, FL, USA*

MARYIA LU • *Department of Integrative Biology and Pharmacology, The University of Texas Health Science Center at Houston, Houston, TX, USA*

YIZHI MENG • *Department of Materials Science and Engineering, Stony Brook University, Stony Brook, NY, USA*

CHRISTOPHER METTER • *Department of Pathology, Stony Brook University Hospital, Stony Brook, NY, USA*

REGINA R. MIFTAKHOVA • *Kazan Federal University, Kazan, Republic of Tatarstan, Russia*

LI MIN • *Department of Biochemistry and Molecular Biology, Peking University Cancer Hospital and Institute, Beijing, People's Republic of China*

SUBHASREE NAG • *Department of Pharmaceutical Sciences, School of Pharmacy, Texas Tech University Health Sciences Center, Amarillo, TX, USA*

KALNISHA NAIDOO • *Institute of Cancer Research, London, UK*

NENGTAI OUYANG • *Department of Pathology, Sun Yat-Sen Memorial Hospital, Sun Yat-Sen University, Guangzhou, China*

MICHAEL L. PEARL • *Division of Gynecologic Oncology, Stony Brook Medicine, Stony Brook, NY, USA*

RICHARD G. PESTELL • *Department of Cancer Biology, Sidney Kimmel Cancer Center, Thomas Jefferson University, Philadelphia, PA, USA; Kazan Federal University, Kazan, Republic of Tatarstan, Russia*

SARAH E. PINDER • *Research Oncology, Division of Cancer Studies, King's College London, Guy's Hospital, London, UK*

ASHLEIGH PULKOSKI-GROSS • *Scientific Affairs Department, AXON Communications, Rye Brook, NY, USA*

DEREK RAMMELKAMP • *Department of Materials Science and Engineering, Stony Brook University, Stony Brook, NY, USA*

ALBERT A. RIZVANOV • *Kazan Federal University, Kazan, Republic of Tatarstan, Russia*

ERIC ROTH • *Division of Cancer Prevention, Department of Medicine, Stony Brook University, Stony Brook, NY, USA*

QING-XIANG AMY SANG • *Department of Chemistry and Biochemistry, Florida State University, Tallahassee, FL, USA; Institute of Molecular Biophysics, Florida State University, Tallahassee, FL, USA*

DAVID SCHMITT • *Mitchell Cancer Institute, Mobile, AL, USA*

CHENGCHAO SHOU • *Department of Biochemistry and Molecular Biology, Peking University Cancer Hospital and Institute, Beijing, People's Republic of China*

MACIEJ J. STAWIKOWSKI • *Department of Chemistry and Biochemistry, Florida Atlantic University, Jupiter, FL, USA*

CHRISTOPHER D. SUAREZ • *Department of Chemistry and Biochemistry, Harper Cancer Research Institute, University of Notre Dame, Notre Dame, IN, USA*

YU SUN • *Department of Chemical Biology, Rutgers University, Piscataway, NJ, USA*

MING TAN • *Mitchell Cancer Institute, Mobile, AL, USA; Department of Biochemistry and Molecular Biology, University of South Alabama, Mobile, AL, USA*

LI WEI RACHEL TAY • *Department of Integrative Biology and Pharmacology, The University of Texas Health Science Center at Houston, Houston, TX, USA*

SHAUN TULLEY • *Vitatex Incorporated, Stony Brook, NY, USA*

KAI WANG • *Division of Cancer Prevention, Department of Medicine, Stony Brook University, Stony Brook, NY, USA; Department of Hepatobiliary Surgery, Nanfang Hospital, Southern Medical University, Guangzhou, China*

LIN WANG • *Department of Pathology, Sun Yat-Sen Memorial Hospital, Sun Yat-Sen University, Guangzhou, China*

WEI WANG • *Department of Pharmaceutical Sciences, School of Pharmacy, Texas Tech University Health Sciences Center, Amarillo, TX, USA; Cancer Biology Center, School of Pharmacy, Texas Tech University Health Sciences Center, Amarillo, TX, USA*

ZIQING WANG • *Department of Integrative Biology and Pharmacology, The University of Texas Health Science Center at Houston, Houston, TX, USA*

QIAN ZHANG • *Division of Cancer Prevention, Department of Medicine, Stony Brook University, Stony Brook, NY, USA*

RUIWEN ZHANG • *Department of Pharmaceutical Sciences, School of Pharmacy, Texas Tech University Health Sciences Center, Amarillo, TX, USA; Cancer Biology Center, School of Pharmacy, Texas Tech University Health Sciences Center, Amarillo, TX, USA*

QIANG ZHAO • *Vitatex Incorporated, Stony Brook, NY, USA; Division of Gynecologic Oncology, Stony Brook Medicine, Stony Brook, NY, USA*

JIE ZHOU • *Department of Hepatobiliary Surgery, Nanfang Hospital, Southern Medical University, Guangzhou, China*

XIAOXIA ZHU • *Department of Radiation Oncology, Nanfang Hospital, Southern Medical University, Guangzhou, China*

WEI-XING ZONG • *Department of Chemical Biology, Rutgers University, Piscataway, NJ, USA*

Part I

Detection of Molecular Markers of Breast Cancer

Chapter 1

Basic Histopathological Methods and Breast Lesion Types for Research

Nengtai Ouyang and Lin Wang

Abstract

The in situ observation on the tissues, such as histopathology, immunohistochemistry (IHC), immuno-fluorescence (IF), and in situ hybridization (ISH), is one of the most important methods in the biomedical scientific research. In this chapter we introduce the most often used methods—hematoxylin and eosin (H&E) and double IF staining. H&E staining is used for general morphology by which the different pathological types of breast lesions are identified. The double IF staining is often used to study the protein–protein interaction on tissues for signaling mechanisms. This chapter also includes the histopathology of primary or simplified breast lesion types that is essential for applying the above methods and the reclassification of breast cancers by molecular markers.

Key words Histopathology, Breast cancer, Hematoxylin & eosin, Immunofluorescence

1 Introduction

The recognition to the morphology of breast and breast lesions is basic for studying breast cancer using in situ approaches such as H&E, immunohistochemistry, IF, and in situ hybridization. The detailed pathological classification of breast lesions is sophisticated to non-pathologist researchers. The following is the most common and simplified types of breast ductal epithelial-originated lesions. Researchers who focus on the specific or comprehensive types of breast cancer may refer to the 2003 [1] or 2012 [2] WHO classification of tumors of the breast.

1.1 Normal Breast and Fibrocystic Breast Conditions

Breast or mammary gland consists of many lobular units that contain parenchyma and connective tissue stroma (Fig. 1). Breast parenchyma is formed from branched ducts and terminal acini. Each lactiferous duct has branches and terminal acini that radially arranged centering on the nipple, resembling a tree-like structure. Acini and ducts are made up of two layers of cells. The inside cells

Jian Cao (ed.), *Breast Cancer: Methods and Protocols*, Methods in Molecular Biology, vol. 1406,
DOI 10.1007/978-1-4939-3444-7_1, © Springer Science+Business Media New York 2016

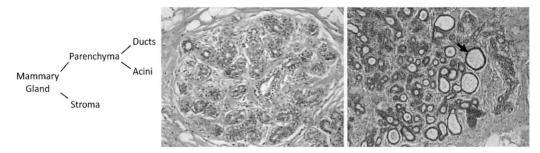

Fig. 1 The normal structure and disorders of the breast. *Left:* Diagram of normal breast lobular unit; *Middle:* Histology of normal breast lobular unit; *Right:* Fibrocystic breast condition. (H&E, 40×)

are columnar or cuboidal glandular epithelial cells and the outside are astroid myoepithelial cells.

Fibrocystic breast condition (also known as fibrocystic breast disease) is the most common cause of "lumpy breasts" in women. It includes three morphological changes: cystic change, fibrosis, and adenosis. As shown in Fig. 1 *right*, some of the acini are enlarged to cysts (arrow indicates), and the number of acini per lobule is increased.

1.2 Intraductal Papilloma and Fibroadenoma

Intraductal papilloma (IDP) and fibroadenoma are the most common benign tumor of the breast. IDP is a benign proliferative lesion consisting of a branching fibrovascular core with overlying epithelial and myoepithelial layers (Fig. 2 *Left*). A fibroadenoma is made up of both breast glandular tissue and stromal tissue (Fig. 2 *Middle*). It is also assumed to be aberrations of normal breast development or the product of hyperplastic processes rather than true neoplasm. Both IDP and fibroadenoma do not pose breast cancer risk unless they undergo other changes, such as atypical hyperplasia.

1.3 Atypical Ductal Hyperplasia

Atypical ductal hyperplasia (ADH) refers to a group of lesions that have special organizational structure and cell morphology. Patients with ADH have a high risk of breast cancer. The diagnostic standard of ADH is as following: (1) Cellular atypia: a single form, uniform distribution, increased nucleus, and increased nucleoplasm ratio, with or without nuclear hyperchromatism; (2) Structural atypia may exist (for example growth pattern with nipple, arch, bridge, solid, or screen structure) as shown in Fig. 2 *Right*.

1.4 Ductal Carcinoma In Situ

Ductal carcinoma in situ (DCIS) is the most common type of non-invasive breast cancer ("in situ" means "in its original place"). DCIS is usually found by mammograms, which detect tiny bits of calcium that develop in dead cancer cells. Cancer cells are surrounded by prominent concentric basement membrane and myoepithelial cells.

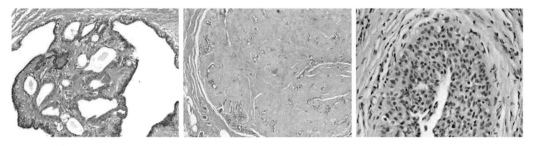

Fig. 2 Benign lesions of the breast. *Left*: Intraductal papilloma of breast (H&E, 40×); *Middle*: Fibroadenoma of breast (H&E, 40×); *Right*: Atypical ductal hyperplasia (H&E, 200×)

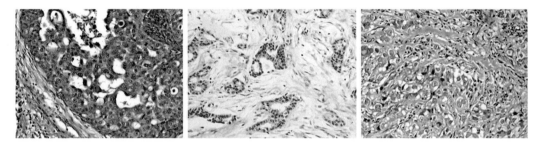

Fig. 3 Malignant tumors of the breast. *Left*: Ductal carcinoma in situ (H&E, 200×); *Middle*: Invasive ductal carcinoma, low grade (H&E, 100×); *Right*: Invasive ductal carcinoma, high grade (H&E &, 200×)

It has not spread beyond the duct into any normal surrounding breast tissue (Fig. 3 *Left*). DCIS is not life-threatening, but it may develop to an invasive breast cancer later on.

1.5 Invasive Ductal Carcinoma

Invasive ductal carcinoma (IDC) is the most common malignant breast tumor. Cancer cells break through the basement membrane and infiltrate in the stroma or the surrounding normal tissue. There are no prominent concentric basement membrane and myoepithelial cells surrounding the cancer cells. IDC can be classified as low grade and high grade based on the differentiation degree of cancer cell.

1. Low degree IDC: cancer cells have an obvious glandular differentiation with central lumens of tubular structure as presented in Fig. 3 *Middle*.

2. High degree IDC: cancer cells show high cell polymorphism or high cell atypia as seen in Fig. 3 *Right*. Pathologists define the IDC grade according to the histological features including the proportion of glands, atypia, and mitotic number of nucleus.

1.6 Molecular Subtypes of Breast Cancer

Four molecular subtypes determined by IHC markers as shown in Fig. 4 including estrogen receptor (ER), progesterone receptor (PR), Her2, Ki67, cytokeratins (CK) 5/6, and epidermal growth factor receptor (EGFR) are widely used in clinical practice [3, 4].

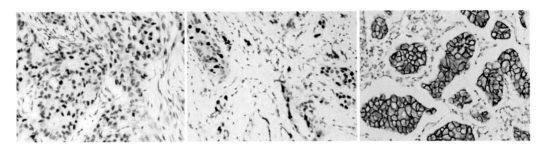

Fig. 4 Immunohistochemistry markers for breast cancer. *Left*: ER positive; *Middle*: PR positive; *Right*: Her2 positive. (IHC, 200×)

1. Luminal A: ER (+), PR (+), Her2 (−), and Ki67 ≤14 %. This subtype makes up about 40 % of all breast cancers, and carries the best prognosis.

2. Luminal B: ER (+), PR (+), Her2 (−), and Ki67 >14 %. About 20 % of breast cancers are luminal B subtype. Compared to luminal A tumors, they tend to lead to a poorer prognosis including poorer tumor grade, larger tumor size, and lymph node metastasis.

3. Her2-enriched: ER (−), PR (−), Her2 (+), and Ki67 (high). This subtype makes up about 10–15 % of breast cancers, is characterized by high expression of Her2 and proliferative gene cluster. ER and PR are usually negative. Her2 subtype tumors have a fairly poor prognosis and are prone to early and frequent recurrence and metastases.

4. Basal-like/triple-negative: ER (−), PR (−), and Her2 (−). It is called "basal-like" because the expression of high molecular weight cytokeratins (e.g., CK5/6) and/or EGFR that usually expressed in basal cells/myoepithelial cells. This subtype is also known as "triple negative." They cannot be treated with hormone therapies or trastuzumab (Herceptin) because they are ER (−) and Her2 (−). This subtype of breast cancer is often aggressive and has a poorer prognosis compared to the receptor-positive subtypes [5, 6].

2   Materials and Reagents

1. Hematoxylin solution (Mayer's): 50 g aluminum ammonium sulfate, 1.2 g hematoxylin crystals, 0.2 g sodium iodate, 1 g citric acid, 50 g chloral hydrate, add distilled water up to final volume of 1 l.

2. Eosin solution: 2.5 g of eosin Y, 0.5 ml of glacial acetic acid, mixed in 495 ml of distilled water.

3. 1 % HCl solution: 2.6 ml of concentrated HCl is diluted in 97.4 ml of distilled water.

4. Fluorescence microscope: with at least three colors of filters, including blue (for DAPI), green, and red.

5. PBS (0.01 M, pH 7.2): Na_2HPO_4 1.09 g, NaH_2PO_4 0.32 g, NaCl 9 g, dissolved in 1 l of distilled water.

6. PBST solution: 0.01 M PBS contains 0.2 % of Tween-20.

7. Citrate buffer (0.01 M, pH 6.0): Tri-sodium citrate 2.94 g, dissolved in 1 l of distilled water, adjust to pH 6.0 with 1 N HCl solution.

8. Mounting medium: anti-fade medium contains 4′, 6-diamidino-2-phenylindole (DAPI), which binds to DNA and shows blue fluorescence for nuclear counterstaining.

3 Methods

3.1 H&E Staining

1. Xylene (1), 5 min.

2. Xylene (2), 5 min.

3. 100 % ethanol (1), 5 min.

4. 100 % ethanol (2), 5 min.

5. 90 % ethanol, 5 min.

6. 70 % ethanol, 5 min (*See* **Note 1** for frozen or smear slides).

7. Rinse with tap water for 3 min and distilled water once.

8. Stain with hematoxylin solution for 1–3 min (Filter Hematoxylin before use).

9. Wash with tap water, change three times and keep in tap water for 5 min.

10. 1 % HCl solution is used for differentiation.

11. Wash with tap water, change three times and keep in tap water for 5 min.

12. Rinse with distilled water once.

13. Stain in eosin solution, 10–30 s.

14. Wash with tap water, change three times and keep in tap water until the red color appropriate (check under microscope).

15. Rinse with distilled water once.

16. Dry slides overnight in the oven (or dehydrate with ascending ethanol).

17. Mount with coverslips using mounting medium.

3.2 Immunofluo-rescence Double Staining

3.2.1 Preparation of Slides

1. Cell lines.
   - Grow cultured cells on sterile glass coverslips or slides overnight at 37 °C.
   - Wash with PBS.
   - Fix as desired. Possible procedures include:
     10 min with 10 % formalin in PBS (keep wet).
     5 min with ice-cold methanol, allow to air-dry.
     5 min with ice-cold acetone, allow to air-dry.

2. Frozen Sections.
   - Snap-freeze fresh tissues in liquid nitrogen or isopentane precooled in liquid nitrogen, embed in OCT compound in cryomolds, and store frozen blocks at –80 °C.
   - Cut 4–8 μm thick cryostat sections, and mount on Superfrost Plus slides or gelatin coated slides. Store slides at –80 °C until needed.
   - Before staining, warm slides at room temperature for 30 min and fix in ice-cold acetone for 5 min. Air-dry for 30 min.

3. Paraffin Sections.
   - Xylene, 2 × 5 min.
   - 100 % ethanol, 2 × 5 min.
   - 90 % ethanol, 5 min.
   - 70 % ethanol, 5 min.
   - Rinse in distilled water.
   - Pretreatment: perform antigen retrieval using 0.01 M citrate buffer (pH 6.0) in a microwave oven at high power up to boiling, then low power for 10 min to keep the temperature; Cool down at room temperature for at least 30 min; rinse twice with distilled water (*See* **Note 2** for frozen or smear slides).

3.2.2 Staining

1. Pre-incubate sections with PBST for 5 min.

2. Incubate sections with 10 % normal serum (from the species that the secondary antibody was raised in) for 30 min to block unspecific binding of the antibodies in a humidified chamber (alternative blocking solutions: 1 % BSA or 1 % gelatin).

3. Incubate sections with the mixture of two primary antibodies (e.g., rabbit against target-1 and mouse against target-2) in PBST (contains 1 % BSA) in a humidified dark chamber for 1 h at room temperature (RT) or overnight at 4 °C.

4. Decant the mixture solution and wash the sections in PBST, 3 × 5 min.

5. Incubate sections with the mixture of two secondary antibodies with two different fluorochromes, i.e., Texas Red-conjugated against rabbit and FITC-conjugated against mouse) in PBST for 1 h at RT in a humidified dark chamber.

6. Decant the mixture of the secondary antibody solution and wash with PBST for 3 × 5 min in the dark.

7. Mount the slides with coverslips using an anti-fade mounting media aqueously and observe the results immediately (*See* **Note 3**).

4   Notes

1. **Steps 1–6** in Subheading 3.1 are for de-wax and rehydration; for frozen section or smear slide start from **step 7**.

2. Antigenic determinants masked by formalin-fixation and paraffin-embedding often may be exposed by epitope unmasking, enzymatic digestion, or saponin. Do not use this pretreatment with frozen sections or cultured cells that are not paraffin-embedded.

3. The immunofluorescence-stained slides should be observed for the results immediately or stored in 4 °C for maximum 24 h.

References

1. World Health Organization (2003) Tumours of the breast and female genital organs. Oxford University Press, Oxford [Oxfordshire]

2. Lakhani SR et al (2012) WHO classification of tumours of the breast. IARC, Lyon

3. Prat A, Perou CM (2011) Deconstructing the molecular portraits of breast cancer. Mol Oncol 5:5–23

4. Geyer FC et al (2009) The role of molecular analysis in breast cancer. Pathology 41:77–88

5. Perou CM (2011) Molecular stratification of triple-negative breast cancers. Oncologist 16:61–70

6. Ross JS (2009) Multigene classifiers, prognostic factors, and predictors of breast cancer clinical outcome. Adv Anat Pathol 16:204–215

Chapter 2

Clinical Applications for Immunohistochemistry of Breast Lesions

Kester Haye, Rajarsi Gupta, Christopher Metter, and Jingxuan Liu

Abstract

Immunohistochemical analysis has been a key clinical tool that shows the protein expression of molecular markers. Expression of molecular markers in breast pathology has been used to distinguish breast cancers from benign lesions, classify subtypes of breast cancers, and determine therapeutic intervention. It is a relatively fast and efficient option in stratifying breast lesions to assist in both determining pathology diagnosis and offer strategies to the best course of clinical action. In this chapter, we discuss the use of immunohistochemistry testing for some of the key molecular markers involved in breast pathology that are crucial for classifying breast cancers and the guidelines for the interpretation of testing results that assist in clinical management.

Key words Immunohistochemistry, Myoepithelial cells, Cytokeratin, E-cadherin, ER, PR, HER2, Ki-67

1 Introduction

From the histopathologic examination of biopsies and resection specimens (lumpectomies, mastectomies, and metastatic lesions), useful prognostic information such as lesion type (ductal vs. lobular), tumor differentiation (well, moderately, and poorly differentiated), invasiveness, lymphovascular invasion, lymph node status, and when applicable, tumor size is derived. Most of this information can be derived from the cytological and histological morphology observed by hematoxylin and eosin (H & E) staining.

However, the utility of these traditional morphology-based parameters can be limited in providing accurate risk assessment per patient, both in terms of local or distant recurrence and in terms of providing the best options for treatment. Cytological and histological analyses by themselves give little information about the specific expression of proteins that are tightly associated to prognosis. In fact, multiple molecular markers have been identified to help differentiate one type of breast lesion from another, determine

Jian Cao (ed.), *Breast Cancer: Methods and Protocols*, Methods in Molecular Biology, vol. 1406,
DOI 10.1007/978-1-4939-3444-7_2, © Springer Science+Business Media New York 2016

invasiveness of the tumor, help define lymphovascular invasion, and allow breast cancers to be stratified into different groups associated with variable degrees of survivability. Here, we will discuss how immunohistochemistry techniques can be used towards these applications.

2   Materials

2.1   Tissue Processing

10 % Formalin, 100 % ethanol, 95 % ethanol, xylene, paraffin.

2.2   Immunohistochemistry Slide Prepping

EZPrep, cell conditioning buffer #1 are pre-made solution reagents created by Ventana Medical Systems, Inc. (Ventana) used in immunohistochemistry (IHC) reactions carried out on VENTANA BenchMark XT automated slide staining systems (*see* **Note 1**).

1. Deparaffinization fluid: 1× EZPrep. EZPrep is an aqueous-based detergent. 10× EZPrep is diluted with nine parts deionized H_2O.

2. Cell conditioning buffer #1 (CC1): This is a slightly basic, Tris-based buffer.

3. Reaction buffer: Tris-based buffer at pH 7.6 used for rinsing slides. 10× Reaction buffer is diluted with nine parts deionized H_2O.

4. Wash buffer: 1× SSC buffer. This is a sodium chloride/sodium citrate buffer which acts as a stringent aqueous wash buffer. 10× SSC buffer is diluted with nine parts deionized H_2O.

5. Rinse buffer: 1× phosphate buffer saline (PBS).

6. Liquid coverslip (LCS; a combination of low-density, paraffinic hydrocarbon and mineral oil).

2.3   Immunohistochemistry Staining Via the UltraView Universal DAB Detection Kit by Ventana

This detection system (including primary antibodies unless otherwise specified) is created by Ventana Medical Systems, Inc. (Ventana) used in immunohistochemistry (IHC) reactions performed on VENTANA BenchMark XT automated slide staining platforms (*see* **Note 1**).

1. Primary antibodies—mouse monoclonal p63 (clone 4A4) (*see* **Note 2**), mouse monoclonal SMMHC (clone SMMS-1) (*see* **Note 2**), mouse monoclonal E-cadherin (clone 36) (*see* **Note 4**), mouse monoclonal Pancytokeratin (clones AE1/AE3/PCK26) (*see* **Note 6**), rabbit monoclonal CDX2 (clone EPR2764Y) (*see* **Note 7**), rabbit monoclonal ER (clone SP1) (*see* **Notes 8** and **9**), rabbit monoclonal PR (clone 1E2) (*see* **Notes 8** and **9**), rabbit monoclonal HER2 (clone 4B5) (*see* **Notes 8** and **9**), rabbit monoclonal Ki67 (clone 30-9) (*see* **Note 10**).

2. UV INHIBITOR—3 % H_2O_2.

3. UV HRP UNIV MULT—Cocktail of goat anti-mouse IgG/IgM and goat anti-rabbit IgG that are conjugated to horse-radish peroxidase (HRP) @ a concentration of ~50 μg/mL.

4. UV DAB chromogen—0.2 % aqueous solution of 3, 3′-diaminobenzidine tetrahydrochloride.

5. UV H_2O_2—0.04 % H_2O_2 in 1× phosphate buffer solution (PBS).

6. UV Copper—Aqueous copper sulfate solution @ 5 g/L in acetate buffer.

7. Hematoxylin—48 % Hematoxylin dye in glycol and acetic acid.

8. Bluing reagent—Contains 0.1 M lithium carbonate in 0.5 M sodium carbonate aqueous solution.

2.4 Mammaglobin and GATA3 Immunohistochemistry

Mammaglobin immunohistochemistry was performed at Quest laboratories (*see* **Note 7**). GATA3 immunohistochemistry was performed by Clarient Diagnostic Services (*see* **Note 7**). Interpretations of these studies were performed at Stony Brook University Hospital.

3 Methods

3.1 Tissue Processing

Tissue sections are processed according to the automated processing protocol used at Stony Brook Hospital Histopathological Laboratories.

Tissue processing for tissue blocks from lumpectomies and mastectomies:

1. Two-cycle incubation in 10 % formalin for 1½ h @ 42 °C, 15 mmHg.

2. One-cycle incubation in 60 % ethanol for 1 h @ 42 °C, 15 mmHg.

3. Two-cycle incubation in 95 % ethanol for 1 h @ 42 °C, 15 mmHg.

4. Three-cycle incubation in 100 % ethanol for 1 h @ 42 °C, 15 mmHg.

5. Two-cycle incubation in Xylene for 1 h @ 42 °C, 15 mmHg.

6. Two-cycle incubation in Paraffin for 1½ h @ 60 °C, 15 mmHg.

Tissue Processing for Tissue Blocks from Biopsies

1. Two-cycle incubation in 10 % formalin for 15 min @ 42 °C, 15 mmHg.

2. One-cycle incubation in 60 % ethanol for 15 min @ 42 °C, 15 mmHg.

3. Two-cycle incubation in 95 % ethanol for 15 min @ 42 °C, 15 mmHg.

4. Three-cycle incubation in 100 % ethanol for 15 min @ 42 °C, 15 mmHg.

5. One-cycle incubation in xylene for 10 min @ 42 °C, 15 mmHg.

6. One-cycle incubation in xylene for 15 min @ 42 °C, 15 mmHg.

7. One-cycle incubation in paraffin for 10 min @ 42 °C, 15 mmHg.

8. One-cycle incubation in paraffin for 15 min @ 42 °C, 15 mmHg.

Immunohistochemistry slide prepping (*see* Subheading 3.2) and staining (*see* Subheading 3.3) are performed according to the BenchMArk XT IHC/ISH Staining Module protocols by Ventana.

3.2 Immunohistochemistry Slide Prepping

1. Using a microtome, obtain tissue section from block 4 µM in thickness.

2. Place sections of formalin-fixed paraffin-embedded (FFPE) tissue on positively charged glass slides.

3. Warm slide to 75 °C, and incubate for 4 min.

4. Apply EZPrep and rinse with 1× PBS. Repeat twice.

5. Apply Liquid coverslip (LCS), warm slide to 76 °C, and incubate for 4 min.

6. Rinse slide with 1× PBS, and apply Liquid coverslip (LCS).

7. Wash with 1× SSC wash buffer, warm slide to 95 °C, and incubate for 8 min.

8. Apply cell conditioner #1 and LCS.

9. Warm slide to 100 °C, and incubate for 4 min.

10. Apply LCS and cell conditioner #1. Repeat four times.

11. Apply LCS and incubate for 8 min.

12. Rinse slide with reaction buffer.

13. Apply LCS. Rinse slide with reaction buffer.

3.3 Immunohistochemistry Staining

1. Obtain prepped slide (from Subheading 3.2).

2. Warm slide to 37 °C, and incubate for 4 min.

3. Rinse with reaction buffer.

4. Add one drop of UV INHIBITOR, apply LCS and incubate for 4 min.

5. Rinse slide with reaction buffer, and warm slide to 37 °C for 4 min.

6. Add LCS, then one drop of primary antibody and incubate for 8 min.

7. Rinse slide with reaction buffer, add LCS, and warm to 37 °C for 4 min. Apply one drop of UV HRP UNIV MULT, add coverslip and incubate for 8 minutes. Rinse with reaction buffer.

8. Apply reaction buffer, add one drop of UV DAB and one drop of UV DAB H_2O_2.

9. Rinse with reaction buffer.

10. Apply one drop of UV COPPER, apply LCS, and incubate for 4 min.

11. Rinse with reaction buffer.

12. Apply one drop of HEMATOXYLIN, LCS, and incubate for 4 min.

13. Rinse with reaction buffer, and apply LCS. Repeat once.

14. Add one drop of BLUING REAGENT, apply LCS, and incubate for 4 min.

15. Rinse with reaction buffer, and then wash with 1× SSC.

16. Add one drop of mounting solution, cover the slide with a glass coverslip and allow drying before histological examination.

3.4 Hematoxylin and Eosin (H & E) Staining   Performed according to the BenchMArk XT H & E Staining Module protocol by Ventana.

4 Notes

1. Principles of immunohistochemistry.

   The UltraView Universal DAB Detection Kit by Ventana is a detection system used in immunohistochemistry (IHC) reactions carried out on VENTANA BenchMark XT automated slide staining platforms. This system is based on a biotin-free method for staining antigens bound by mouse or rabbit IgG antibodies on formalin-fixed, paraffin-embedded tissue sections. Together, the automated platform allows for efficient, staining of multiple slides in real time with high efficiency.

   Tissue sections are processed with incubations in different solutions (formalin, ethanol) which freeze cellular functions and preserves cellular components through crosslinking carboxy and amino groups (formalin) or by protein coagulation (ethanol). Incubation with xylene perforates cell membranes for easier staining of cellular components while preserving cellular integrity. Incubation with paraffin allows for long term storage. However, this "fixed" state must be partially reversed for adequate staining of tissue sections. The immunostaining prepping process achieves this by using the EZPrep detergent solution, along with heating, to deparaffinize tissue. Cell conditioning buffer reverses covalent bonds formed during the formalin fixation process, thus renaturing proteins and preserving epitope antigenicity for proper antibody affinity. Reaction buffer (a Tris-based buffer at an appropriate pH 7.6) supplies an adequate aqueous medium for the antibodies to bind their respective targets.

Liquid coverslip (LCS; a combination of low density, paraffinic hydrocarbon and mineral oil) provides a semipermeable liquid barrier allowing reagents to contact the tissue section, but preventing excessive evaporation of water.

The immunohistochemistry staining assaying is based on an indirect immune complex reaction incorporating the protein target of interest, a respective primary antibody and a secondary antibody conjugated to horse-radish peroxidase (HRP) to label the protein of interest via a chromogen precipitate reaction. To reduce background signal from nonspecific reactions, endogenous tissue peroxidases are inactivated with a high dose of hydrogen peroxide (UV INHIBITOR). The tissue section is then incubated with the primary antibody (in most cases a mouse or rabbit IgG) to label the specific protein of interest. This antibody/antigen complex is then incubated with UV HRP UNIV MULT containing secondary antibodies goat anti-mouse or goat anti-rabbit IgG conjugated to HRP. The secondary antibody binds to the primary antibody. The HRP motif, in the presence of copper and low concentration of hydrogen peroxide, drives an oxidation reaction of the DAB chromogen, generating a brown precipitate at the site of the antibody/antigen complex, which can be visualized on the tissue section with the cellular background highlighted by hematoxylin and bluing reagent.

Using the above techniques, detection of specific protein expression or lack thereof can help pathologists assess invasiveness of breast lesions, origin of breast lesions, or lymph node metastasis, determine metastasis of mammary origin and help predict patient prognosis by detection of prognostic factors.

2. Invasiveness: absence of myoepithelial cells (MECs).

Breast lobules and ducts are composed of two cell types: An outer myoepithelial cell (MEC) layer and an inner luminal epithelial cell layer [1]. The existence of these two cell layers is an important feature in separating precancerous (in situ) lesions from invasive cancers, as most invasive cancers can be characterized by their loss of association to MECs. Each layer expresses a different combination of cytokeratins (CKs) [2]. Studies have demonstrated that MECs express cytokeratins CK5, CK14, and CK17 [3], neuroendocrine marker S100 [4], and cytoskeletal elements such as smooth muscle actin (SMA), smooth muscle heavy chain myosin (SMMHC), and calponin [5]. MECs also express p63, a homologue to the tumor suppressor protein p53 [6].

Although these proteins are not exclusive to MECs, they are not expressed in intraductal luminal cells. In fact, p63, SMMHC, and calponin have been described as the most sensitive and specific combination of targets for immunohistochemical staining [5]. The nuclear staining of p63 and the cytoplasmic staining of SMMHC and calponin highlight the absence of MECs, providing evidence of malignancy. These immunohistochemical markers are very useful in distinguishing benign entities with histological patterns similar to invasive carcinoma, such as in a benign lesion like sclerosing adenosis [3] (Fig. 1), and also in other cases to demonstrate areas of invasion at sites of in situ carcinoma [7] (Fig. 1).

3. Immunohistochemistry studies for differential diagnosis of ductal intraepithelial proliferations.

Ductal intraepithelial proliferations include the lesions usual ductal hyperplasia (UDH), atypical ductal hyperplasia (ADH), and ductal carcinoma in situ (DCIS) [8, 9]. Histologically, usual ductal hyperplasia can be described as an

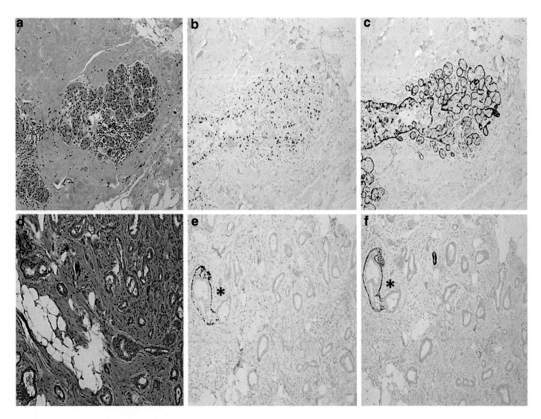

Fig. 1 Myoepithelial markers. Sclerosing adenosis (**a–c**). (**a**) Hematoxylin–eosin staining with (**b**) corresponding p63 (clone 4A4) and (**c**) smooth muscle myosin heavy chain (SMMHC, clone SMMS-1) highlighting the myoepithelial cells at 100×. Invasive ductal carcinoma (**d–f**). Note the *asterisk* marking the benign duct highlighted by p63 (**e**) and SMMHC (**f**) staining juxtaposed to carcinoma glands, absent of p63 and SMMHC staining at 100×

intraductal proliferation of cells with pleomorphic nuclei, haphazard arrangement, and irregular slit-like spaces or fenestrations. Frequently, the cells and nuclei show overlapping with rare to no mitoses. Conversely, with atypical ductal hyperplasia, cells begin to have monomorphic nuclei with nuclear enlargement and less cellular overlap and begin to show more regular, symmetrical fenestrations with possible cellular bridges (roman-arch bridges). In DCIS, the duct is filled with cells with monomorphic nuclei, without cellular overlap and with possible symmetrical fenestrations. This spectrum of lesions is suggested to be the precursor lesions for invasive ductal carcinoma, with increased gain of function mutations in cell proliferation genes, as well as increased loss-of-function mutations in cell cycle regulatory and apoptosis genes, paving a pathway to malignancy [10–12]. Consistent with this observation, the presence of UDH confers a 1.9 times relative risk of cancer development, ADH has a four to five times risk, and DCIS has an eight to ten times risk of cancer [13].

As a result, there are different approaches to clinical management where the detection of ADH or DCIS on biopsy requires obligate excision of the lesion. Therefore, effort has been placed on identifying molecular markers to help distinguish between UDH, ADH, and DCIS. Studies have demonstrated that the cytokeratin family of proteins, a type of intermediate filaments used for cell structural elements, are variably expressed in ductal proliferative lesions. High molecular weight cytokeratins (CK903 and CK5/6) have higher and diffuse expression in UDH compared to ADH/DCIS lesions with reduced staining [14–16]. Though this technique has improved the diagnostic agreement among pathologists [17], there is evidence that across the spectrum of progression from a hyperplastic lesion to in situ carcinomatous lesion, there is variability with respect to expression of CK5/6 and CK903 among ADH and DCIS lesions [18], suggesting that these cytokeratin combinations are most useful for distinguishing ADH and low-grade DCIS lesions from UDH. However, they are not applicable for studying high-grade DCIS, as CK5/6 may be expressed in some high grade DCIS type lesions.

4. Ductal vs. lobular neoplasias.

There are several subtypes of invasive breast lesions (including tubular, mucinous, micropapillary, papillary, cribriform, and medullary), where the most clinically significant patterns include ductal and lobular patterns since these two entities compromise most of the breast lesions encountered on a daily basis [8]. At the level of precursor lesions (DCIS and lobular neoplasia), lobular neoplasias, which include atypical lobular hyperplasia (ALH) and lobular carcinoma in situ (LCIS), are histologically different in appearance to ductal lesions [19].

Although these cells are monomorphic, with large nuclei and do not overlap in a similar fashion to atypical ductal proliferations, they have a more discohesive appearance [19]. These cells can fill the acini of the lobules without expansion (as in ALH), with expansion in a lobular pattern (as in LCIS), and involve the ducts described as pagetoid spread [19].

Even though lobular neoplasias are associated with invasive cancers (more so with invasive lobular carcinoma), there has been debate as to whether they are precursor lesions for invasive carcinoma [20]. Lobular lesions, when compared to ductal-type lesions, display different clinicopathological behaviors. The presence of lobular neoplasias is associated with increased diffuse disease and involvement of the bilateral breasts [21–23]. However, in terms of the presence of noninvasive lobular neoplasias at specimen margins, the cancer recurrence rate in patients with positive margins is comparable to the rate in patients with negative margins [24]. Therefore, unlike the management of DCIS, lobular neoplasias (ALH and LCIS) at specimen margins do not require re-excision.

Invasive lobular carcinomas (ILCs) and invasive ductal carcinomas (IDCs) also demonstrate different clinical behavior. Though the rate of lymph node metastasis is similar to that of invasive ductal carcinomas, invasive lobular carcinomas tend to metastasize to the skin and visceral organs, whereas invasive ductal carcinomas tend to metastasize to the lungs [25, 26]. However, some clinical differences are debatable. For instance, one study showed patients with invasive lobular carcinomas have similar prognosis to those with invasive ductal carcinomas [27]. Conversely, another study showed using multivariate analysis that patients with ILC had worse survival than patients with IDC [25]. Nevertheless, distinguishing lobular from ductal-type lesions is of clinical consequence. To that end, molecular markers to distinguish these two histotypes have been identified.

Consistent with the discohesive cellular morphology of lobular-type neoplasias, it has been demonstrated that the cell adhesion signaling pathway mediated by E-cadherin, p120 catenin, and β-catenin is altered in these lesions [28]. E-cadherin is a transmembrane cell adhesion molecule that interacts with intracellular proteins of the catenin family (p120, α and β-catenin) that associates with actin and other cytoskeletal elements to regulate cell integrity and cell proliferation [29, 30]. Immunohistochemical analysis of these proteins in normal tissue shows strong membranous localization for E-cadherin and p120, and membranous staining for β-catenin [29, 31]. Somatic mutations, genetic deletions, loss of heterozygosity, and epigenetic changes that silence the expression of the gene that encodes for E-cadherin (*CDH1)* have been routinely detected in lobular neoplasias [32, 33]. As a result, immuno-histochemistry of lobular-type lesions for E-cadherin shows a reduction in protein expression [34]. In conjunction with these perturbations, there is a shift in p120 localization from the cell membrane to the cytoplasm with scant cytoplasmic β-catenin expression [31, 35]. In our laboratory, E-cadherin IHC staining is routinely used to differentiate the two lesions (Fig. 2), and has been a very useful tool in classifying histologically ambiguous lesions.

5. Lymph-vascular invasion.

Lymph-vascular invasion (LVI) is an important prognostic parameter that is used to determine the risk of local recurrence and distant metastases [36, 37]. In patients without nodal involvement, the subset without lymph-vascular invasion has lower rates of future metastasis and increased disease-free and overall survival [37, 38]. Lymphatic invasion is routinely considered in therapeutic decision-making for patients with a borderline tumor size and negative lymph node status. There are four criteria used in the definition of lymph-vascular invasion derived from Rosen et al. [39]: (1) invasion must be detected outside the border of invasive carcinoma, (2) tumor emboli should not fit exactly within the confines of the enclosing space, (3) endothelial cells should line the confining space, and (4) lymphatics are found nearby to blood vessels. In situations where it is difficult to ascertain LVI histologically, immunohistochemical methods have been employed to highlight the endothelial cells of vascular and lymphatic spaces [40]. Vascular and lymphatic endothelial cells express CD31 and CD34, whereas lymphatic vascular cells express D2-40 and podoplanin [40, 41]. Using these markers to determine LVI, pathologists are able to increase the quantity and accuracy of detection of LVI in breast specimens [40, 41].

6. Lymph node metastasis.

Metastasis to regional lymph nodes is associated with decreased disease-free and overall survival [42–44]. Pathological assessment of regional lymph nodes is key in staging a patient's cancer [45]. Sentinel lymph nodes (SLNs) are the first series of lymph nodes that drain a particular region of the breast and have been observed to be the first type of lymph nodes that contain metastasis. Clinically, the

Fig. 2 (continued) proliferation highlighted by strong E-cadherin staining, with areas of lobular-type proliferation showing absent E-cadherin expression marked by *arrows*. (**c**) H & E and (**d**) E-cadherin immunohistochemistry of LCIS at 200×. (**e**) H & E and (**f**) E-cadherin Immunohistochemistry of pagetoid spread of lobular neoplasia at 100×. Note the lobular lesions beneath the ductal epithelium highlighted by E-cadherin staining

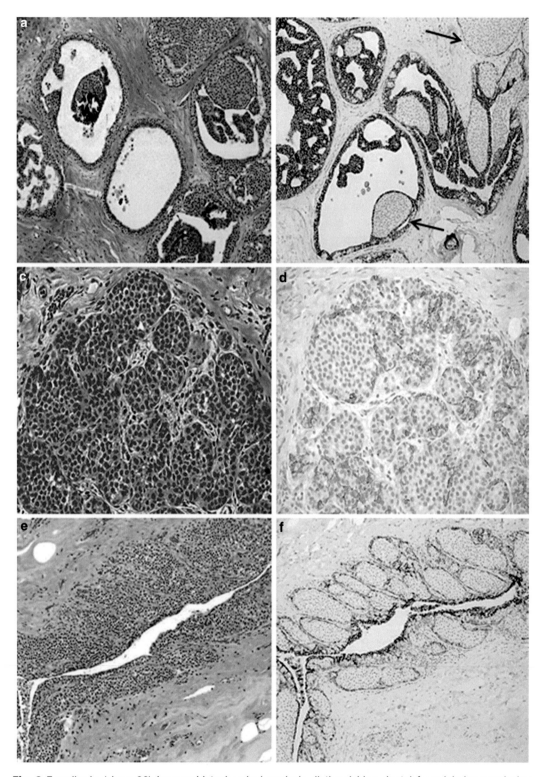

Fig. 2 E-cadherin (clone 36) immunohistochemical analysis distinguishing ductal from lobular neoplasias. (**a**, **b**) at 200×. Hematoyxylin-eosin (H & E) staining of micropapillary ductal carcinoma in situ with pagetoid spread of (**a**) lobular carcinoma in situ (LCIS) with corresponding E-cadherin staining (**b**). Note the ductal

SLNs are identified by highlighting them with intraoperative gamma radiotracer detected by a Geiger counter or visually detected blue-colored dye. Once highlighted as being "hot" and/or "blue," a surgical biopsy is performed. An intraoperative assessment of SLNs can be performed on H & E slides of SLN frozen sections. If the SLNs are positive for metastatic tumor by frozen section evaluation, the practice for the surgeon in certain clinical situations is to perform an axillary dissection to obtain more axillary nodes. This would provide more accurate staging as positive SLNs correlate with axillary lymph node metastasis in a proportion of patients [46].

However, it has been demonstrated that there can be variability in diagnosing the presence and size of lymph node metastasis [47, 48]. Thus, measuring the size of metastasis or quantification of the number of cells is important in establishing prognosis and adjuvant treatment [48, 49]. Three main categories have been used: isolated tumor cells ($\leq$0.2 mm or 200 cells), micrometastasis (more than 0.2 mm but $\leq$2.0 mm and/or $\geq$200 cells), and macrometastases (>2.0 mm). These categories correlate with nonsentinel axillary lymph node metastasis and prognosis [43, 46, 49]. To assist pathologists to this effect, IHC using the antibody combination AE1/AE3 for pancytokeratin expression found in breast tumor cells [50] has been widely used in determining lymph node metastasis, and size if applicable (Fig. 3) [43, 46, 48, 51].

7. Determination of breast metastasis from neoplasms of uncertain origins.

When metastatic lesions of unknown origin are encountered, it is prudent to include metastatic breast cancer as a major differential among possible sources, especially in female patients. Lineage-specific expression of proteins is

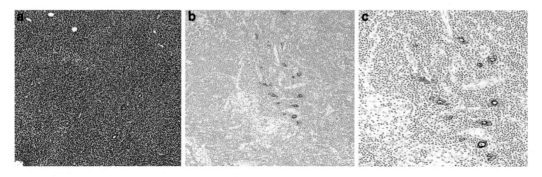

Fig. 3 Immunohistochemistry of sentinel lymph node for pancytokeratin revealing metastasis (isolated tumor cells) of Invasive Ductal Carcinoma. (**a**) Hematoxylin–eosin at 100×. Image of pancytokeratin staining (clone AE1/AE3/PCK26) of sentinel lymph node at 100× (**b**), and at 200× (**c**)

useful in classifying metastasis by tissue of origin [52]. To help differentiate among various origins of cancer, a panel of IHC studies can be employed. As the majority of metastatic breast cancers are epithelial in origin, AE1/AE3 pancytokeratin antibodies are useful to validate epithelial differentiation. CK7 and CK20 stains are also performed as breast lesions are mostly CK7+ and CK20−, and this combination can distinguish them from other cell lineages such as colon and urothelial. To further distinguish from other CK7+/CK20− tumors such as nonmucinous lung adenocarcinomas, other combinations of mammary specific markers such as GATA3, estrogen receptor (ER), gross cystic disease fluid protein 15 (GCDFP-15), and mammaglobin can also be utilized (Fig. 4).

8. Predictive and prognostic molecular factors and subclassification.

Estrogen is the primary hormone that regulates the proliferation of breast cancer cells through the interaction with its receptor, ER [53]. There are two known isoforms of ER, designated as ER-α and ER-β, where ER-α is the dominant regulator of estrogen signaling in breast cancer pathogenesis. ER expression is a strong predictive factor in terms of determining the potential benefits from adjuvant hormonal therapy.

Progesterone receptor (PR) is a superfamily of nuclear receptors, where a single copy of the PR gene has separate promoters and translational start sites to produce two isoforms, PR-α and PR-β [54]. PR is also considered important in cancer pathogenesis as PR is the codependent partner of ER in terms of the biological behavior of breast cancer at the molecular level, since the interaction of progesterone and PR is an essential component of physiology.

Human Epidermal Growth Factor Receptor 2 (HER2), expressed by the gene ERBB2, is a member of a family of transmembrane growth factor receptors that play pivotal roles in regulating normal cell proliferation and transmitting signals for cell growth and survival [55]. The HER2 receptor tyrosine kinase plays a very important role in both the biological behavior and the clinical course of breast cancer. Although the identity of the high-affinity ligand for HER2 remains unclear, HER2 is thought to dimerize with other HER receptors within the family, leading to activation of cytoplasmic tyrosine kinase thus initiating downstream signaling for cell proliferation, migration, and survival of tumor cells overexpressing HER receptors. When the HER2 receptor tyrosine kinase is activated, multiple cellular signaling pathways are initiated that include both the mitogen-activated protein kinase (MAPK) and phosphatidylinositol 3-kinase (PI3K) signaling pathways. Normally, cells contain a single copy of the HER2 gene on each copy of chromosome 17 [56]. Breast epithelial

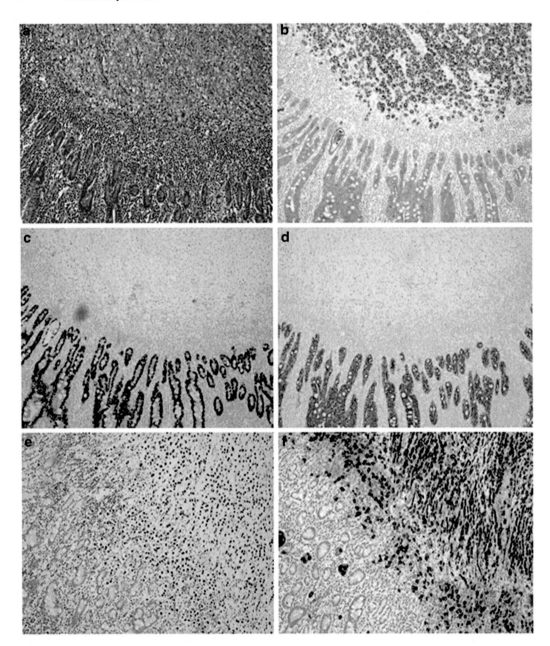

Fig. 4 Immunohistochemical determination of tissue origin of metastasis. Hematoxylin–eosin (H & E) staining at 100× (**a**) of metastatic breast lobular carcinoma located in small bowel submucosa. (**b**) Pancytokeratin immunostaining at 100× highlighting the benign small bowel epithelium (*bottom half*), and the submucosal metastatic breast lesion (*top half*). CDX2 (**c**), and E-cadherin (**d**) immunohistochemistry at 100× with positive staining of the small bowel epithelium (*bottom half*) and negative staining of the lobular breast lesion (*top half*). (**e**) GATA3 immunohistochemistry at 100× showing positive nuclear staining of the metastatic breast lesion (*upper right*). The cells stained in the lower left are lymphocytes within the small bowel lamina propria. (**f**) Mammaglobin immunohistochemistry at 100× showing positive cytoplasmic staining of the metastatic breast lesion (*top right*), as well as infiltrating breast carcinoma cells within the small bowel mucosa (*lower left*)

cells express the HER2 gene, which is translated into a 185 kDa transmembrane growth factor receptor with cytoplasmic tyrosine kinase activity. HER2 genes can be amplified from twofold to greater than 20-fold in each tumor cell nucleus relative to chromosome 17 in approximately 15–25 % of breast cancer cases, resulting in the expression of cell surface HER2 receptors with up to 100 times the normal number of receptors found in normal breast epithelial cells.

Based on gene expression profiling from cDNA microarray analysis of clinically-acquired breast lesions, subgroup classifications with associated biological and clinical behaviors can be determined by similar expression of ER, PR, and HER2 [57]. A first group was described as having a transcriptome profile similar to luminal cells with expression of the hormone receptors ER and PR. A second group demonstrated high expression of the ERBB2 coding for the HER2. A third was demonstrated to have an expression profile similar to basal cells with triple negative expression of ER, PR, and HER2. And a fourth group was described as having an expression profile close to that of normal breast tissue. It was subsequently demonstrated that luminal and basal subtypes could be further divided. In the luminal subgroup, a portion of tumors co-express HER2, thus generating the subtype luminal A (ER+/HER2−) and luminal B (ER+/HER2+) [58, 59]. Furthermore, based on the expression of cytokeratin CK5/6 and the Epidermal Growth Factor receptor 1 (EGFR), the basal-type triple negative subgroup could be further divided into two groups that co-express or lack expression of these proteins [60, 61].

Altogether, four major subclasses are clinically recognized: Luminal A (ER+ HER2−), Luminal B (ER+ HER2+), HER2 (ER− HER2+), and Triple Negative (TNC) (ER− PR− HER−; CK5/6±, EGFR±). Assays based on qRT-PCR methods for measuring the gene expression of select genes including ER, PR, and HER2 simulate the early cDNA gene array profiling studies, and can be performed for clinical prognostication by classifying the various breast subtypes [62, 63]. However, a more economic form of breast subtype classification can be performed via IHC-based methods staining for the ER, PR and HER2 combinations [60] (Fig. 5).

Molecular-derived classification has displayed a hierarchy of particular clinical behaviors. TNCs and HER2 subgroups have been demonstrated to have worse overall survival compared to the luminal subgroups [64–66]. TNCs have the highest mitotic activity, followed by HER2, luminal B, and luminal A group cancers in descending order [61, 64]. HER2 subgroup cancers are associated with the highest rate of lymph node metastases, followed by luminal B, TNC, and luminal A lesions [64].

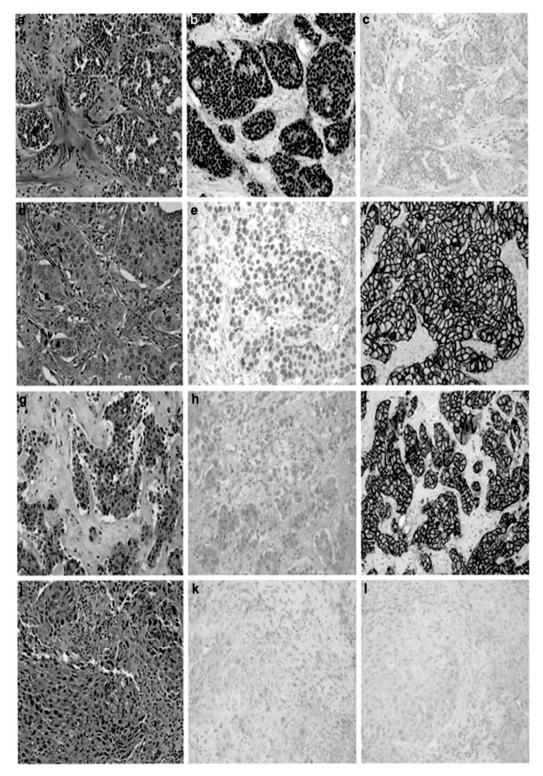

Fig. 5 Molecular subtyping of breast invasive carcinoma. Hematoxylin–eosin (H & E), and immunohistochemistry of ER (clone SP1) and HER2 (clone 4B5) at 200×, classifying Luminal A (**a**, H & E; **b**, ER; **c**, HER2), Luminal B (**d**, H & E; **e**, ER; **f**, HER2), HER2 (**g**, H & E; **h**, ER; **i**, HER2), and Triple negative cancer (**j**, H & E; **k**, ER; **l**, HER2) subgroups. PR staining not shown

9. Scoring systems for ER, PR, and HER2 expression by IHC.

The quantity of biomarker expression, based on molecular classification of the breast cancers, governs clinical management. Luminal cancers, with their higher level of expression of ER and ER-related genes, are considered for hormonal-based therapy such as adjuvant tamoxifen or aromatase inhibitors as a main stay for therapeutic intervention. Interestingly, luminal subgroups show differential response to anti-hormonal therapy [62]. Luminal A cancers have a higher response to hormonal-based therapy compared to luminal B cancers, perhaps in part to its dual HER2 expression. Conversely, luminal B cancers show increased susceptibility to anthracycline-based chemotherapy [67]. Similarly, in the HER2 subgroup, clinical treatment is based on targeting HER2 overexpression by combination of adjuvant chemotherapy and anti-HER2 antibodies, like Trastuzumab, with increased clinical benefit over adjuvant chemotherapy alone [68, 69]. With respect to TNCs, patients within this subgroup have been shown to gain benefit in disease-free and overall survival from neoadjuvant anthracycline-based chemotherapy [70, 71], as well as high-dose adjuvant chemotherapy compared to conventional doses of chemotherapy [72]. Therefore, a standardized protocol for quantifying IHC staining is important to determining positivity of biomarker expression as it strongly influences clinical management.

In daily practice, the presence of ER and PR in breast tissue is measured on formalin-fixed and paraffin-embedded (FFPE) breast tissue containing the maximal amount of viable tumor cells. IHC analysis of hormone receptor expression has been described as a more superior detection compared to previously utilized ligand binding methods [73, 74]. Increased levels of ER and PR expression by IHC correlate with response to anti-hormonal therapies [74, 75]. However, other studies have shown that tumors with even 1 % of nuclei showing nuclear expression of hormone receptor can respond to hormonal therapy [76]. Based on these observations, a consensus was reached by the American Society of Clinical Oncology (ASCO) and the College of American Pathologists (CAP) on testing interpretation criteria that include the definition of positive and negative status for hormone receptor by IHC studies and the reporting of these corresponding results [76, 77], where any nuclear immunoreactivity ≥1 % be reported as "positive" along with average intensity and extent of staining.

ASCO and CAP also recommend that HER-2 status can be determined by IHC for protein overexpression [78]. Therefore, all newly diagnosed breast cancer cases are tested for the HER-2 molecular marker using FFPE tissue sections of invasive breast cancer. Evaluations of breast cancer cell membrane HER2 protein expression by IHC are semi-quantitatively reported in pathology reports, where an absence of membranous staining or incomplete, faint membranous staining in ≤10 % of invasive tumor cells is scored as '0', incomplete, faint membranous staining in >10 % of invasive tumor cells is scored as '1+', incomplete and/or weak circumferential membrane staining in >10 % of invasive tumor cells or complete, intense, circumferential membranous staining in ≤10 % of invasive tumor cells is scored as '2+', and circumferential membranous staining of at least 10 % of the tumor cells with a thick staining ring and refractile quality is scored as '3+' (Fig. 6). Only cases with strong circumferential membrane staining, scored as "3+," show clinically relevant concordance with HER2 gene amplification by FISH. The HER2+ breast cancers that are IHC 3+ by IHC staining are candidates for targeted Trastuzumab treatment that will provide the most benefit to patients [79].

10. Ki-67 coupling histopathologic and molecular factors to predict prognosis and treatment.

Pathologists generally use the Nottingham Combined Histologic Grade for standardized grading for breast tumors [80]. This overall grading methodology is based on the sum of the individual assessments of the degree of tubular formation, nuclear pleomorphism, and mitotic activity, where the combined score puts the tumor in a tiered system that characterizes the tumor as either low-, intermediate-, or high-grade. Even though grading is qualitative and dependent on observer variability, histologic grade is still an important parameter in terms of predicting clinical outcome [81–83]. Of the grading schema, cell proliferation has gained particular attention as it can be used to further differentiate cancer groups into high and low categories and influencing prognosis [84, 85]. And even though increased cell proliferation is associated with poor prognosis, it predicts increased response to certain chemotherapeutic treatments [70, 71, 84]. Therefore, Ki-67 has become the surrogate marker for cell proliferation as it is increasingly and specifically expressed in the nuclei of all stages of active cellular division except G0 and early G1 [86].

Fig. 6 (continued) staining, grade "0"—Negative. (**c**) H & E and corresponding HER2 immunohistochemistry (**d**) showing faint, incomplete membranous staining of >10 % tumor cells, grade "1"—Negative. (**e**) H & E and (**f**) and corresponding HER2 immunohistochemistry with weak to moderate, incomplete, circumferential staining of >10 % tumor cells, grade "2"—Equivocal. (**g**) H & E and corresponding HER2 immunohistochemistry (**h**) with strong, complete circumferential membranous staining, grade "3"—Positive

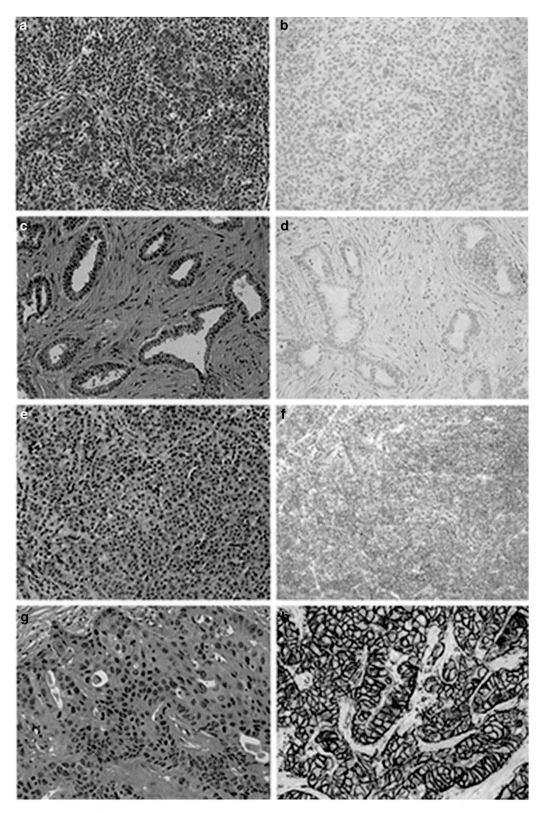

Fig. 6 Grading of HER2 immunohistochemistry staining of invasive carcinoma (clone 4B5) at 200×. (**a**) Hematoxylin–eosin (H & E) and corresponding HER2 immunohistochemistry (**b**) showing absence of membranous

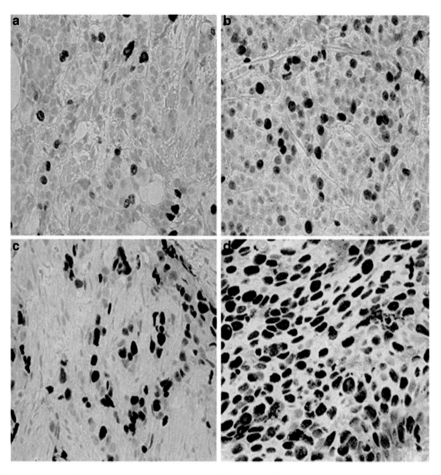

Fig. 7 Cell proliferation by Ki-67 immunohistochemistry at 400×. Nuclear staining of carcinoma cells by Ki-67 (clone 30-9) immunohistochemistry with various quantities of expression. (**a**) 10–15 %, (**b**) 35–40 %, (**c**) 60–65 %, (**d**) >90 % of tumor cells present are positive for Ki-67 nuclear staining.

To assist in quantification of cell proliferation, IHC staining for Ki-67 is routinely used in assessing cancers, coupling the histological grading system to the molecular-based subgroups of breast cancer [85]. Recently, the International Ki-67 in Breast Cancer Working Group generated recommendations for the application of Ki-67 staining and quantitative scoring in breast cancers. Cells suitable for scoring should have complete nuclear expression, suggestive of good specimen quality (Fig. 7). For a quantitative measurement of Ki-67 expression, at least three fields of cells at 40× objective located at the tumor's invasive edge should be analyzed. However, if there are heterogeneous regions of Ki-67 expression throughout the lesion, described as "hot spots," an overall average of the Ki-67 expression should be calculated [85].

Scoring of Ki-67 expression also has a predictive role in treatment of breast cancers. In luminal cancers, increased Ki-67 expression was detected in the luminal B subgroup, and can be used to differentiate from the luminal A subgroup with a cutoff point of 13.25 % [84]. This has been illustrated in the IMPACT and the P024 clinical trials, where expression of Ki-67 status post neoadjuvant chemotherapy with combined hormonal and anthracycline-based therapy has a linear correlation with recurrence [87, 88]. These findings have been supported by other studies showing increased Ki-67 expression associated with increased complete pathological response to anthracycline-based neoadjuvant chemotherapy [70, 71, 84]. This observation is crucial with the clinical management of the triple negative cancer (TNC) subgroup, since these lesions have been demonstrated to have high-grade disease and high Ki-67 expression with poor clinical outcomes [61, 70, 71, 84].

11. Quality assurance of breast specimens for accurate assessment of biomarker expression.

Proper handling and care of breast specimens is necessary for accurate quantification and assessment of biomarker expression by IHC. Once the breast specimen is removed from the patient, the length of time before placement in formalin, known as the cold ischemic time, is a key factor in affecting integrity of biomarker expression. It has been demonstrated that refrigerated specimens with cold ischemic times over 4 h and unrefrigerated specimens with cold ischemic times over 2 h begin to show drastic decreases in ER, PR, and HER2 expression [89]. This has led to the ASCP/CAP recommendation of keeping ischemic times no longer than 1 h [76, 89]. Formalin times are another important factor in maintaining quality of biomarker expression. Formalin times below 6 h have been documented to decrease detectable expression of ER, PR, and HER2 [90]. Conversely, formalin times greater than 72 h also result in decreased expression for ER, PR, and HER2 expression [91–93]. Combining these observations with quality assurance of immunohistochemistry results for ER, PR, HER2, and Ki-67 have led to established CAP/ASCO guidelines that have recommended that specimens must be incubated in formalin no less than 6 h and no more than 72 h [76, 78, 85].

As a result of rapidly advancing biotechnologies, especially that of gene expression studies, our knowledge of breast pathology has been greatly expanded. With this enhanced knowledge, we have implemented the use of immunohistochemical analysis of diagnostic and predictive markers as an invaluable tool in modern clinical practice to help determine both the clinical management of breast disease and to improve overall patient care. In this chapter, we have discussed the ways in which IHC analysis can enhance H & E examinations to further distinguish benign from malignant breast lesions, differentiate subtypes of breast neoplasias, and to quantify cancer cells and biomarker expression to support various clinical interventions. With the ongoing work into the discovery and development of better prognostic biomarkers, this invaluable tool will ultimately be refined, to the benefit of pathologists, clinicians, and most importantly, our patients.

Acknowledgement

The authors would like to thank Ms. Laura Birney, Lucille Camille Kutcher, and Julie Elder of the Histology Laboratory at the Stony Brook University Hospital for their technical assistance.

References

1. Collins LC, Schnitt SJ (2007) Histology for pathologists. Lippincott Williams & Wilkins, Philadelphia

2. Bocker W, Bier B, Freytag G, Brommelkamp B, Jarasch ED, Edel G, Dockhorn-Dworniczak B, Schmid KW (1992) An immunohistochemical study of the breast using antibodies to basal and luminal keratins, alpha-smooth muscle actin, vimentin, collagen IV and laminin. Part II: epitheliosis and ductal carcinoma in situ. Virchows Arch A Pathol Anat Histopathol 421(4):323–330

3. Jarasch ED, Nagle RB, Kaufmann M, Maurer C, Bocker WJ (1988) Differential diagnosis of benign epithelial proliferations and carcinomas of the breast using antibodies to cytokeratins. Hum Pathol 19(3):276–289

4. Egan MJ, Newman J, Crocker J, Collard M (1987) Immunohistochemical localization of S100 protein in benign and malignant conditions of the breast. Arch Pathol Lab Med 111(1):28–31

5. Werling RW, Hwang H, Yaziji H, Gown AM (2003) Immunohistochemical distinction of invasive from noninvasive breast lesions: a comparative study of p63 versus calponin and smooth muscle myosin heavy chain. Am J Surg Pathol 27(1):82–90

6. Barbareschi M, Pecciarini L, Cangi MG, Macri E, Rizzo A, Viale G, Doglioni C (2001) p63, a p53 homologue, is a selective nuclear marker of myoepithelial cells of the human breast. Am J Surg Pathol 25(8):1054–1060

7. Damiani S, Ludvikova M, Tomasic G, Bianchi S, Gown AM, Eusebi V (1999) Myoepithelial cells and basal lamina in poorly differentiated in situ duct carcinoma of the breast. An immunocytochemical study. Virchows Arch 434(3):227–234

8. Lakhani S, Ellis I, Schnitt S (2012) WHO classification of tumours of the breast. IARC Press, Lyon

9. Tavassoli F, Devilee P (eds) (2003) Pathology and genetics of tumours of the breast and female genital organs. IARC Press, Lyon

10. Mommers EC, Poulin N, Sangulin J, Meijer CJ, Baak JP, van Diest PJ (2001) Nuclear cytometric changes in breast carcinogenesis. J Pathol 193 (1):33–39. doi:10.1002/1096-9896(2000) 9999:9999<::AID-PATH744>3.0.CO;2-Q

11. Mommers EC, van Diest PJ, Leonhart AM, Meijer CJ, Baak JP (1998) Expression of proliferation and apoptosis-related proteins in usual ductal hyperplasia of the breast. Hum Pathol 29(12):1539–1545

12. Mommers EC, van Diest PJ, Leonhart AM, Meijer CJ, Baak JP (1999) Balance of cell proliferation and apoptosis in breast carcinogenesis. Breast Cancer Res Treat 58(2):163–169

13. Dupont WD, Page DL (1985) Risk factors for breast cancer in women with proliferative breast disease. N Engl J Med 312(3):146–151. doi:10.1056/NEJM198501173120303

14. Moinfar F, Man YG, Lininger RA, Bodian C, Tavassoli FA (1999) Use of keratin 35betaE12 as an adjunct in the diagnosis of mammary intraepithelial neoplasia-ductal type – benign and malignant intraductal proliferations. Am J Surg Pathol 23(9):1048–1058

15. Otterbach F, Bankfalvi A, Bergner S, Decker T, Krech R, Boecker W (2000) Cytokeratin 5/6 immunohistochemistry assists the differential diagnosis of atypical proliferations of the breast. Histopathology 37(3):232–240

16. Boecker W, Moll R, Dervan P, Buerger H, Poremba C, Diallo RI, Herbst H, Schmidt A, Lerch MM, Buchwalow IB (2002) Usual ductal hyperplasia of the breast is a committed stem (progenitor) cell lesion distinct from atypical ductal hyperplasia and ductal carcinoma in situ. J Pathol 198(4):458–467. doi:10.1002/path.1241

17. Jain RK, Mehta R, Dimitrov R, Larsson LG, Musto PM, Hodges KB, Ulbright TM, Hattab EM, Agaram N, Idrees MT, Badve S (2011) Atypical ductal hyperplasia: interobserver and intraobserver variability. Mod Pathol 24(7):917–923. doi:10.1038/modpathol.2011.66

18. Lacroix-Triki M, Mery E, Voigt JJ, Istier L, Rochaix P (2003) Value of cytokeratin 5/6 immunostaining using D5/16 B4 antibody in the spectrum of proliferative intraepithelial lesions of the breast. A comparative study with 34betaE12 antibody. Virchows Arch 442(6):548–554. doi:10.1007/s00428-003-0808-0

19. Haagensen CD, Lane N, Lattes R, Bodian C (1978) Lobular neoplasia (so-called lobular carcinoma in situ) of the breast. Cancer 42(2):737–769

20. Jorns J, Sabel MS, Pang JC (2014) Lobular neoplasia: morphology and management. Arch Pathol Lab Med 138(10):1344–1349. doi:10.5858/arpa.2014-0278-CC

21. Chuba PJ, Hamre MR, Yap J, Severson RK, Lucas D, Shamsa F, Aref A (2005) Bilateral risk for subsequent breast cancer after lobular carcinoma-in-situ: analysis of surveillance, epidemiology, and end results data. J Clin Oncol 23(24):5534–5541. doi:10.1200/JCO.2005.04.038

22. Zengel B, Yararbas U, Duran A, Uslu A, Eliyatkin N, Demirkiran MA, Cengiz F, Simsek C, Postaci H, Vardar E, Durusoy R (2013) Comparison of the clinicopathological features of invasive ductal, invasive lobular, and mixed (invasive ductal + invasive lobular) carcinoma of the breast. Breast Cancer. doi:10.1007/s12282-013-0489-8

23. Hofmeyer S, Pekar G, Gere M, Tarjan M, Hellberg D, Tot T (2012) Comparison of the subgross distribution of the lesions in invasive ductal and lobular carcinomas of the breast: a large-format histology study. Int J Breast Cancer 2012:436141. doi:10.1155/2012/436141

24. Ciocca RM, Li T, Freedman GM, Morrow M (2008) Presence of lobular carcinoma in situ does not increase local recurrence in patients treated with breast-conserving therapy. Ann Surg Oncol 15(8):2263–2271. doi:10.1245/s10434-008-9960-8

25. Korhonen T, Kuukasjarvi T, Huhtala H, Alarmo EL, Holli K, Kallioniemi A, Pylkkanen L (2013) The impact of lobular and ductal breast cancer histology on the metastatic behavior and long term survival of breast cancer patients. Breast 22(6):1119–1124. doi:10.1016/j.breast.2013.06.001

26. Ferlicot S, Vincent-Salomon A, Medioni J, Genin P, Rosty C, Sigal-Zafrani B, Freneaux P, Jouve M, Thiery JP, Sastre-Garau X (2004) Wide metastatic spreading in infiltrating lobular carcinoma of the breast. Eur J Cancer 40(3):336–341

27. Arpino G, Bardou VJ, Clark GM, Elledge RM (2004) Infiltrating lobular carcinoma of the breast: tumor characteristics and clinical outcome. Breast Cancer Res 6(3):R149–R156. doi:10.1186/bcr767

28. Chan JK, Wong CS (2001) Loss of E-cadherin is the fundamental defect in diffuse-type gastric carcinoma and infiltrating lobular carcinoma of the breast. Adv Anat Pathol 8(3):165–172

29. Ohkubo T, Ozawa M (1999) p120(ctn) binds to the membrane-proximal region of the E-cadherin cytoplasmic domain and is involved in modulation of adhesion activity. J Biol Chem 274(30):21409–21415

30. Nelson WJ, Nusse R (2004) Convergence of Wnt, beta-catenin, and cadherin pathways. Science 303(5663):1483–1487. doi:10.1126/science.1094291

31. Dabbs DJ, Kaplai M, Chivukula M, Kanbour A, Kanbour-Shakir A, Carter GJ (2007) The spectrum of morphomolecular abnormalities of the E-cadherin/catenin complex in pleomorphic lobular carcinoma of the breast. Appl Immunohistochem Mol Morphol 15(3):260–266. doi:10.1097/01.pai.0000213128.78665.3c

32. Lerwill MF (2006) The evolution of lobular neoplasia. Adv Anat Pathol 13(4):157–165

33. Mastracci TL, Tjan S, Bane AL, O'Malley FP, Andrulis IL (2005) E-cadherin alterations in atypical lobular hyperplasia and lobular carcinoma in situ of the breast. Mod Pathol 18(6):741–751. doi:10.1038/modpathol.3800362

34. Handschuh G, Candidus S, Luber B, Reich U, Schott C, Oswald S, Becke H, Hutzler P, Birchmeier W, Hofler H, Becker KF (1999) Tumour-associated E-cadherin mutations alter

cellular morphology, decrease cellular adhesion and increase cellular motility. Oncogene 18(30):4301–4312. doi:10.1038/sj.onc.1202790

35. Sarrio D, Perez-Mies B, Hardisson D, Moreno-Bueno G, Suarez A, Cano A, Martin-Perez J, Gamallo C, Palacios J (2004) Cytoplasmic localization of p120ctn and E-cadherin loss characterize lobular breast carcinoma from preinvasive to metastatic lesions. Oncogene 23(19):3272–3283. doi:10.1038/sj.onc.1207439

36. Davis BW, Gelber R, Goldhirsch A, Hartmann WH, Hollaway L, Russell I, Rudenstam CM (1985) Prognostic significance of peritumoral vessel invasion in clinical trials of adjuvant therapy for breast cancer with axillary lymph node metastasis. Hum Pathol 16(12):1212–1218

37. Lauria R, Perrone F, Carlomagno C, De Laurentiis M, Morabito A, Gallo C, Varriale E, Pettinato G, Panico L, Petrella G et al (1995) The prognostic value of lymphatic and blood vessel invasion in operable breast cancer. Cancer 76(10):1772–1778

38. Colleoni M, Rotmensz N, Peruzzotti G, Maisonneuve P, Mazzarol G, Pruneri G, Luini A, Intra M, Veronesi P, Galimberti V, Torrisi R, Cardillo A, Goldhirsch A, Viale G (2005) Size of breast cancer metastases in axillary lymph nodes: clinical relevance of minimal lymph node involvement. J Clin Oncol 23(7):1379–1389. doi:10.1200/JCO.2005.07.094

39. Rosen PP (1983) Tumor emboli in intramammary lymphatics in breast carcinoma: pathologic criteria for diagnosis and clinical significance. Pathol Annu 18(Pt 2):215–232

40. Van den Eynden GG, Van der Auwera I, Van Laere SJ, Colpaert CG, van Dam P, Dirix LY, Vermeulen PB, Van Marck EA (2006) Distinguishing blood and lymph vessel invasion in breast cancer: a prospective immunohistochemical study. Br J Cancer 94(11):1643–1649. doi:10.1038/sj.bjc.6603152

41. Mohammed RA, Martin SG, Gill MS, Green AR, Paish EC, Ellis IO (2007) Improved methods of detection of lymphovascular invasion demonstrate that it is the predominant method of vascular invasion in breast cancer and has important clinical consequences. Am J Surg Pathol 31(12):1825–1833. doi:10.1097/PAS.0b013e31806841f6

42. Treseler P (2006) Pathologic examination of the sentinel lymph node: what is the best method? Breast J 12(5 Suppl 2):S143–S151. doi:10.1111/j.1075-122X.2006.00328.x

43. Nasser IA, Lee AK, Bosari S, Saganich R, Heatley G, Silverman ML (1993) Occult axillary lymph node metastases in "node-negative" breast carcinoma. Hum Pathol 24(9):950–957

44. Mullenix PS, Brown TA, Meyers MO, Giles LR, Sigurdson ER, Boraas MC, Hoffman JP, Eisenberg BL, Torosian MH (2005) The association of cytokeratin-only-positive sentinel lymph nodes and subsequent metastases in breast cancer. Am J Surg 189(5):606–609. doi:10.1016/j.amjsurg.2005.01.031, discussion 609

45. Connolly JL (2006) Changes and problematic areas in interpretation of the AJCC Cancer Staging Manual, 6th Edition, for breast cancer. Arch Pathol Lab Med 130(3):287–291. doi:10.1043/1543-2165(2006)130[287:CAPAII]2.0.CO;2

46. Dabbs DJ, Fung M, Landsittel D, McManus K, Johnson R (2004) Sentinel lymph node micrometastasis as a predictor of axillary tumor burden. Breast J 10(2):101–105

47. Carcoforo P, Bergossi L, Basaglia E, Soliani G, Querzoli P, Zambrini E, Pozza E, Feggi L (2002) Prognostic and therapeutic impact of sentinel node micrometastasis in patients with invasive breast cancer. Tumori 88(3):S4–S5

48. Tan LK, Giri D, Hummer AJ, Panageas KS, Brogi E, Norton L, Hudis C, Borgen PI, Cody HS 3rd (2008) Occult axillary node metastases in breast cancer are prognostically significant: results in 368 node-negative patients with 20-year follow-up. J Clin Oncol 26(11):1803–1809. doi:10.1200/JCO.2007.12.6425

49. Kahn HJ, Hanna WM, Chapman JA, Trudeau ME, Lickley HL, Mobbs BG, Murray D, Pritchard KI, Sawka CA, McCready DR, Marks A (2006) Biological significance of occult micrometastases in histologically negative axillary lymph nodes in breast cancer patients using the recent American Joint Committee on Cancer breast cancer staging system. Breast J 12(4):294–301. doi:10.1111/j.1075-122X.2006.00267.x

50. Czerniecki BJ, Scheff AM, Callans LS, Spitz FR, Bedrosian I, Conant EF, Orel SG, Berlin J, Helsabeck C, Fraker DL, Reynolds C (1999) Immunohistochemistry with pancytokeratins improves the sensitivity of sentinel lymph node biopsy in patients with breast carcinoma. Cancer 85(5):1098–1103

51. Clare SE, Sener SF, Wilkens W, Goldschmidt R, Merkel D, Winchester DJ (1997) Prognostic significance of occult lymph node metastases in node-negative breast cancer. Ann Surg Oncol 4(6):447–451

52. Lin F, Liu H (2014) Immunohistochemistry in undifferentiated neoplasm/tumor of uncertain origin. Arch Pathol Lab Med 138(12):1583–1610. doi:10.5858/arpa.2014-0061-RA

53. Deroo BJ, Korach KS (2006) Estrogen receptors and human disease. J Clin Invest 116(3):561–570. doi:10.1172/JCI27987

54. Jacobsen BM, Horwitz KB (2012) Progesterone receptors, their isoforms and progesterone regulated transcription. Mol Cell Endocrinol 357(1–2):18–29. doi:10.1016/j.mce.2011.09.016

55. Harari D, Yarden Y (2000) Molecular mechanisms underlying ErbB2/HER2 action in breast cancer. Oncogene 19(53):6102–6114. doi:10.1038/sj.onc.1203973

56. Shah SS, Wang Y, Tull J, Zhang S (2009) Effect of high copy number of HER2 associated with polysomy 17 on HER2 protein expression in invasive breast carcinoma. Diagn Mol Pathol 18(1):30–33. doi:10.1097/PDM.0b013e31817c1af8

57. Perou CM, Sorlie T, Eisen MB, van de Rijn M, Jeffrey SS, Rees CA, Pollack JR, Ross DT, Johnsen H, Akslen LA, Fluge O, Pergamenschikov A, Williams C, Zhu SX, Lonning PE, Borresen-Dale AL, Brown PO, Botstein D (2000) Molecular portraits of human breast tumours. Nature 406(6797):747–752. doi:10.1038/35021093

58. Sorlie T, Perou CM, Tibshirani R, Aas T, Geisler S, Johnsen H, Hastie T, Eisen MB, van de Rijn M, Jeffrey SS, Thorsen T, Quist H, Matese JC, Brown PO, Botstein D, Lonning PE, Borresen-Dale AL (2001) Gene expression patterns of breast carcinomas distinguish tumor subclasses with clinical implications. Proc Natl Acad Sci U S A 98(19):10869–10874. doi:10.1073/pnas.191367098

59. Sorlie T, Tibshirani R, Parker J, Hastie T, Marron JS, Nobel A, Deng S, Johnsen H, Pesich R, Geisler S, Demeter J, Perou CM, Lonning PE, Brown PO, Borresen-Dale AL, Botstein D (2003) Repeated observation of breast tumor subtypes in independent gene expression data sets. Proc Natl Acad Sci U S A 100(14):8418–8423. doi:10.1073/pnas.0932692100

60. Nielsen TO, Hsu FD, Jensen K, Cheang M, Karaca G, Hu Z, Hernandez-Boussard T, Livasy C, Cowan D, Dressler L, Akslen LA, Ragaz J, Gown AM, Gilks CB, van de Rijn M, Perou CM (2004) Immunohistochemical and clinical characterization of the basal-like subtype of invasive breast carcinoma. Clin Cancer Res 10(16):5367–5374. doi:10.1158/1078-0432.CCR-04-0220

61. Livasy CA, Karaca G, Nanda R, Tretiakova MS, Olopade OI, Moore DT, Perou CM (2006) Phenotypic evaluation of the basal-like subtype of invasive breast carcinoma. Mod Pathol 19(2):264–271. doi:10.1038/modpathol.3800528

62. Paik S, Shak S, Tang G, Kim C, Baker J, Cronin M, Baehner FL, Walker MG, Watson D, Park T, Hiller W, Fisher ER, Wickerham DL, Bryant J, Wolmark N (2004) A multigene assay to predict recurrence of tamoxifen-treated, node-negative breast cancer. N Engl J Med 351(27):2817–2826. doi:10.1056/NEJMoa041588

63. van de Vijver MJ, He YD, van't Veer LJ, Dai H, Hart AA, Voskuil DW, Schreiber GJ, Peterse JL, Roberts C, Marton MJ, Parrish M, Atsma D, Witteveen A, Glas A, Delahaye L, van der Velde T, Bartelink H, Rodenhuis S, Rutgers ET, Friend SH, Bernards R (2002) A gene-expression signature as a predictor of survival in breast cancer. N Engl J Med 347(25):1999–2009. doi:10.1056/NEJMoa021967

64. Carey LA, Perou CM, Livasy CA, Dressler LG, Cowan D, Conway K, Karaca G, Troester MA, Tse CK, Edmiston S, Deming SL, Geradts J, Cheang MC, Nielsen TO, Moorman PG, Earp HS, Millikan RC (2006) Race, breast cancer subtypes, and survival in the Carolina Breast Cancer Study. JAMA 295(21):2492–2502. doi:10.1001/jama.295.21.2492

65. Rakha EA, El-Sayed ME, Green AR, Lee AH, Robertson JF, Ellis IO (2007) Prognostic markers in triple-negative breast cancer. Cancer 109(1):25–32. doi:10.1002/cncr.22381

66. Carey LA, Dees EC, Sawyer L, Gatti L, Moore DT, Collichio F, Ollila DW, Sartor CI, Graham ML, Perou CM (2007) The triple negative paradox: primary tumor chemosensitivity of breast cancer subtypes. Clin Cancer Res 13(8):2329–2334. doi:10.1158/1078-0432.CCR-06-1109

67. Parker JS, Mullins M, Cheang MC, Leung S, Voduc D, Vickery T, Davies S, Fauron C, He X, Hu Z, Quackenbush JF, Stijleman IJ, Palazzo J, Marron JS, Nobel AB, Mardis E, Nielsen TO, Ellis MJ, Perou CM, Bernard PS (2009) Supervised risk predictor of breast cancer based on intrinsic subtypes. J Clin Oncol 27(8):1160–1167. doi:10.1200/JCO.2008.18.1370

68. Slamon DJ, Leyland-Jones B, Shak S, Fuchs H, Paton V, Bajamonde A, Fleming T, Eiermann W, Wolter J, Pegram M, Baselga J, Norton L (2001) Use of chemotherapy plus a monoclonal antibody against HER2 for metastatic breast cancer that overexpresses HER2. N Engl J Med 344(11):783–792. doi:10.1056/NEJM200103153441101

69. Seidman AD, Berry D, Cirrincione C, Harris L, Muss H, Marcom PK, Gipson G, Burstein H, Lake D, Shapiro CL, Ungaro P, Norton L, Winer E, Hudis C (2008) Randomized phase III trial of weekly compared with every-3-weeks paclitaxel for metastatic breast cancer,

with trastuzumab for all HER-2 overexpressors and random assignment to trastuzumab or not in HER-2 nonoverexpressors: final results of Cancer and Leukemia Group B protocol 9840. J Clin Oncol 26(10):1642–1649. doi:10.1200/JCO.2007.11.6699

70. Yoshioka T, Hosoda M, Yamamoto M, Taguchi K, Hatanaka KC, Takakuwa E, Hatanaka Y, Matsuno Y, Yamashita H (2013) Prognostic significance of pathologic complete response and Ki67 expression after neoadjuvant chemotherapy in breast cancer. Breast Cancer. doi:10.1007/s12282-013-0474-2

71. Horimoto Y, Arakawa A, Tanabe M, Sonoue H, Igari F, Senuma K, Tokuda E, Shimizu H, Kosaka T, Saito M (2014) Ki67 expression and the effect of neo-adjuvant chemotherapy on luminal HER2-negative breast cancer. BMC Cancer 14:550. doi:10.1186/1471-2407-14-550

72. Gluz O, Nitz UA, Harbeck N, Ting E, Kates R, Herr A, Lindemann W, Jackisch C, Berdel WE, Kirchner H, Metzner B, Werner F, Schutt G, Frick M, Poremba C, Diallo-Danebrock R, Mohrmann S, West German Study Group (2008) Triple-negative high-risk breast cancer derives particular benefit from dose intensification of adjuvant chemotherapy: results of WSG AM-01 trial. Ann Oncol 19(5):861–870. doi:10.1093/annonc/mdm551

73. Barnes DM, Harris WH, Smith P, Millis RR, Rubens RD (1996) Immunohistochemical determination of oestrogen receptor: comparison of different methods of assessment of staining and correlation with clinical outcome of breast cancer patients. Br J Cancer 74(9):1445–1451

74. Elledge RM, Green S, Pugh R, Allred DC, Clark GM, Hill J, Ravdin P, Martino S, Osborne CK (2000) Estrogen receptor (ER) and progesterone receptor (PgR), by ligand-binding assay compared with ER, PgR and pS2, by immuno-histochemistry in predicting response to tamoxifen in metastatic breast cancer: a Southwest Oncology Group Study. Int J Cancer 89(2):111–117

75. Stendahl M, Ryden L, Nordenskjold B, Jonsson PE, Landberg G, Jirstrom K (2006) High progesterone receptor expression correlates to the effect of adjuvant tamoxifen in premenopausal breast cancer patients. Clin Cancer Res 12(15):4614–4618. doi:10.1158/1078-0432.CCR-06-0248

76. Hammond ME, Hayes DF, Dowsett M, Allred DC, Hagerty KL, Badve S, Fitzgibbons PL, Francis G, Goldstein NS, Hayes M, Hicks DG, Lester S, Love R, Mangu PB, McShane L, Miller K, Osborne CK, Paik S, Perlmutter J, Rhodes A, Sasano H, Schwartz JN, Sweep FC, Taube S, Torlakovic EE, Valenstein P, Viale G, Visscher D, Wheeler T, Williams RB, Wittliff JL, Wolff AC (2010) American Society of Clinical Oncology/College of American Pathologists guideline recommendations for immunohistochemical testing of estrogen and progesterone receptors in breast cancer. Arch Pathol Lab Med 134(6):907–922. doi:10.1043/1543-2165-134.6.907

77. Harris L, Fritsche H, Mennel R, Norton L, Ravdin P, Taube S, Somerfield MR, Hayes DF, Bast RC Jr, American Society of Clinical Oncology (2007) American Society of Clinical Oncology 2007 update of recommendations for the use of tumor markers in breast cancer. J Clin Oncol 25(33):5287–5312. doi:10.1200/JCO.2007.14.2364

78. Wolff AC, Hammond ME, Hicks DG, Dowsett M, McShane LM, Allison KH, Allred DC, Bartlett JM, Bilous M, Fitzgibbons P, Hanna W, Jenkins RB, Mangu PB, Paik S, Perez EA, Press MF, Spears PA, Vance GH, Viale G, Hayes DF, American Society of Clinical Oncology; College of American Pathologists (2013) Recommendations for human epidermal growth factor receptor 2 testing in breast cancer: American Society of Clinical Oncology/College of American Pathologists clinical practice guideline update. J Clin Oncol 31(31):3997–4013. doi:10.1200/JCO.2013.50.9984

79. Menard S, Casalini P, Campiglio M, Pupa SM, Tagliabue E (2004) Role of HER2/neu in tumor progression and therapy. Cell Mol Life Sci 61(23):2965–2978. doi:10.1007/s00018-004-4277-7

80. Ellis IO, Elston CW (2006) Histologic grade. Elsevier, Philadelphia

81. Rakha EA, El-Sayed ME, Lee AH, Elston CW, Grainge MJ, Hodi Z, Blamey RW, Ellis IO (2008) Prognostic significance of Nottingham histologic grade in invasive breast carcinoma. J Clin Oncol 26(19):3153–3158. doi:10.1200/JCO.2007.15.5986

82. Harris GC, Denley HE, Pinder SE, Lee AH, Ellis IO, Elston CW, Evans A (2003) Correlation of histologic prognostic factors in core biopsies and therapeutic excisions of invasive breast carcinoma. Am J Surg Pathol 27(1):11–15

83. Pinder SE, Murray S, Ellis IO, Trihia H, Elston CW, Gelber RD, Goldhirsch A, Lindtner J, Cortes-Funes H, Simoncini E, Byrne MJ, Golouh R, Rudenstam CM, Castiglione-Gertsch M, Gusterson BA (1998) The importance of the histologic grade of invasive breast carcinoma and response to chemotherapy. Cancer 83(8):1529–1539

84. Cheang MC, Chia SK, Voduc D, Gao D, Leung S, Snider J, Watson M, Davies S, Bernard PS, Parker JS, Perou CM, Ellis MJ, Nielsen TO (2009) Ki67 index, HER2 status, and prognosis of patients with luminal B breast cancer. J Natl Cancer Inst 101(10):736–750. doi:10.1093/jnci/djp082

85. Dowsett M, Nielsen TO, A'Hern R, Bartlett J, Coombes RC, Cuzick J, Ellis M, Henry NL, Hugh JC, Lively T, McShane L, Paik S, Penault-Llorca F, Prudkin L, Regan M, Salter J, Sotiriou C, Smith IE, Viale G, Zujewski JA, Hayes DF, International Ki-67 in Breast Cancer Working Group (2011) Assessment of Ki67 in breast cancer: recommendations from the International Ki67 in Breast Cancer working group. J Natl Cancer Inst 103(22):1656–1664. doi:10.1093/jnci/djr393

86. Yu CC, Woods AL, Levison DA (1992) The assessment of cellular proliferation by immunohistochemistry: a review of currently available methods and their applications. Histochem J 24(3):121–131

87. Dowsett M, Smith IE, Ebbs SR, Dixon JM, Skene A, Griffith C, Boeddinghaus I, Salter J, Detre S, Hills M, Ashley S, Francis S, Walsh G, Trialists I (2005) Short-term changes in Ki-67 during neoadjuvant treatment of primary breast cancer with anastrozole or tamoxifen alone or combined correlate with recurrence-free survival. Clin Cancer Res 11(2 Pt 2):951s–958s

88. Ellis MJ, Coop A, Singh B, Tao Y, Llombart-Cussac A, Janicke F, Mauriac L, Quebe-Fehling E, Chaudri-Ross HA, Evans DB, Miller WR (2003) Letrozole inhibits tumor proliferation more effectively than tamoxifen independent of HER1/2 expression status. Cancer Res 63(19):6523–6531

89. Yildiz-Aktas IZ, Dabbs DJ, Bhargava R (2012) The effect of cold ischemic time on the immunohistochemical evaluation of estrogen receptor, progesterone receptor, and HER2 expression in invasive breast carcinoma. Mod Pathol 25(8):1098–1105. doi:10.1038/modpathol.2012.59

90. Kalkman S, Barentsz MW, van Diest PJ (2014) The effects of under 6 hours of formalin fixation on hormone receptor and HER2 expression in invasive breast cancer: a systematic review. Am J Clin Pathol 142(1):16–22. doi:10.1309/AJCP96YDQSTYBXWU

91. Arber DA (2002) Effect of prolonged formalin fixation on the immunohistochemical reactivity of breast markers. Appl Immunohistochem Mol Morphol 10(2):183–186

92. Oyama T, Ishikawa Y, Hayashi M, Arihiro K, Horiguchi J (2007) The effects of fixation, processing and evaluation criteria on immunohistochemical detection of hormone receptors in breast cancer. Breast Cancer 14(2):182–188

93. Tong LC, Nelson N, Tsourigiannis J, Mulligan AM (2011) The effect of prolonged fixation on the immunohistochemical evaluation of estrogen receptor, progesterone receptor, and HER2 expression in invasive breast cancer: a prospective study. Am J Surg Pathol 35(4):545–552. doi:10.1097/PAS.0b013e31820e6237

Chapter 3

Immunohistochemistry for Triple-Negative Breast Cancer

Kalnisha Naidoo and Sarah E. Pinder

Abstract

Triple-negative breast cancers are a heterogeneous group of tumors that are, as yet, not entirely understood. Although triple-negative carcinomas are strictly defined as invasive carcinomas lacking expression of estrogen receptor, progesterone receptor, and HER2, some use the terms triple-negative and basal-like cancer synonymously. It should be noted that these are not entirely equivalent. Nevertheless, it has been shown that a panel of immunohistochemical markers can be used as a surrogate for genomic profiling and thus to identify basal-like breast cancers. We describe the panels of immunohistochemical markers that can be applied and how to interpret these markers herein.

Key words Triple-negative breast cancer, ER, HER2, Basal-like breast carcinoma, Immunohistochemistry

1 Introduction

The onset of the molecular age has transformed the way in which we classify, treat, and predict outcomes for breast cancer. At a genetic level, at least five subtypes exist: luminal A; luminal B; human epidermal growth factor 2 (HER-2) enriched; normal breast-like; and basal-like [1, 2]. Reconciling the basal-like group of cancers to traditional morphological subtypes, however, has proven difficult. Indeed, triple-negative breast cancers (i.e., cancers lacking estrogen receptor (ER), progesterone receptor (PR), and HER-2 expression) overlap with the basal-like subtype, but are not entirely the same entity, with approximately 75 % equivalence [3].

Distinguishing triple-negative carcinomas from basal-like lesions may be of clinical value; it has been shown that epidermal growth factor receptor (EGFR) and/or basal cytokeratin (CK) 5/6 expression within this triple-negative group allows for the identification of a "Core Basal" subtype [4, 5]. These "Core Basal" cancers are reported to not respond as well to adjuvant anthracycline-based chemotherapy in some series and thus, have a significantly poorer prognosis [5–7]. The identification of such cancers,

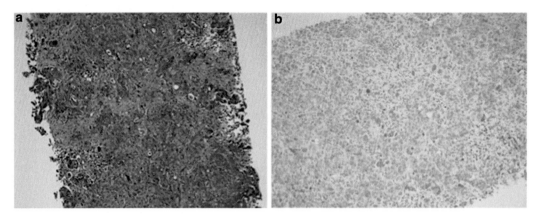

Fig. 1 Core of ER-negative grade 3 invasive breast carcinoma. A hematoxylin and eosin-stained section (H&E) is shown in (**a**). ER staining of the same tumor is shown in (**b**); no nuclear expression is seen

therefore, has potential clinical implications although this is not universally accepted, and, in our opinion, more evidence and research is required.

It is also clear that both triple-negative and basal-like breast cancers are not homogeneous entities and that these are not truly individual tumor types but groups of lesions. Although many are grade 3 malignant cancers with high mitotic frequency (Fig. 1), irregular areas of necrosis (Fig. 2) or fibrosis and a pushing margin, perhaps even with a lymphoid infiltrate, it must be remembered that even these are a heterogeneous spectrum [5]. The microscopic appearance of triple-negative carcinoma varies enormously and includes the salivary gland-like lesions, such as adenoid cystic carcinomas that have a good prognosis (Fig. 3a–c), and spindled cell/metaplastic carcinomas (both low (Fig. 3d, e) and high grade, with good and poor prognosis respectively); all these lesions typically also express basal markers immunohistochemically. Thus the prognostic relevance of subtyping according to genomic or immunohistochemical assays cannot be accepted in isolation of the histology of the lesion. It should also be remembered that none of the additional biomarkers available as yet have proven predictive value for a specific targeted therapy, above and beyond ER and HER2 in this group of lesions.

Immunohistochemistry (IHC) is an invaluable, relatively inexpensive, widely used antibody-dependent technique that evaluates protein expression and localization in tissue sections. Although the technique can be used on fresh, frozen, and/or formalin-fixed tissue, an in-depth analysis of tissue fixation and its effects on antigen retrieval is beyond the scope of this chapter. The method described herein therefore applies to formalin-fixed paraffin-embedded tissue, which is the diagnostic standard in laboratories worldwide.

IHC can be divided into two stages: staining (i.e., the technical process of applying an antibody to a tissue section so that a protein

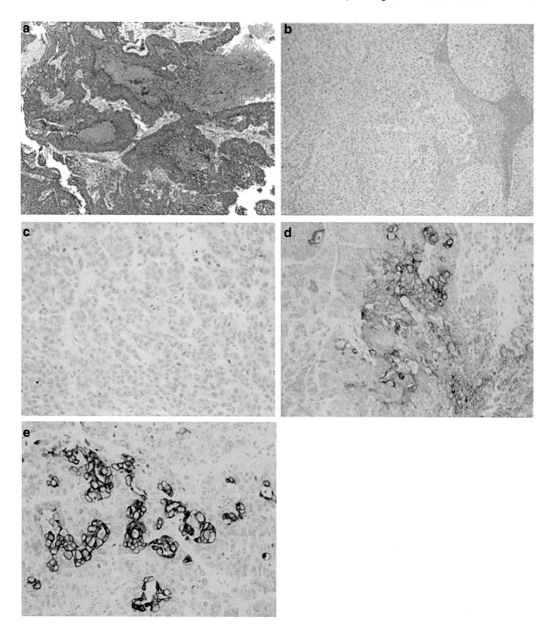

Fig. 2 "Core Basal" grade 3 invasive breast cancer with geographic irregular areas of necrosis. A hematoxylin and eosin-stained section (H&E) is shown in (**a**). ER negativity in the nodal metastasis seen in (**b**) as an absence of nuclear reactivity. HER2 negativity in the same primary breast cancer is seen in (**c**) as an absence of any membrane reactivity. Cytokeratin 5 and cytokeratin 14 cytoplasmic positivity, respectively, in the same tumor is shown in (**d, e**)

can be visualized easily) and interpretation (i.e., semi-quantitative analysis of protein expression and localization). With regards to the former, protocols need to be optimized in-house, as although the principles of staining (discussed below) are generic, the differences in tissue processing and reagents between laboratories will affect

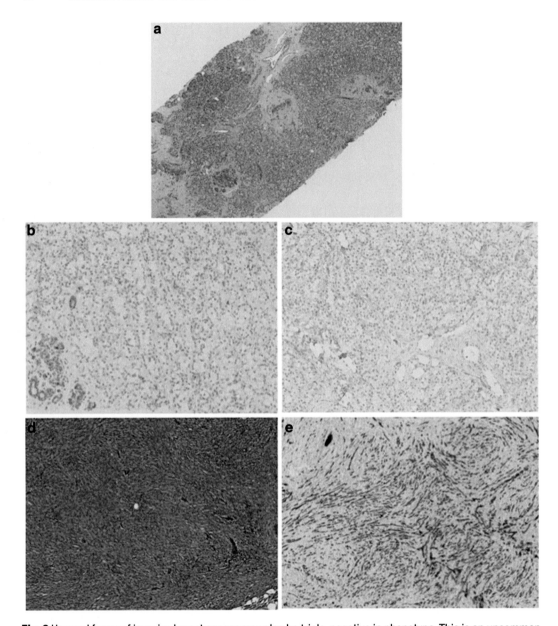

Fig. 3 Unusual forms of invasive breast cancer may also be triple-negative in phenotype. This is an uncommon, acinic cell (variant of salivary gland-like), carcinoma (**a**; H&E), which is ER negative (**b**) and HER2 negative (**c**). Normal breast glands act as an internal control and show ER reactivity, as seen in (**b**). Other forms of triple-negative carcinoma include metaplastic (spindle cell) carcinomas (**d**; H&E), which retain cytokeratin expression (**e**; PanK) confirming the epithelial nature of the lesion

the quality of staining achieved. Thus, we have focused on a conceptual explanation of technique rather than a proscriptive one.

Of note, it is particularly important that all possible quality assurance (internal and external) protocols are applied in the assessment of receptor status of invasive breast cancers as these are intimately linked with the selection of the most appropriate treatments for patients [8, 9]. Any errors, either technical or interpretive, may result in receipt of treatments that are unlikely to be effective, or conversely in omission of therapies that may be highly beneficial [8, 9]. Standardization of both technique and interpretation, and regular, ongoing participation in external quality assurance is essential. Appropriate controls (negative and positive) are also mandatory.

2   Materials

By far, the most important determinant of staining quality is the choice of antibody.

1. Fully automated staining system with relevant reagents.
2. ER antibody (SP1 clone) at 1:100 dilution.
3. PR antibody (1A6 clone) at 1:300 dilution.
4. HER-2 ready-to-use kit (Leica Microsystems, Milton Keynes, UK).
5. EGFR antibody (384 clone) at 1:100 dilution.
6. CK5/6 antibody (D5/16 B4 clone) at 1:100 dilution.
7. Slides.
8. Coverslips.

3   Methods

3.1 Staining

Standardization of immunohistochemical staining in the recommended automated staining platforms obviates a detailed description of technique. Thus, a brief summary of the rationale behind the machination is described below.

3.1.1 Removal of Paraffin and Tissue Rehydration

Following fixation, tissue is dehydrated to permit paraffin penetration for storage. This process needs to be reversed gradually prior to staining so that the antibody can permeate the tissue entirely.

3.1.2 Reducing Background Staining and Nonspecific Antibody Binding

This step in essence serves to reduce "noise" so that only the protein of interest is ultimately lit up. "Noise" can result from endogenous peroxidases, nonspecific (non-immunological) binding between the tissue and the antibody due to hydrophobic and electrostatic forces, and endogenous biotin.

3.1.3 Antigen Retrieval

Formalin fixation crosslinks proteins, and masks epitopes. For antibodies to be able to bind, these crosslinks must be broken. This can be done using heat, enzymes, or both. Rarely, if the epitope has managed to avoid the effects of fixation, this step can be omitted entirely.

3.1.4 Antibody Binding

As mentioned, the most important component of this entire process is the quality of the antibody used.

Potential antibodies for the evaluation of triple-negative and basal-like breast cancers include the SP1 clone for ER, 1A6 for PR, 384 for EGFR, and D5/16 B4 for CK5/6. The UK National External Quality Assurance Scheme for Immunocytochemistry (NEQAS ICC) and in situ hybridization provides information on the mostly widely used primary antibodies and their successful application. The Ventana Pathway 790-2991 (4B5) and the Dako A0485 C-erB-2 (poly) are the most commonly used at present (http://www.ukneqasicc.ucl.ac.uk/journal_run_107.pdf), whilst Ventana 790-4324 (SP1) is the most commonly used in ER assays (http://www.ukneqasicc.ucl.ac.uk/run_106.pdf) and Ventana 790-2223 (1E2) (A&B) for PR assessment (http://www.ukneqasicc.ucl.ac.uk/journal_run_107.pdf).

Many of these antibodies are monoclonal. This relates to the importance of antibody specificity. A great deal of structural and functional homology exists between the steroid receptors (ER and PR) [10]; the ErbB family of tyrosine kinase receptors (HER2 and EGFR) [11, 12]; and the various cytokeratins that form the intermediate filaments of the cytoskeleton [13]. Expression of both the steroid and tyrosine kinase receptors in breast cancer, through various mechanisms, results in autonomous cell growth and proliferation [14]. Targeted therapies towards ER (tamoxifen and aromatase inhibitors) and HER2 (such as trastuzumab), by blocking this aberrant process, have significantly improved survival in a subset of patients [15–17].

Cytokeratin expression is used as a marker of cellular differentiation. In the breast, CK 5/6 is normally present in myoepithelial and basal epithelial cells, whilst CK7 and CK8 are present in luminal cells [18]. Thus, cytokeratin expression can help define tumor subtype.

3.1.5 Antibody Detection

Once the primary antibody has been allowed sufficient time to bind, the next step is to amplify the signal produced for visualization. For this protocol, an indirect method is used in which a labeled secondary antibody is added to the tissue. An enzyme reporter is then added with a chromogenic substrate.

3.1.6 Counterstain, Dehydrate, and Mount

Sections are then counterstained to highlight the nuclei, and dehydrated. Finally, they are mounted and a coverslip is added.

3.2 Interpretation and Analysis

The stained slides are analyzed to localize and quantitate protein expression.

The accurate assessment of estrogen receptor alpha is vital in predicting the response of patients with invasive breast cancer to hormone therapies (tamoxifen and/or aromatase inhibitors) and immunohistochemistry is the "gold standard" for determining ER status (an ER-positive tumor is shown in Fig. 4). It is thus crucial

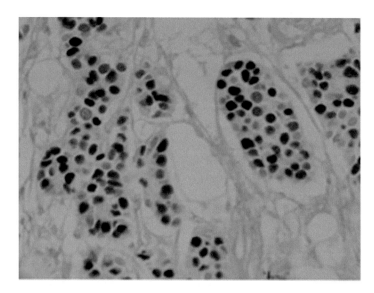

Fig. 4 Strongly ER-positive tumor, seen as brown nuclear reactivity in invasive breast carcinoma

that correctly validated antibodies are used in a standardized way and that appropriate control tissues are included with each run [8].

A separate matter is the interpretation of ER and PR immunostaining and defining positive/negative breast cancers. Evaluation most commonly includes an interpretation of the frequency and the intensity of reactivity of the tumor nuclei in a system such as the Allred score (Table 1). An Allred score of 0 or 2 is widely regarded as defining a breast cancer as ER negative [19]. However, there is a relative paucity of data on response to hormone therapies of patients with invasive breast cancer according to degree of ER reactivity. The most recent recommendations (ASCO CAP) are that a cut-off of 1 % or more of tumor cells is used to define a lesion as ER positive [8] and this is now also the recommendation in the UK. Nevertheless, as elsewhere, in our own Unit the Multidisciplinary Team still desires a more semi-quantified score be included in histopathology reports in order to make a recommendation for individual patient management; those with very weakly ER-positive tumors (Fig. 5) would not, for example, be considered ideal candidates for primary endocrine therapy to downsize a tumor prior to surgery. Such cases are relatively rare, as the distribution of estrogen receptor score in breast cancer is bimodal; most invasive carcinomas are either completely negative (0 % cells) or extensively (>66 % of cells) positive, the latter with typically moderate or strong intensity of reactivity. However, accurately and consistently defining a cut-off of clinical relevance is difficult as the most appropriate threshold may vary according to clinical context, i.e., adjuvant vs. neoadjuvant/pre-surgical/primary therapy vs. metastatic settings.

Table 1

Allred scoring system for ER and PR immunohistochemistry

Allred score	Score for proportion	Score for intensity	Result
0	No staining	No staining	Negative
1	<1 % Nuclear staining	Weak	Positive if total score is >2
2	1–10 % Nuclear staining	Moderate	Positive
3	11–33 % Nuclear staining	Strong	Positive
4	34–66 % Nuclear staining		Positive
5	67–100 % Nuclear staining		Positive

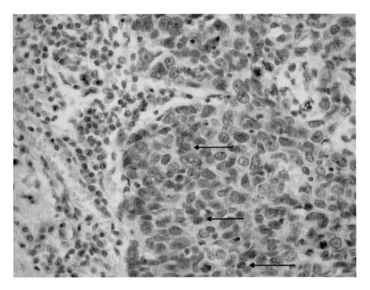

Fig. 5 Percentage ER positivity directly informs therapeutic management. Although ER positivity is defined as 1 % or more of cells within a tumor, cases with very scanty weakly ER reactive tumor nuclei (*arrows*) are not generally considered ideal candidates for preoperative/neoadjuvant/primary hormone therapy

HER2 immunohistochemistry has been routinely used as a predictive marker in breast cancer for more than 10 years. Whilst patients with HER2-positive tumors generally have a poor overall prognosis, those with both early and metastatic disease have a markedly improved survival when anti-HER2 systemic therapy, such as trastuzumab, is given [11, 12]. Guidelines for testing for HER2 [9, 20] provide advice on methodology, quality assurance, and interpretation of HER2 assays (Table 2 and Fig. 6). However, as HER2 assessment, as well as ER, forms an integral part of the pathology minimum dataset and is a requirement for optimum

Table 2

Scoring system for HER2 immunohistochemistry

IHC score	Criteria	Result
0	No staining is observed, or <10 % of tumor cells	Negative
1+	A faint/barely perceptible membrane staining is detected in >10 % of tumor cells; the cells are only stained in part of the membrane	Negative
2+	A weak to moderate complete membrane staining is observed in >10 % of tumor cells	Borderline
3+	A strong complete membrane staining is observed in >10 % of tumor cells	Positive

A similar system may be used for EGFR scoring

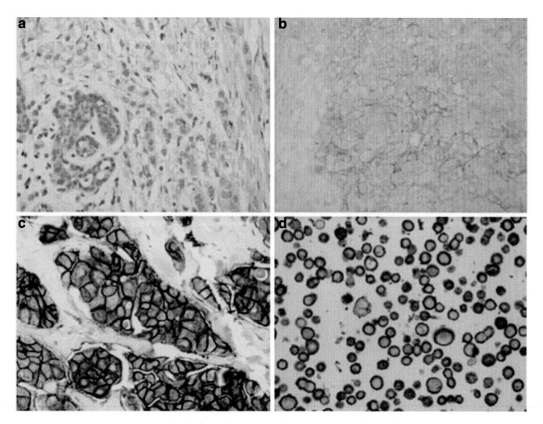

Fig. 6 Range of possible staining with HER2. An invasive breast cancer and adjacent normal glands showing no reactivity with HER2 is seen in (**a**). An invasive breast cancer showing weak to moderate, but complete, membrane reactivity regarded as borderline, 2+ is shown in (**b**). FISH assay demonstrated no amplification of the *Her2* gene in this case. An invasive breast cancer showing 3+, strong, complete membrane reactivity regarded as HER2 positive is seen in (**c**), while (**d**) represents a cell line control for HER2 showing strong complete membrane reactivity

patient management it is imperative that standardized protocols and methodologies are followed.

Whilst there are established methods described for interpretation of ER, PR, and HER2 biomarkers in invasive cancer, there is no global agreement on the assessment of other "basal" markers such as CK5, CK5/6, CK14, or EGFR. Numerous other markers are also described as more frequently expressed in basal-like breast cancers, e.g. nestin, caveolin, NGFR, and others [21]. However, there is a lack of uniformity in cut-offs used for defining positivity for these assays in the literature, as well as in finding the right combination of markers to accurately recapitulate the gene expression profile of this group of tumors.

Subsequent to the recognition that a surrogate IHC profile could be used to identify basal-like breast cancers with a more widely available technique than gene expression profiling by Nielsen et al [4], the same group described that tumors can be separated into a "Core Basal" phenotype, with reactivity for EGFR and/or CK 5/6, and a "Five Marker Negative Phenotype" (5NP) negative for all five markers (ER, PR, HER2, CK5/6, and EGFR). These authors define reactivity for CK5/6 and EGFR as those cases showing any reactivity but others have used a 10 % cut off for the basal cytokeratins with lesions showing less than 10 % of cells defined as negative [22]. It remains unclear, however, whether there is significance in the proportion of a tumor that shows reactivity for these markers with regard to classification or prognosis [23]. It is clear that invasive breast cancers frequently show heterogeneous expression of, for example, basal cytokeratin markers, and further investigation is required regarding the underlying biology and clinical significance of this (Fig. 7).

4 Notes

1. We have shown that morphological examination alone, including the identification of features such as histological grade 3 and geographic necrosis, on hematoxylin and eosin-stained sections is insufficient for the accurate recognition of entities as basal breast cancer by gene profiling techniques [5]. Similarly, there is not complete overlap between basal-like lesions with genomic and IHC profiling, and present methods of surrogate profiling using IHC must therefore be regarded as imperfect. Indeed other literature suggests that rather than the five intrinsic subgroups widely described, there are at least ten groupings of breast cancers when both genome and transcriptome aberrations are integrated [24]. However, most centers are not equipped to run such an analysis on every patient's cancer, whereas tumor immunoprofiling can be performed routinely. Nevertheless, the value of the assessment of panels of immuno-

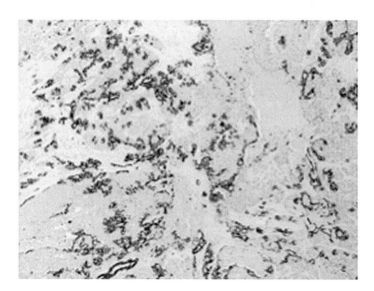

Fig. 7 Heterogeneous expression of the basal cytokeratins is frequently seen. Although some series have defined expression as any positivity, others have applied a cut-off of 10 % or more of tumor cells to regard as case as expressing these markers. Low power view of the cytokeratin 14 immunohistochemically stained slides from the same case as seen in Fig. 2, showing patchy nature of reactivity

markers for any tumor is ultimately is governed by and in particular of the clinical relevance, for example whether the biomarker serves as a "target" for a therapeutic agent. This will also be influenced by local and national guidelines, resource availability, and individual preference.

2. In the context of this complexity and uncertainty, the diagnosis of triple-negative breast cancer is, more straightforward and is conversely, defined by IHC alone; as a minimum, ER and HER2 receptor status should be evaluated on all primary invasive breast carcinomas, and on all metastatic and recurrent breast cancers that have tissue available for assessment. For ER, we recommend that all cancers with 1 % or more ER positivity be reported as ER positive as per ASCO CAP and UK guidelines but, in addition, a semi-quantified score (such as the Allred score) should be included. These data may prove useful in elucidating how therapy correlates to staining intensity in the future. HER2 status is reported as shown in Table 2. Borderline cases should be sent for in situ hybridization studies. At present this is undertaken in our Unit by dual probe fluorescence and both the ratio of *Her2* to chromosome 17 centromeric probe number and average copy number is reported. A ratio of <2.0 is regarded as non-amplified and *Her2* negative, as is copy number less than 6. There is conflicting data as to whether progesterone receptor status adds to ER assessment or not; therefore

routine analysis is optional at present in the UK minimum dataset for invasive breast cancer pathology.

3. With regards to EGFR and CK5/6, our laboratory has shown that additional staining for these two markers better identifies those patients with the worst outcome (i.e., those patients with a 5NP phenotype) [5]. However, in the absence of any targeted therapy for patients with triple-negative tumors, the presence of expression of EGFR or CK5/6 has relatively little impact on clinical recommendations regarding choice of chemotherapy in the routine setting at present. There is also a paucity of data regarding selection of antibody clone, standardization or technical methodology and a lack of consensus on how to interpret these markers. It remains to be seen whether these, or other novel markers, truly provide additional clinically valuable information in the routine setting in triple-negative breast cancer and which additional biomarkers or indeed molecular techniques will be used for classifying these heterogeneous tumors in the future.

References

1. Sorlie T, Perou CM, Tibshirani R, Aas T, Geisler S, Johnsen H, Hastie T, Eisen MB, van de Rijn M, Jeffrey SS, Thorsen T, Quist H, Matese JC, Brown PO, Botstein D, Lonning PE, Borresen-Dale AL (2001) Gene expression patterns of breast carcinomas distinguish tumor subclasses with clinical implications. Proc Natl Acad Sci U S A 98(19):10869–10874. doi:10.1073/pnas.191367098

2. Sorlie T, Tibshirani R, Parker J, Hastie T, Marron JS, Nobel A, Deng S, Johnsen H, Pesich R, Geisler S, Demeter J, Perou CM, Lonning PE, Brown PO, Borresen-Dale AL, Botstein D (2003) Repeated observation of breast tumor subtypes in independent gene expression data sets. Proc Natl Acad Sci U S A 100(14):8418–8423. doi:10.1073/pnas.0932692100

3. Badve S, Dabbs DJ, Schnitt SJ, Baehner FL, Decker T, Eusebi V, Fox SB, Ichihara S, Jacquemier J, Lakhani SR, Palacios J, Rakha EA, Richardson AL, Schmitt FC, Tan PH, Tse GM, Weigelt B, Ellis IO, Reis-Filho JS (2011) Basal-like and triple-negative breast cancers: a critical review with an emphasis on the implications for pathologists and oncologists. Mod Pathol 24(2):157–167. doi:10.1038/modpathol.2010.200

4. Nielsen TO, Hsu FD, Jensen K, Cheang M, Karaca G, Hu Z, Hernandez-Boussard T, Livasy C, Cowan D, Dressler L, Akslen LA, Ragaz J, Gown AM, Gilks CB, van de Rijn M, Perou CM (2004) Immunohistochemical and clinical characterization of the basal-like subtype of invasive breast carcinoma. Clin Cancer Res 10(16):5367–5374. doi:10.1158/1078-0432.CCR-04-0220

5. Gazinska P, Grigoriadis A, Brown JP, Millis RR, Mera A, Gillett CE, Holmberg LH, Tutt AN, Pinder SE (2013) Comparison of basal-like triple-negative breast cancer defined by morphology, immunohistochemistry and transcriptional profiles. Mod Pathol 26(7):955–966. doi:10.1038/modpathol.2012.244

6. Cheang MC, Voduc D, Bajdik C, Leung S, McKinney S, Chia SK, Perou CM, Nielsen TO (2008) Basal-like breast cancer defined by five biomarkers has superior prognostic value than triple-negative phenotype. Clin Cancer Res 14(5):1368–1376. doi:10.1158/1078-0432.CCR-07-1658

7. Dent R, Trudeau M, Pritchard KI, Hanna WM, Kahn HK, Sawka CA, Lickley LA, Rawlinson E, Sun P, Narod SA (2007) Triple-negative breast cancer: clinical features and patterns of recurrence. Clin Cancer Res 13(15 Pt 1):4429–4434. doi:10.1158/1078-0432.CCR-06-3045

8. Hammond ME, Hayes DF, Dowsett M, Allred DC, Hagerty KL, Badve S, Fitzgibbons PL, Francis G, Goldstein NS, Hayes M, Hicks DG, Lester S, Love R, Mangu PB, McShane L,

Miller K, Osborne CK, Paik S, Perlmutter J, Rhodes A, Sasano H, Schwartz JN, Sweep FC, Taube S, Torlakovic EE, Valenstein P, Viale G, Visscher D, Wheeler T, Williams RB, Wittliff JL, Wolff AC, American Society of Clinical Oncology; College of American Pathologists (2010) American Society of Clinical Oncology/ College of American Pathologists guideline recommendations for immunohistochemical testing of estrogen and progesterone receptors in breast cancer (unabridged version). Arch Pathol Lab Med 134(7):e48–e72. doi:10.1043/1543-2165-134.7.e48

9. Wolff AC, Hammond ME, Schwartz JN, Hagerty KL, Allred DC, Cote RJ, Dowsett M, Fitzgibbons PL, Hanna WM, Langer A, McShane LM, Paik S, Pegram MD, Perez EA, Press MF, Rhodes A, Sturgeon C, Taube SE, Tubbs R, Vance GH, van de Vijver M, Wheeler TM, Hayes DF, American Society of Clinical Oncology; College of American Pathologists (2007) American Society of Clinical Oncology/ College of American Pathologists guideline recommendations for human epidermal growth factor receptor 2 testing in breast cancer. J Clin Oncol 25(1):118–145. doi:10.1200/JCO.2006. 09.2775

10. Mangelsdorf DJ, Thummel C, Beato M, Herrlich P, Schutz G, Umesono K, Blumberg B, Kastner P, Mark M, Chambon P, Evans RM (1995) The nuclear receptor superfamily: the second decade. Cell 83(6):835–839

11. Arteaga CL, Sliwkowski MX, Osborne CK, Perez EA, Puglisi F, Gianni L (2012) Treatment of HER2-positive breast cancer: current status and future perspectives. Nat Rev Clin Oncol 9(1):16–32. doi:10.1038/nrclinonc.2011.177

12. Baselga J, Swain SM (2009) Novel anticancer targets: revisiting ERBB2 and discovering ERBB3. Nat Rev Cancer 9(7):463–475. doi:10.1038/nrc2656

13. Chung BM, Rotty JD, Coulombe PA (2013) Networking galore: intermediate filaments and cell migration. Curr Opin Cell Biol 25(5):600–612. doi:10.1016/j.ceb.2013.06.008

14. Hanahan D, Weinberg RA (2000) The hallmarks of cancer. Cell 100(1):57–70

15. Davies C, Godwin J, Gray R, Clarke M, Cutter D, Darby S, McGale P, Pan HC, Taylor C, Wang YC, Dowsett M, Ingle J, Peto R, Early Breast Cancer Trialists' Collaborative Group (2011) Relevance of breast cancer hormone receptors and other factors to the efficacy of adjuvant tamoxifen: patient-level meta-analysis of randomised trials. Lancet 378(9793):771–784. doi:10.1016/S0140-6736(11)60993-8

16. Rabaglio M, Aebi S, Castiglione-Gertsch M (2007) Controversies of adjuvant endocrine treatment for breast cancer and recommendations of the 2007 St Gallen conference. Lancet Oncol 8(10):940–949. doi:10.1016/ S1470-2045(07)70317-0

17. Smith IE, Dowsett M (2003) Aromatase inhibitors in breast cancer. N Engl J Med 348(24):2431–2442. doi:10.1056/NEJMra0 23246

18. Abd El-Rehim DM, Pinder SE, Paish CE, Bell J, Blamey RW, Robertson JF, Nicholson RI, Ellis IO (2004) Expression of luminal and basal cytokeratins in human breast carcinoma. J Pathol 203(2):661–671. doi:10.1002/ path.1559

19. Harvey JM, Clark GM, Osborne CK, Allred DC (1999) Estrogen receptor status by immunohistochemistry is superior to the ligand-binding assay for predicting response to adjuvant endocrine therapy in breast cancer. J Clin Oncol 17(5):1474–1481

20. Walker RA, Bartlett JM, Dowsett M, Ellis IO, Hanby AM, Jasani B, Miller K, Pinder SE (2008) HER2 testing in the UK: further update to recommendations. J Clin Pathol 61(7):818–824. doi:10.1136/jcp.2007. 054866

21. Won JR, Gao D, Chow C, Cheng J, Lau SY, Ellis MJ, Perou CM, Bernard PS, Nielsen TO (2013) A survey of immunohistochemical biomarkers for basal-like breast cancer against a gene expression profile gold standard. Mod Pathol 26(11):1438–1450. doi:10.1038/ modpathol.2013.97

22. Rakha EA, El-Sayed ME, Green AR, Paish EC, Lee AH, Ellis IO (2007) Breast carcinoma with basal differentiation: a proposal for pathology definition based on basal cytokeratin expression. Histopathology 50(4):434–438. doi:10.1111/j.1365-2559.2007.02638.x

23. Fulford LG, Reis-Filho JS, Ryder K, Jones C, Gillett CE, Hanby A, Easton D, Lakhani SR (2007) Basal-like grade III invasive ductal carcinoma of the breast: patterns of metastasis and long-term survival. Breast Cancer Res 9(1):R4. doi:10.1186/bcr1636

24. Curtis C, Shah SP, Chin SF, Turashvili G, Rueda OM, Dunning MJ, Speed D, Lynch AG, Samarajiwa S, Yuan Y, Graf S, Ha G, Haffari G, Bashashati A, Russell R, McKinney S, Group M, Langerod A, Green A, Provenzano E, Wishart G, Pinder S, Watson P, Markowetz F, Murphy L, Ellis I, Purushotham A, Borresen-Dale AL, Brenton JD, Tavare S, Caldas C, Aparicio S (2012) The genomic and transcriptomic architecture of 2,000 breast tumours reveals novel subgroups. Nature 486(7403):346–352. doi:10.1038/ nature10983

Chapter 4

In Situ Hybridization of Breast Cancer Markers

Li Min and Chengchao Shou

Abstract

In situ hybridization is an important technique in breast cancer research, which is widely applied in detection of specific nucleic acid sequences. Here, we describe the detailed protocol of fluorescence in situ hybridization and chromogenic in situ hybridization in detection of gene HER2/*neu* amplification in breast cancer tissues.

Key words In situ hybridization, Breast cancer, HER2, FISH, CISH

1 Introduction

Recently, in situ hybridization is widely applied in detection of specific nucleic acid sequences [1–3].

The mechanism of in situ hybridization is based on complementarity of single-strand nucleic acid [1]. The procedure of in situ hybridization can be summarized as three main steps: preparing radioactive or nonradioactive targeted exogenous nucleic acid (i.e. probe), hybridizing pretreated sample DNA or RNA on the tissue, cell, or chromosome with probes, detecting and mapping the specific hybrid nucleic acid [4, 5].

In breast cancer research, many types of hybridization methods have been used, such as Dot blot, Southern blot, Northern blot, Tissue in situ hybridization, Genome in situ hybridization [2, 6]. Based on different probes and visualization reagents, in situ hybridization can be classified into fluorescence in situ hybridization (FISH), chromogenic in situ hybridization (CISH), radiolabeled in situ hybridization, dual-color silver-enhanced in situ hybridization and et al [7–10]. FISH and CISH are the methods being most widely used in breast cancer research.

Human epidermal growth factor receptor 2 (HER2) is one of the most three important breast cancer markers [11, 12]. HER2 gene amplification occurs in more than 20 % breast cancer cases and was identified as an important prognostic biomarker [13]. HER2 gene amplification can be identified by many different methods, and FISH has become the "Gold Standard" of HER2 gene amplification detection in clinical practice [14, 15]. Recently, CISH was gradually widely studied and used as a viable alternative to FISH in HER2 gene amplification detection [16–18]. In this chapter, we summarized regular protocol of FISH and CISH in detection of HER2, as representative protocols of in situ hybridization of breast cancer markers.

2   Materials

Use ultrapure water (purified deionized water, sensitivity: 18 MΩ·cm at 25 °C) to prepare all buffers and other solutions. Store all solutions at room temperature (about 25 °C) unless otherwise indicated. All reagents used in the experiment should be analytical grade pure unless otherwise indicated.

2.1   Equipment and Reagent

1. 100-W Epifluorescence microscope with appropriate filters (*see* **Note 1**).

2. Water baths set.

3. Microwave oven with appropriate size (for heat pretreatment).

4. Thermal plate set.

5. Silanized microscope slides.

6. "Probe check" quality control slides (Abbott Inc., UK), (*see* **Note 2**).

7. Reveal target unmasking solution (Biocare Inc., Walnut Creek, CA).

8. Staining dishes with 100 % xylene.

9. Staining dishes with 100 %, 90 %, 80 %, 70 % ethanol (*see* **Note 3**).

10. Phosphate buffered saline (PBS), pH 7.4: 0.01 M phosphate buffer, 0.0027 M KCl, 0.137 M NaCl. Dissolve 8 g NaCl, 0.2 g KCl, 1.42 g $Na_2HPO_4 \cdot 2H_2O$, 0.27 g KH_2PO_4 in 800 mL ultrapure water, adjust pH to 7.4 with 10 M NaOH, make up to 1 L with ultrapure water and then autoclave.

11. 2× SSC solution, pH 7.0: 3 M NaCl, 0.3 M sodium citrate. Dissolve 175.3 g NaCl and 88.2 g sodium citrate ($C_6H_5Na_3O_7 \cdot 2H_2O$) in 800 mL ultrapure water, adjust pH to 7.0 with 10 M NaOH, make up to 1 L with ultrapure water and then autoclave, dilute 1:10 with ultrapure water for 2× SSC.

12. Vectashield DAPI solution (Vectorlabs, UK), with 200 ng/mL of 4,6-diamidino-2 phenylindole-2 hydrochloride (Sigma, UK) added.

13. Omnislide hybridization platform (Thermo-Hybaid) with dark plastic lid.

14. 8 % (w/v) Sodium thiocyanate (*see* **Note 4**).

15. Pepsin solution (Digest-All III; Zymed Inc.), (*see* **Note 5**).

2.2 Fluorescence In Situ Hybridization Components

1. HER2/chromosome 17 probe mixture (Pathvysion™ kit).

2. Denaturing solution, pH 7.0: Mix 49 mL Ultrapure formamide (Fluka, UK), 7 mL 2× SSC, with 14 mL distilled water, and then adjust pH to 7.0 with 10 M NaOH.

3. 0.5× SSC: Autoclave 2× SSC solution, dilute 1:4 with ultrapure water for 0.5× SSC.

4. Temporary "coverslips": Cut Parafilm to make temporary coverslips. Temporary "coverslips" are made by cutting regular parafilm into appropriate size, and then add the denaturation solution.

5. 22×22 mm slip.

6. Rubber cement.

2.3 Chromogenic In Situ Hybridization Components

1. A ready-to-use probe mixture for HER2 (SpotLight probe series; Zymed Inc., CA).

2. CISH-UnderCover Slips (Zymed Inc.).

3. 0.5× SSC: Autoclave2× SSC solution, dilute 1:4 with ultrapure water for 0.5× SSC.

4. Hematoxylin solution: There are various methods of hematoxylin dye preparation. The method of Heidenhain [19] is preferred for CISH.

5. Mouse anti-digoxigenin antibody (Roche Biochemicals).

6. Powervision+detection kit (ImmunoVision Inc., CA).

3 Methods

All procedures should be performed at room temperature unless expressly indicated.

3.1 Basic Sample Handling

1. Cut tissue to 5-μm sections on silanized slides and then bake at 60 °C overnight. Store sections at room temperature (about 25 °C) until required (*see* **Note 6**).

2. Immerse slides in xylene for 10 min (*see* **Note 7**), twice.

3. Transfer slides into 100 % ethanol for 5 min, twice.

4. Rehydrate slides in graded ethanol (90 %, 80 %, 70 % ethanol, each for 1 min).

5. Place slides in 0.2 N HCl for 20 min (*see* **Note 8**).

6. Wash slides in distilled water for 5 min.

7. Wash slides in PBS wash buffer for 5 min, twice (*see* **Note 9**).

8. Place slides in 8 % sodium thiocyanate (Vysis, UK or Sigma, UK) in distilled water at 80 °C for 30 min (*see* **Note 4**).

9. Wash slides in distilled water for 1 min.

10. Wash slides in PBS wash buffer for 5 min, twice (*see* **Note 9**).

11. Place in protease buffer at 37 °C for 20 min (*see* **Note 10**).

12. Immerse slides in 2× SSC buffer for 5 min, twice.

13. Dehydrate slides in graded ethanol (70 %, 80 %, 90 %, 100 % ethanol, each for 1 min), (*see* **Note 3**).

14. Air-dry slides in an oven at 50 °C.

15. Apply 10 μL DAPI in mountant and cover the slides by temporary coverslips.

16. Assess digestion of tissue with a filter specific for DAPI (*see* **Note 11**).

17. Place all slides in 2× SSC buffer until the temporary coverslips fall off.

18. Air-dry slides in an oven at 50 °C.

3.2 Fluorescence In Situ Hybridization

All procedures in this section should be performed after completed Basic Sample Handling procedures.

1. Apply 100 μL denaturing solution to each slide in a fume hood. Cover with temporary coverslips.

2. Denature slides for 5 min at 94 °C.

3. Remove temporary coverslips in a fume hood.

4. Dehydrate slides in graded ethanol (70 %, 80 %, 90 %, 100 % ethanol, each for 1 min), (*see* **Note 3**).

5. Air-dry slides in an oven at 50 °C.

6. Apply 10 μL of HER2/chromosome 17 probe mixture to a 22 × 22 mm slip.

7. Invert slides and gently lower them onto 22 × 22 mm slips with probe mixture.

8. Seal slides with rubber cement.

9. Hybridize slides overnight at 37 °C, shielded from light (*see* **Note 12**).

10. Remove rubber cement, place slides in 0.5× SSC buffer at room temperature to make the cover slip fall off.

11. Place slides into 0.5× SSC buffer at 72 °C for 2 min.

12. Air-dry slides in an oven shielded from light at 50 °C.

13. Mount slide in mountant with 0.2 ng/mL DAPI and seal with resin glue (*see* **Note 13**).

14. Carefully remove excess glue with a soft paper, and ready to evaluate the slide by microscopy.

3.3 Chromogenic In Situ Hybridization

All procedures in this section should be performed after completed Basic Sample Handling procedures.

1. Apply 10 μL probe solution (Zymed Spot-Light series) and seal the slides under UnderCover Slips (*see* **Note 14**).

2. Denature probe and target DNA, 94 °C for 5 min (*see* **Note 15**).

3. Hybridize slides overnight at 37 °C, shielded from light (*see* **Note 12**).

4. Remove UnderCover Slips.

5. Wash slides with 0.5× SSC buffer for 5 min at 75 °C.

6. Wash slides with PBS buffer, 5 min.

7. Incubate slides for 45 min at room temperature, using mouse antidigoxigenin antibody (diluted 1: 500 in the Powervision + blocking solution).

8. Wash slides with PBS buffer, 5 min.

9. Incubate slides for 20 min at room temperature, using Powervision + post antibody blocking solution (*see* **Note 16**).

10. Wash slides with PBS buffer, 5 min.

11. Incubate slides for 30 min at room temperature, using Powervision + poly-HRP-goat-x-mouse polymer (*see* **Note 16**).

12. Wash slides with PBS buffer, 5 min.

13. Incubate slides for 5 min at room temperature, using the Powervision + DAB solution (1000 μL distilled water + 50 μL DAB reagent A + 50 μL DAB reagent B, prepared ready for use).

14. Wash slides with distilled water, 1–2 min.

15. Stain slides with standard hematoxylin, 1–5 s.

16. Wash slides with distilled water, 5–10 s.

17. Dehydrate slides in graded ethanol (70 %, 80 %, 90 %, 100 % ethanol, each for 1 min), (*see* **Note 3**).

18. Immerse slides in xylene for 10 min, twice.

19. Mount slides in 0.2 ng/mL DAPI.

20. Cover with a cover slip and seal with resin glue.

21. Carefully remove excess glue with a soft paper, and ready to evaluate the slide by microscopy.

4 Notes

1. Filters specific for DAPI and different spectrums are required and a 100× objective application was preferred.

2. Regularly, both internal and external controls are required for FISH. For large-scale sample scanning, both control sections should be included within each diagnostic run.

3. It is best to prepare this series of graded ethanol fresh each time before use. When used for 10 batches of slides, it is also best to replace them with fresh ones.

4. Sodium thiocyanate is used to break the protein–protein disulfide bonds, which could facilitate the subsequent process of digestion.

5. Fresh pepsin should be added to digestion buffers before starting digestion, and additional pepsin should also be added each 30 min during the whole process of digestion.

6. Even though nearly no attenuation of FISH signals was found when stored for prolonged periods, use of slides within 4 months of cutting is strongly preferred.

7. A fume hood should be used when handling xylene. Some nonorganic solutions (Hemo-de solution for example) may be used as an alternative.

8. As a pretreatment permeabilization process, the use of 0.2 N HCl to acid deproteination allows preservation of better tissue morphology.

9. After washing slides in wash buffer, gently touch the slide edge by absorbent paper to remove remaining fluid.

10. This step is very crucial to achieve clear images, because exposure of slides to protease buffer could ensure adequate pretreatment of tissues prior to application of probes.

11. If digestion of tissue is optimal, proceed to next steps. Otherwise, proceed to **steps 17** and **18** and then repeat the protease treating steps for 1–15 min digestion depending on the detailed condition (*see* **step 11–16**).

12. To ensure the hybridization efficiency, duration of this step should be more than 12 h.

13. Transparent resin glue could be used to prevent slides from drying out.

14. Conventional repeat-containing DNA probes also work well in CISH, as an alternative to Zymed probes.

15. Denaturation of probe and the target DNA is carried out shortly after applying the probes on slides.

16. Probe detection is crucial in all CISH steps. To increase sensitivity, a two-layer antibody approach was applied.

Acknowledgement

This work was supported by National Nature Science Foundation of China (No. 81230046) and 973 Program of National Basic Research Program of China (No. 2015CB553906).

References

1. John MSB (2004) Fluorescence in situ hybridization: technical overview. Molecular Diagnosis of Cancer Methods and Protocols, Second edition 1, 71–77

2. Wolfe KQ, Herrington CS (1997) Interphase cytogenetics and pathology: a tool for diagnosis and research. J Pathol 181:359–361

3. Bell SM, Zuo J, Myers RM et al (1996) Fluorescence in situ hybridization deletion mapping at 4p16.3 in bladder cancer cell lines refines the localisation of the critical interval to 30 kb. Genes Chromosomes Cancer 17:108–117

4. John MSB, Amanda F (2004) HER2 FISH in breast cancer. Molecular Diagnosis of Cancer Methods and Protocols, second edition 1, 89–103

5. Dale P, Jaana LD, Aparna K et al (2003) Molecular markers in ductal carcinoma in situ of the breast. Mol Cancer Res 1:362–375

6. Cole KD, He HJ, Wang L (2013) Breast cancer biomarker measurements and standards. Proteomics Clin Appl 7:17–29

7. Bartlett JMS, Going JJ, Mallon EA et al (2001) Evaluating HER2 amplification and overexpression in breast cancer. J Pathol 195:422–428

8. Young WK, Hee JL, Jong WL et al (2011) Dual-color silver-enhanced in situ hybridization for assessing HER2 gene amplification in breast cancer. Mod Pathol 24:794–800

9. Nitta H, Hauss-Wegrzyniak B, Lehrkamp M et al (2008) Development of automated brightfield double in situ hybridization (BDISH) application for HER2 gene and chromosome 17 centromere (Cen 17) for breast carcinomas and an assay performance comparison to manual dual color HeR2 fluorescence in situ hybridization (FISH). Diagn Pathol 3:41

10. Ulla H, Sven M, Andreas S (2011) Dual color CISH and FISH to CISH conversion. IHC Staining Methods, Fifth edition 14, 97–101

11. Howlader N, Chen VW, Ries LA et al (2014) Overview of breast cancer collaborative stage data items-their definitions, quality, usage, and clinical implications: a review of SEER data for 2004–2010. Cancer 120(Suppl 23):3771–3780

12. Poulsen TS, Espersen ML, Kofoed V et al (2013) Comparison of fluorescence in situ hybridization and chromogenic in situ hybridization for low and high throughput HER2 genetic testing. Int J Breast Cancer 2013:368731

13. Iqbal N, Iqbal N (2014) Human epidermal growth factor receptor 2 (HER2) in cancers: overexpression and therapeutic implications. Mol Biol Int 2014:852748

14. Mitchell MS, Press MF (1999) The role of immunohistochemistry and fluorescence in situ hybridization for HER-2/neu in assessing the prognosis of breast cancer. Semin Oncol 26:108–116

15. Hanna WM, Rüschoff J, Bilous M et al (2014) HER2 in situ hybridization in breast cancer: clinical implications of polysomy 17 and genetic heterogeneity. Mod Pathol 27:4–18

16. Jorma I, Minna T (2004) Chromogenic in situ hybridization in tumor pathology. Molecular Diagnosis of Cancer Methods and Protocols, second edition 1, 133–144

17. Tanner M, Gancberg D, Di Leo A et al (2000) Chromogenic in situ hybridization (CISH): a practical new alternative for FISH in detection of HER-2/neu oncogene amplification. Am J Pathol 157:1467–1472

18. Jacquemier J, Spyratos F, Esterni B et al (2014) SISH/CISH or qPCR as alternative techniques to FISH for determination of HER2 amplification status on breast tumors core needle biopsies: a multicenter experience based on 840 cases. BMC Cancer 13:351

19. Shen EY (1969) Two-solution iron hematoxylin staining (Heidenhain) with HCl differentiation and tris-buffer neutralization. Stain Technol 44:209–210

Chapter 5

Evaluation of Human Epidermal Growth Factor Receptor 2 (HER2) Gene Status in Human Breast Cancer Formalin-Fixed Paraffin-Embedded (FFPE) Tissue Specimens by Fluorescence In Situ Hybridization (FISH)

Harry C. Hwang and Allen M. Gown

Abstract

Current standard of care requires that HER2 gene testing be performed on all newly diagnosed invasive breast cancers in order to determine eligibility for anti-HER2 antibody therapy and should be performed in accordance with current ASCO-CAP guidelines (Hammond et al., J Clin Oncol 29(15):e458, 2011; Wolff et al., J Clin Oncol 31(31):3997–4013, 2013). Here we describe a HER2 FISH methodology to evaluate HER2 gene status in FFPE breast tumor specimens.

Key words Breast cancer, Human epidermal growth factor receptor 2, HER2, Fluorescence in situ hybridization, FISH, Herceptin, PathVysion, DNA probe

1 Introduction

The human epidermal growth factor receptor 2 gene (HER2) is a receptor tyrosine kinase oncogene that is amplified or overexpressed in 15–20 % of human breast cancer [1, 2]. HER2 gene status predicts response to immunotherapies directed at the HER2 (ERBB2) protein, is an independent prognostic marker of adverse clinical outcome in breast cancer, and can predict response to doxorubicin (Adriamycin)-based adjuvant chemotherapy, as well as resistance to tamoxifen, even in the setting of ER/PR expression [3–5]. In addition to trastuzumab (Herceptin™), other HER2-targeted therapies that are in use clinically include pertuzumab (Perjeta™), lapatinib (Tykerb™), and T-DMI (ado-trastuzumab emtansine, Kadcyla™) [6–8].

By current ASCO-CAP guidelines, HER2 testing must be performed on all newly diagnosed invasive breast cancer and is typically evaluated by a reflex testing algorithm of immunohistochemistry followed by FISH, when necessary. HER2 immunostaining is

Jian Cao (ed.), *Breast Cancer: Methods and Protocols*, Methods in Molecular Biology, vol. 1406,
DOI 10.1007/978-1-4939-3444-7_5, © Springer Science+Business Media New York 2016

scored on a 0 to 3+ scale. Cases that exhibit 3+ immunostaining for HER2 are deemed positive for HER2 overexpression, and cases that are 0 or 1+ are deemed as negative for HER2 overexpression. HER2 FISH is employed on cases that show equivocal protein overexpression (2+) [1]. For HER2 FISH, a fluorescently labeled DNA probe to the HER2 gene is hybridized to tumor nuclei, and HER2 copy number is enumerated using fluorescence microscopy along with a reference gene probe.

HER2 FISH testing is typically performed using the FDA-approved Abbott Molecular® PathVysion™ HER2 DNA Probe Kit (IVD) [9], although other HER2 probes are commercially available. The PathVysion™ HER2 (ERBB2) DNA Probe Kit is a dual-color probe kit that contains one probe (SpectrumOrange™ labeled) specific for the HER2 (ERBB2) gene and a second probe (SpectrumGreen™ labeled) specific for chromosome 17 (CEP17 or chromosome enumeration probe 17). The underlying alteration is an amplification of the HER2 gene, resulting in more than the usual two copies of HER2 (ERBB2) (one on each copy of chromosome 17) typically present in normal breast epithelium and the majority of breast cancers without HER2 (ERBB2) alterations. Gene amplification is defined as a ratio of HER2 (ERBB2) to chromosome 17 signals of ≥2.0 or absolute HER2 copy number ≥6.0 [1]. Amplification of the HER2 (ERBB2) gene is almost always associated with overexpression of the HER2 protein, as assessed by immunohistochemistry (IHC) [1, 10, 11].

HER2 FISH testing is best performed using a 2-day workflow on FFPE tumor tissue cut on 4 μm section charged glass slides. On Day 1 of the workflow, specimen pretreatment is performed which entails deparaffinization and enzymatic digestion. The purpose of the enzymatic digestion step is to break down cross linkages that may impede accessibility of DNA-binding sites to the HER2 FISH probe. After pretreatment, HER2 probe is applied, followed by denaturation and hybridization of slides overnight allowing for a ~16 h hybridization time. On Day 2, slides are washed in high-stringency buffer, DAPI-stained to visualize nuclear morphology, and coverslipped. The HER2 FISH slides are then scored using fluorescence microscopy under high power. Scoring of HER2 FISH slides is commonly done under oil objective lens fluorescence microscopy using appropriate filters to visualize the labeled probe signals [9]; however, in our laboratory, we use a semiautomated fluorescence image analysis platform to capture high-power digital images, which are then scored by a technologist and pathologist on a computer screen. Cases are resulted as POSITIVE, NEGATIVE, or EQUIVOCAL for HER2 gene amplification in accordance with current ASCO-CAP guidelines as detailed below [1].

Day-to-day quality control is the main challenge of consistently performing an accurate HER2 FISH clinical assay. A tissue digestion control, which typically is a previously tested breast cancer, should be used to allow for the in-assay DAPI evaluation of enzymatic digestion effectiveness. If deemed adequate, then probing of slides with the HER2 probe can proceed. In addition, all runs should include a positive control, and the HER2 status of this control is recorded and tracked for changes that may occur in the analytic sensitivity of the assay over time. Cell lines such as the MDA-MB-453 cell line can be used as a positive HER2 control; however, other HER2-positive cell lines can also be used depending on availability [12].

2   Materials

2.1   Reagents and Cell Line: Note That Solutions Should Be Prepared in Ultrapure Water ddH₂O

1. Alcohol (100 % ethanol).
2. Fluoroguard Antifade mounting medium (*see* **Note 5**).
3. DAPI (0.3 μg/mL) in Fluoroguard Antifade (*see* **Note 6**).
4. Formalin, 10 % neutral buffered (10 % NBF).
5. 0.2 N HCl (*see* **Note 7**).
6. Pepsin (from porcine gastric mucosa, powder, ≥250 units/mg) 0.8 % in 0.2 N HCl solution.
7. 2× SSC (*see* **Note 8**).
8. Sodium thiocyanate (NaSCN, 8.1 %) (*see* **Note 9**).
9. Post-hybridization wash buffer (2× SSC/0.3 % NP-40) (*see* **Note 10**).
10. Xylene.
11. PathVysion HER2 DNA IVD Probe Kit (Abbott Molecular, Des Plaines, IL).

2.2   Equipment and Supplies

1. VP2000 Processor (Abbott Molecular).
2. Water bath.
3. Pipettes and pipette tips.
4. Charged glass slides and glass coverslips.
5. Glass Coplin jars.
6. Plastic slide trays.
7. Plastic "Tupperware-type" bins.
8. Ovens (recommend use of two-jacketed laboratory grade ovens).
9. Fluorescence microscope with appropriate filters to detect labeled probes.

3　Methods

3.1　Day 1: Pretreatment and Probe Application

HER2 FISH is performed on 4 µm FFPE tissue sections cut on charged glass slides that have been previously baked at 65 °C for 30 min.

Charged glass slides as well as slide baking promote adherence of tissue sections to the slide through the tissue digestion and probing steps. The baking step is also useful in melting away excess paraffin from cut sections, and for this reason, slides should be baked upright so melted paraffin drips off the slide. For pretreatment, tissue section slides are deparaffinized, rehydrated, then incubated in HCl and NaSCN, and then enzymatically digested with pepsin. The purpose of the HCl and NaSCN incubation steps is to break disulfide bonds within the formalin-fixed tissue and dissociate histone proteins from the genomic DNA. This allows for more effective enzymatic digestion of peptide bonds and increases the accessibility of DNA to the labeled probe. This protocol uses pepsin for the enzymatic digestion step. Pepsin, a carboxyl protease, is readily available and inexpensive and provides a gentler enzymatic pretreatment in comparison to proteinase K. Optimal enzymatic pretreatment greatly promotes FISH signal evaluation while at the same time reduces background autofluorescence.

Pretreatment can be done using a VP2000 or a similar automated tissue processor, or manually using Coplin jars in water baths. The advantage of automated tissue processors is that they standardize and automate slide pretreatment, thereby reducing the chance for errors.

3.1.1　Pretreatment Protocol (see **Note 1**)

Pretreatment steps for slides are at room temperature unless otherwise noted. The pretreatment protocol is as follows:

1. Deparaffinize in xylene for 5 min; repeat three times.
2. Rehydrate in 100 % ETOH for 5 min; repeat three times.
3. Air-dry for 5 min.
4. Acid treat in 0.2 N HCl for 20 min.
5. Rinse in water bath for 5 min.
6. Wash in 2× SSC for 3 min.
7. Pretreat in NaSCN for 35 min at 80 °C.
8. Rinse in water bath for 5 min.
9. Wash in 2× SSC for 5 min; repeat two times.
10. Digest in 0.8 % Pepsin for 10 min at 37 °C.
11. Rinse in water bath for 5 min.
12. Wash in 2× SSC for 5 min; repeat two times.
13. Postfix in 10 % NBF for 4 min.

14. Rinse in water bath for 5 min.

15. Wash in 2× SSC for 5 min; repeat two times.

16. Rinse in water bath for 2 min.

17. Air-dry for 15 min.

3.1.2 Quality Control Tissue Check of Pretreatment Steps

The purpose of the tissue check Qc is to provide an in-assay assessment of the previous tissue pretreatment steps.

1. Apply DAPI (0.3 µg/mL) in Fluoroguard Antifade to each slide, coverslip, and incubate at room temperature for 10 min.

2. Review slides under the fluorescent microscope with the DAPI filter. Look for solid strongly blue nuclei; the rest of the tissue should look dark. If the tissue is inadequately digested, it will look whitish and opaque, and nuclei will be difficult to see. If it is over-digested, the nuclei will look dark and hollow, and the rest of the tissue will also look diminished or be absent.

3. If the tissue check QC is acceptable, proceed with general probing.

3.1.3 Preparation and Application of HER2 Probe to FISH Slides

1. Warm PathVysion HER2 probe, which is in hybridization buffer, to room temperature so that the viscosity decreases sufficiently to allow accurate pipetting. Vortex to mix contents and then spin down in a benchtop microcentrifuge for ~5 s to bring contents to the bottom of the tube.

2. Determine the amount of probe solution needed to cover the entire target tumor area; for example, 10 µL is sufficient for the area under a 22 × 22 mm coverslip.

3. Pipette probe solution onto tissue and gently coverslip (without creating bubbles).

3.1.4 Denature Tissues and Probe

1. Place slides in slide trays and set denaturation oven to 79 °C.

2. Place slide trays in 79 °C oven, working as quickly as possible to minimize the amount of time the oven door is open. Place thermometer probe on top of slide trays and then close the door firmly.

3. The actual temperature in the oven will drop significantly (~15–30 °C) at this point, and it will take anywhere from 10 to 25 min for it to come back up to 79 °C, depending on the number of trays and type of oven. When the temperature reaches 79 °C again, incubate in oven for 16 min.

3.1.5 Hybridize Tissues in Hybridization Oven

1. Set temperature of hybridization oven to 37 °C.

2. Place slide trays in plastic "Tupperware-type" bins with moistened liner to create a humid environment. If needed, add water to moisten liner.

3. Place plastic bins with slides into 37 °C oven.

4. Hybridize overnight (~16 h).

3.2 Day 2: Washing and Coverslipping

3.2.1 Preparation of Slides for Stringent Wash

1. Remove slides from hybridization oven.

2. Soak off coverslips in 2× SSC. Allow approximately 10–15 min. If the coverslip does not come off easily, place back into 2× SSC for another 10–15 min.

3. Retain de-coverslipped slides in 2× SSC until ready for next step.

3.2.2 Stringent Wash

1. Wash slides in Post-Hybridization Wash Buffer (2× SSC/0.3 % NP-40) for 3 min at 79 °C.

2. Remove slides from the water bath and place in 2× SSC.

3. Wash slides in 2× SSC for 3 min. Each boat of 2× SSC may be used twice before disposal.

4. Rinse slides in deionized water, using one to two quick dips.

5. Stand slides on end on the counter, propped up against racks, water bath, etc., to air-dry.

3.2.3 Counterstain/Coverslip

1. Bring DAPI in Fluoroguard and Antifade Fluoroguard to room temperature. Vortex and centrifuge. Mix in equal parts. Use mixture to coverslip slides.

2. Clean off the back of the slides as needed.

3. Apply enough of the DAPI/Fluoroguard mixture to cover the probed area, and coverslip gently, avoiding bubbles and moving the coverslip.

3.3 Day 2: Analysis of HER2 FISH Slides

3.3.1 Pre-scan Slide Review

1. Review and mark H&E or HER2-immunostained slide for areas of invasive carcinoma to be samples.

2. Score the HER2-immunostained slide (0, 1+, 2+, or 3+) as per ASCO-CAP guidelines and record results.

3.3.2 Fluorescence Microscopy (see Notes 2-4)

1. Visualize tumor nuclei using a fluorescence microscope at high power in the areas of invasive carcinoma.

2. Engage fluorescence filters to visualize orange-labeled HER2 signal, green-labeled CEP17 signal, and DAPI-labeled blue nuclei. Score nonoverlapping tumor nuclei for the absolute number of HER2 and CEP17 signals per nuclei and record data for each case. Count 60 nuclei and only count nuclei having at least one orange HER2 signal and one green CEP17 signal. Calculate HER2:CEP17 ratio.

3.3.3 Scoring and Resulting of Cases

Cases are scored in accordance with ASCO-CAP guidelines as follows [1]:

1. NEGATIVE: HER2:CEP17 ratio <2.0 AND absolute HER2 count <4.0 signals/nuclei.

2. POSITIVE: HER2:CEP17 ratio ≥2.0 OR absolute HER2 count ≥6.0 signals/nuclei.

3. EQUIVOCAL: HER2:CEP17 ratio <2.0 AND absolute HER2 count between 4.0 and 6.0.

4 Notes

1. Pretreatment Optimization: Empirical pretreatment optimization for the FFPE tissue sections in one's laboratory is critical to the success of this protocol. Because pepsin is a crude laboratory reagent that may vary in enzymatic activity depending on source, we recommend testing a variety of pepsin concentrations (e.g., 0.2, 0.8, 1.2 %, etc.) and pepsin incubation times (e.g., 5, 10, 15, and 20 min) that bracket the conditions detailed in this protocol. For each tested condition, review the resulting FISH slide and evaluate tissue sections for extent of digestion and FISH signal strength. After performing a series of optimization experiments, the best combination of pepsin concentration and pepsin incubation time can be chosen. Particular care should be taken to accurately prepare dilute acid solutions that are used for tissue section incubation as over acidification can lead to damaged DNA and poor FISH signal strength. For this reason we recommend preparing such dilute acid solutions from an intermediate-strength acid stock of no greater than 1.0 N.

2. Microscopic review and scoring: With the PathVysion HER2 IVD probe kit, successful execution of this protocol will yield crisp and distinct FISH signals in uniformly stained DAPI nuclei. The green FISH signals are typically brighter and have a "chunkier" appearance in comparison to the orange HER2 probe, and both should be easily enumerated. Care should be taken to count only invasive tumor nuclei and not nuclei associated with in situ disease.

3. Troubleshooting.

    (a) Weak or no FISH signal: The situation where weak or no FISH signal is seen is not an uncommon occurrence, and one must determine in a systematic way whether the lack of FISH signal is due to pre-analytic versus analytic factors. Causes of pre-analytic-associated failure of a FFPE FISH assays include inadequate formalin fixation, suboptimal tissue processing, or over-decalcification in low-pH decalcification solutions. Common causes of analytic or post-analytic failures for FFPE FISH include errors in performing tissue pretreatment steps (i.e., wrong solutions or solutions used in the wrong order), errors in pHing acid solutions, and problems with the fluorescence filters.

(b) High background: High background artifactual signal in either the orange or green filter channels can impede microscopic evaluation and enumeration of true FISH signals. Most commonly high background can be caused by intrinsic tissue factors such as under-digestion leading to excess intact cytoplasmic and nuclear protein or exogenous factors such as specimen ink. In the former case, optimization of pretreatment incubation digestion times should be performed empirically. Typically under-digestion can be addressed by empirically testing two or three increasing incubation times for the pepsin pretreatment step.

(c) Tissue falling off the slide: This is a common situation and impedes microscopic evaluation. For well-fixed and processed tissue, the majority of the tissue sample will remain adherent to the slide after pretreatment and probing. If excess tissue falls off during pretreatment, the slides can be baked overnight in a humidified chamber at 65 °C to promote better adherence. If this does not correct the problem, then one should examine the fixation and processing parameters of the sample, as under-fixation or improper tissue processing may be the source of the problem.

4. EQUIVOCAL case handling.

By current 2013 ASCO-CAP guidelines, if a case is found to be EQUIVOCAL, then the testing laboratory should do further evaluation in an attempt to obtain a non-equivocal result [1]. The first step in this process is to perform an additional FISH signal count, which can be used over the first count if a NEGATIVE or POSITIVE result is found. If other tumor-containing tissue is available for the case, HER2 testing can be performed on such tissue as a repeat test. Finally, alternative chromosome 17 enumeration probes such as SMS, RARA, and P53 can be performed on the case, and the FISH signal counts for these loci can be used instead of CEP17 to calculate the HER2:CEP17 ratio [1, 13, 14]. If the HER2:Chromosome 17 ratio (by alternative probe) is ≥ 2.0, then the case is deemed POSITIVE, and if the ratio is <2.0, the case is scored as NEGATIVE. Such use of alternative chromosome 17 assessment is recommended in current ASCO-CAP guidelines and has been shown in published studies to be useful in converting EQUIVOCAL cases to either POSITIVE or NEGATIVE [13, 14].

5. Fluoroguard Antifade is a liquid-based mounting medium that is applied to slides when coverslipping. This mounting medium contains proprietary chemicals to slow the quenching of the probe dyes. We use Insitus Fluoroguard Antifade

(Insitus Cat. No. F001), but Fluoroguard Antifade mounting medium is available from other lab chemical and reagent manufacturers.

6. DAPI (4′,6-diamidino-2-phenylindole dihydrochloride) is a fluorescent DNA-binding dye. Make a 300 µg/mL stock solution for DAPI by adding 3 mg of DAPI to 10 mL dH$_2$O. Then add 1 µL of this DAPI stock per 1 mL Fluoroguard Antifade to make the DAPI in Fluoroguard Antifade solution. Alternatively premade DAPI in Fluoroguard Antifade reagent solution can also be purchased from commercial manufacturers such as Insitus.

7. To make 0.2 N HCl, first prepare 1 N HCl from concentrated HCl (37 % or 12 N HCl) by adding 83.5 mL of concentrated HCl to 916.5 mL of deionized dH$_2$O and mix well. Then make a 1:5 dilution of 1 N HCl by adding 200 mL of 1 N HCl to 800 mL dH$_2$O to make 0.2 N HCl. Mix well.

8. To make 2× SSC, first prepare 20× SSC by adding 175.3 g NaCl and 88.2 g Sodium Citrate (Na$_3$C$_6$H$_5$O$_7$) to 800 mL dH$_2$O. After dissolving all the NaCl and sodium citrate, bring volume to 1 L. Make 2× SSC by adding 100 mL of 20× SSC to 900 mL of dH$_2$O. Mix well and pH to between 7.5 and 8.0.

9. To make Sodium Thiocyanate (NaSCN, 8.1 %) solution, add 81 g of NaSCN to 800 mL of dH$_2$O. After NaSCN dissolves, bring volume up to 1 L. Store in amber glass bottle.

10. To make 2× SSC/0.3 % NP-40, first obtain 100 % NP-40 (nonionic detergent P-40) from a commercial chemical manufacturer (if 100 % is not available, then 90 % can be used). NP-40 (Nonidet P40) is a highly viscous nonionic detergent that can cause skin and eye irritation. Because of its high viscosity and difficulty pipetting small volumes of NP-40, a large volume (4 L) of Post-Hybridization Wash Buffer should be prepared as follows. Add 3388 mL of dH$_2$O to a 4 L container. Add 400 mL of 20× SSC to the container and then add 12 mL of NP-40. Mix well and pH to between 7.0 and 7.5.

Acknowledgments

The authors would like to thank Christopher Tse, Stephanie Rodriguez, and April Carr for their excellent technical assistance in optimizing this FISH protocol.

References

1. Wolff AC, Hammond ME, Hicks DG, Dowsett M, McShane LM, Allison KH, Allred DC, Bartlett JM, Bilous M, Fitzgibbons P, Hanna W, Jenkins RB, Mangu PB, Paik S, Perez EA, Press MF, Spears PA, Vance GH, Viale G, Hayes DF, American Society of Clinical Oncology; College of American Pathologists (2013) Recommendations for human epidermal growth factor receptor 2 testing in breast cancer: American Society of Clinical Oncology/College of American Pathologists clinical practice guideline update. J Clin Oncol 31(31):3997–4013. doi:10.1200/JCO.2013.50.9984

2. Slamon DJ, Clark GM, Wong SG, Levin WJ, Ullrich A, McGuire WL (1987) Human breast cancer: correlation of relapse and survival with amplification of the HER-2/neu oncogene. Science 235(4785):177–182

3. Hammond ME, Hayes DF, Wolff AC (2011) Clinical Notice for American Society of Clinical Oncology-College of American Pathologists guideline recommendations on ER/PgR and HER2 testing in breast cancer. J Clin Oncol 29(15):e458. doi:10.1200/JCO.2011.35.2245

4. Gusterson BA, Gelber RD, Goldhirsch A, Price KN, Save-Soderborgh J, Anbazhagan R, Styles J, Rudenstam CM, Golouh R, Reed R et al (1992) Prognostic importance of c-erbB-2 expression in breast cancer. International (Ludwig) Breast Cancer Study Group. J Clin Oncol 10(7):1049–1056

5. Rayson D, Richel D, Chia S, Jackisch C, van der Vegt S, Suter T (2008) Anthracycline-trastuzumab regimens for HER2/neu-overexpressing breast cancer: current experience and future strategies. Ann Oncol 19(9):1530–1539. doi:10.1093/annonc/mdn292

6. Geyer CE, Forster J, Lindquist D, Chan S, Romieu CG, Pienkowski T, Jagiello-Gruszfeld A, Crown J, Chan A, Kaufman B, Skarlos D, Campone M, Davidson N, Berger M, Oliva C, Rubin SD, Stein S, Cameron D (2006) Lapatinib plus capecitabine for HER2-positive advanced breast cancer. N Engl J Med 355(26):2733–2743. doi:10.1056/NEJMoa064320

7. Baselga J, Cortes J, Kim SB, Im SA, Hegg R, Im YH, Roman L, Pedrini JL, Pienkowski T, Knott A, Clark E, Benyunes MC, Ross G, Swain SM, CLEOPATRA Study Group (2012) Pertuzumab plus trastuzumab plus docetaxel for metastatic breast cancer. N Engl J Med 366(2):109–119. doi:10.1056/NEJMoa1113216

8. Verma S, Miles D, Gianni L, Krop IE, Welslau M, Baselga J, Pegram M, Oh DY, Dieras V, Guardino E, Fang L, Lu MW, Olsen S, Blackwell K, EMILIA Study Group (2012) Trastuzumab emtansine for HER2-positive advanced breast cancer. N Engl J Med 367(19):1783–1791. doi:10.1056/NEJMoa1209124

9. PathVysion HER-2 DNA IVD Probe Kit Package Insert and Quality Assurance Certificate

10. Jacobs TW, Gown AM, Yaziji H, Barnes MJ, Schnitt SJ (1999) Comparison of fluorescence in situ hybridization and immunohistochemistry for the evaluation of HER-2/neu in breast cancer. J Clin Oncol 17(7):1974–1982

11. Yaziji H, Goldstein LC, Barry TS, Werling R, Hwang H, Ellis GK, Gralow JR, Livingston RB, Gown AM (2004) HER-2 testing in breast cancer using parallel tissue-based methods. JAMA 291(16):1972–1977. doi:10.1001/jama.291.16.1972

12. Holliday DL, Speirs V (2011) Choosing the right cell line for breast cancer research. Breast Cancer Res 13(4):215. doi:10.1186/bcr2889

13. Troxell ML, Bangs CD, Lawce HJ, Galperin IB, Baiyee D, West RB, Olson SB, Cherry AM (2006) Evaluation of Her-2/neu status in carcinomas with amplified chromosome 17 centromere locus. Am J Clin Pathol 126(5):709–716. doi:10.1309/9EYM-6VE5-8F2Y-CD9F

14. Tse CH, Hwang HC, Goldstein LC, Kandalaft PL, Wiley JC, Kussick SJ, Gown AM (2011) Determining true HER2 gene status in breast cancers with polysomy by using alternative chromosome 17 reference genes: implications for anti-HER2 targeted therapy. J Clin Oncol 29(31):4168–4174. doi:10.1200/JCO.2011.36.0107

Part II

Genetic Detection for Breast Cancer

Chapter 6

Quantification of mRNA Levels Using Real-Time Polymerase Chain Reaction (PCR)

Yiyi Li, Kai Wang, Longhua Chen, Xiaoxia Zhu, and Jie Zhou

Abstract

Real-time quantitative reverse transcription PCR technique has advanced greatly over the past 20 years. Messenger RNA (mRNA) levels in cells or tissues can be quantified by this approach. It is well known that changes in mRNA expression in disease, and correlation of mRNA expression profiles with clinical parameters, serve as clinically relevant biomarkers. Hence, accurate determination of the mRNA levels is critically important in describing the biological, pathological, and clinical roles of genes in health and disease. This chapter describes a real-time PCR approach to detect and quantify mRNA expression levels, which can be used for both laboratorial and clinical studies in breast cancer research.

Key words mRNA, RNA isolation, Reverse transcription, Real-time PCR

1 Introduction

During physiological and pathological processes in human, such as carcinogenesis, expression levels of mRNA and proteins are highly regulated to orchestrate biological outcomes. Generally, upon stimulation, gene expression may be regulated transcriptionally, resulting in alteration of mRNA levels and followed by the increase or decrease in levels of its coding protein; eventually, this procedure regulates intracellular and/or extracellular biology. Expression levels of mRNA and protein can serve as molecular markers providing accurate prognosis of any disease during corresponding therapy, for predicting response, resistance, and toxicity to therapy in patients with breast cancer [1].

To determine alterations in mRNA levels, traditional methods, including northern blotting, in situ hybridization, ribonuclease protection, or cDNA arrays can be used [2]. However, the requirement of large amounts of RNA and the low sensitivity of detection are the main limitations of these methods [2]. Reverse transcription polymerase chain reaction (RT-PCR) is the most sensitive and

Jian Cao (ed.), *Breast Cancer: Methods and Protocols*, Methods in Molecular Biology, vol. 1406,
DOI 10.1007/978-1-4939-3444-7_6, © Springer Science+Business Media New York 2016

accurate method, but it needs intensive and laborious post-PCR manipulations. Facing to these difficulties, real-time PCR or quantitative PCR came into being. It allows measuring PCR product accumulation during the exponential phase of the reaction, thus quantifying PCR product in a real-time manner [2]. Using small amounts of RNA to determine mRNA expression levels, real-time PCR assays can be completed rapidly since no post-amplification manipulations are required. This technology is very flexible; many alternative instruments and fluorescent probe systems have been developed. In this chapter, we will focus on two-step real-time PCR methodology using SYBR Green I in Bio-Rad iCycler iQ system and describe a step-by-step protocol about RNA amplification and quantification.

2   Materials

2.1   Reagents and Equipment for RNA Isolation

1. RNA isolation kits. There are commercially available kits that are used for isolating total RNA, mostly the procedure starts with organic extraction followed by isolation using silica columns. Kits that we have extensively tested with satisfactory results are given below:

   (a) RNeasy Mini Kit (Qiagen) for cultured cells.

   (b) RNeasy FFPE Kit (Qiagen) for formalin-fixed, paraffin-embedded tissues.

2. NanoDrop 1000 Spectrophotometer (Thermo) for determining the RNA quality and concentration.

2.2   Reagents and Equipment for Reverse Transcription and Real-Time PCR

The scope of this chapter is restricted to the poly(A)-tailed reverse transcription and SYBR Green I real-time PCR. The reagents and equipment used for these assays are as follows:

1. iScript cDNA Synthesis Kit (Bio-Rad), PCR tubes, microcentrifuge, and PCR thermal cycler for reverse transcription.

2. iQ SYBR Green Supermix (Bio-Rad), 96-well PCR plates, 96-well PCR plate centrifuge, and PCR thermal cycler (Bio-Rad iCycler iQ) for real-time PCR.

3. Primers of targeted gene (*see* **Note 1**).

3   Methods

3.1   Isolation of Total RNA Including mRNA

1. Methods of homogenization are varied from sample types. For mammalian tissues, it is necessary to disrupt with a mortar and pestle before adding lysis reagent (Trizol, Qiazol, etc.) and continue with the homogenization by using a syringe and needle (0.9 mm) or homogenizer; for mammalian serum and

plasma, since these samples typically do not contain larger RNA species that are commonly used for normalization of mRNA expression, we suggest to either add a synthetic mRNA into sample after the addition of lysis reagent or use a stably expressed mRNA which is chosen based on literature or preexisting data for normalization; for formalin-fixed, paraffin-embedded tissue sections, which require a different isolation approach with other sample types, we recommend to use laser capture microdissection or manual macrodissection techniques to isolate the regions of interest from tissue sections following with RNA isolation by using special commercially available kit. For cell samples, remove the cell-culture media, disrupt the cells by adding appropriate volume of lysis reagent (Trizol, Qiazol, etc.) which is a monophasic solution of phenol and guanidine thiocyanate to facilitate lysis, and inhibit RNases.

2. Transfer the lysate into a microcentrifuge tube and homogenize by vortexing for 1 min.

3. Place the lysate at room temperature for 5 min to promote dissociation of nucleoprotein complexes.

4. Supplement the lysate with 0.2 volume of chloroform per 1 volume of lysis reagent; shake the tube vigorously for 15 s.

5. Place the tube at room temperature for 3 min; centrifuge for 15 min at $12,000 \times g$ at 4 °C (*see* **Note 2**).

6. Transfer the upper aqueous phase to a new microcentrifuge tube. Add 1.5 volume of 100 % ethanol (or isopropanol) per 1 volume of aqueous phase, and mix thoroughly by pipetting up and down several times to precipitate RNA (*see* **Note 3**).

7. Pipet the appropriate volume of the sample into a silica-membrane spin column in a collection tube, centrifuge at $\geq 8000 \times g$ for 1 min at room temperature, and discard the flow-through (*see* **Note 4**).

8. Add series of washing buffers containing ethanol to the spin column and centrifuge for 1 min at $\geq 8000 \times g$ to wash the spin column membrane, respectively. Discard the flow-through after each time centrifugation.

9. Place the spin column into a new collection tube; centrifuge at maximum speed for 1 min to eliminate any possible carryover of ethanol from washing buffers or residual flow-through remaining on the outside of spin column.

10. Transfer the spin column to a new microcentrifuge tube. Pipet the appropriate volume of RNase-free water directly onto the spin column membrane. Incubate at room temperature for 1 min and centrifuge for 1 min at $\geq 8000 \times g$ to elute the RNA (*see* **Note 5**).

11. Determine the RNA quality and concentration by using spectrophotometer. RNA should be stored at −80 °C if reverse transcription is not performing immediately.

3.2 Reverse Transcription for cDNA Synthesis

The reverse transcription and the following real-time PCR can be achieved as either a one-step or a two-step reaction. In the one-step approach, the entire reaction from cDNA synthesis to real-time-PCR amplification is carried out in a single tube (one tube). On the other hand, by the two-step reaction, the reverse transcriptase reaction and real-time PCR amplification are performed in separate tubes (two tubes). Although the one-step approach is designed to minimize experimental variation, the starting RNA templates are found to be prone to degradation in the one-step reaction, and it is reported to be less accurate compared to the two-step reaction. Two-step reaction is considered as the preferred method of real-time PCR when using DNA-binding dyes such as SYBR Green. In this chapter we will describe the method of two-step reaction starting with reverse transcription.

1. Prepare the reverse transcription master mixture on ice according to kit instructions. The master mix should contain all components required for cDNA synthesis including poly(A) polymerase, reverse transcriptase, dNTPs, ATP, oligo-dT primers, and reaction buffer (*see* **Notes 6** and **7**). Mix the reverse transcription master mixture thoroughly but gently, then dispense appropriate volume into PCR tubes.

2. Add appropriate amount and volume of template RNA to each PCR tube containing reverse transcription master mix. Mix gently and centrifuge briefly (*see* **Note 8**).

3. Incubate at 37 °C for 60 min to proceed the reactions, then incubate at 95 °C for 5 min to inactivate reverse transcriptase mix and place on ice.

4. Cool down the reaction and add 2 U RNAse H, then incubate at 37 °C for 20 min. Or the reaction can be frozen if the RNAse H treatment is not performed immediately.

5. The resulting cDNA can be stored at –20 °C or –80 °C if real-time PCR is not performed immediately (*see* **Note 9**).

3.3 Real-Time PCR Quantification of RNA

To date, two fluorescent technologies have found application to RNA detection: TaqMan probes and SYBR Green I. Taqman probes, which are fluorescent reporter oligonucleotide probes, can release fluorescence signal during amplification process. SYBR Green will emit light upon excitation after it binds to the double-stranded DNA of the PCR products; the intensity of the fluorescence increases as the PCR products accumulate. Compared to Taqman probes, using SYBR Green is more economical and easier. Here we will focus on the quantification of RNA by using SYBR Green I in real-time PCR. cDNA from reverse transcription is the appropriate starting material for this part.

1. Prepare reaction mix containing DNA polymerase, dNTP mix, SYBR Green I, SYBR Green PCR buffer, $MgCl_2$, target-specific forward primer, and universal reverse primer (*see* **Note 10**).

2. Mix the reaction thoroughly but gently, then dispense appropriate volume into each well of 96-well PCR plate.

3. Dispense appropriate amount and volume of template cDNA into the individual plate wells containing reaction mix.

4. Carefully, tightly seal the 96-well PCR plate with film. Centrifuge at room temperature to remove bubbles.

5. Program the real-time PCR thermal cycler: activate hot start DNA polymerase for 15 min at 95 °C followed with 40 cycles of three steps: denaturation for 15 s at 94 °C, annealing for 30 s at 55 °C, and extension for 30 s at 70 °C (*see* **Note 11**).

6. Place the 96-well PCR plate in the real-time PCR thermal cycler and start the cycling program.

7. Perform data normalization and standardization (*see* **Note 12**).

4 Notes

1. Success, including specificity, of real-time PCR assays depends on the optimal primers used. Some of the considerations for optimizing primer design include primer length, GC content, primer self-dimer, or secondary structure formation. Software used for real-time PCR primer design is freely available [3].

2. After centrifugation, the lysate separates into three phases: an upper, colorless, aqueous phase containing RNA, a white interphase containing DNA, and a lower, red phenol–chloroform phase containing proteins. The setting of centrifugation in this step and all the following related steps may be slightly different among kits from different companies.

3. The volume of the aqueous phase should be about 50 % of the volume of lysis reagent. However, the actual volume we take is always less, because we do not want any DNA or proteins contamination from interphase and lower phase.

4. Total RNA can bind to the membrane and other contaminants can be washed away in following step. The appropriate volume of the sample for each time pipetting and the RNA-binding capacity of the silica membrane may be different among kits from different companies.

5. The minimum elution volume and the dead volume of spin column may be different among kits from different companies. Repeating elution with the same elute may increase the RNA yeild. However, repeating elution requires adding more RNase-free water into the elute, this may eventually lower the final RNA concentration.

6. First-strand cDNA synthesis reactions can be performed with sequence-specific (SS), random, or oligo(dT) primers. SSPs

offer the greatest specificity. However, they do not offer the flexibility of oligo(dT) and random primers, and a new cDNA synthesis reaction must be performed for each targeted gene. Random primers may overestimate copy number when used in real-time RT-PCR experiments, and they are used only in two-step real-time PCR reactions. Random primers are preferred for synthesizing large pools of cDNA. They are also used for non-polyadenylated RNA, such as bacterial RNA, and degraded RNA such as FFPE samples. Oligo(dT) primers are preferred for two-step cDNA synthesis reactions, because of their specificity for mRNA, they also allow many different targets to be studied from the same cDNA sample. Oligo(dT) primers are not recommended as the only primer for cDNA synthesis if 18S rRNA is used for normalization in a real-time PCR experiment as the oligo(dT) primer will not anneal.

7. A control reaction that lacks reverse transcriptase can be set up to test if there is any amplification caused by genomic DNA.

8. Total RNA should be used as starting material for reverse transcription reactions. All reactions should be set up on ice to minimize the risk of RNA degradation.

9. SYBR Green I can preferentially bind to double-stranded DNA and can be used as a dye for the quantification in quantitative PCR. $MgCl_2$ can be used with DNA polymerase to improve amplification rate. Due to the hot start, it is not necessary to keep samples on ice during reaction setup.

10. Dilute the cDNA by adding RNase-free water prior to real-time PCR. The dilution depends on abundance of microRNA of interest and the starting amount required from different kits. Reactions should be done in duplicate or triplicate to reduce erroneous results due to pipetting error. Increasing the reaction volume will decrease pipetting error but increase the cost per reaction. A non-template control reaction should also be included, and a dissociation curve should be calculated to determine if nonspecific primer interactions occur.

11. In denaturation step, the DNA template can be briefly heated to between 92 and 95 °C to break the hydrogen bonds that keep DNA in its characteristic double-helix form and separate it into two single strands. After denaturation, the mixture is cooled to a temperature as low as 55 °C; this allows hydrogen bonds to form between complementary nucleotide base pairs of the DNA template and primers. Once annealing has occurred, a heat-stable DNA polymerase steps in to begin rapid, sequential addition of the nucleotides to the primer strands; to kick this into gear, the mixture is again heated to 72 °C. The fluorescence data collection can be performed during extension.

12. The estimation of the amount of reference genes such as β-actin across samples is useful to correct for sample-to-sample variation. Once the running is completed, the relative quantity of each microRNA can be calculated by the $\Delta\Delta Ct$ method. ΔCt is calculated using

$$\Delta Ct = Ct(\text{target}) - Ct(\text{reference}) \tag{1}$$

where ΔCt (target) is the Ct value of targeted gene and Ct (reference) is Ct value of reference gene. $\Delta\Delta Ct$ values are calculated using

$$\Delta\Delta Ct = Ct(\text{sample}) - Ct(\text{calibrator}) \tag{2}$$

where ΔCt (sample) represents the expression value of the targeted gene calculated using Eq. 1 and ΔCt (calibrator) is the expression value of the sample (control sample) to which other samples in the data set are normalized. The relative expression value was obtained from $\Delta\Delta Ct$ values by using

$$\text{relative expression} = 2^{-\Delta\Delta Ct} \tag{3}$$

References

1. Bernard PS, Wittwer CT (2002) Real-time PCR technology for cancer diagnostics [J]. Clin Chem 48(8):1178–1185
2. Sluijter JP, Smeets MB, Pasterkamp G et al (2003) Methods in molecular cardiology: quan-titative real-time polymerase chain reaction [J]. Neth Heart J 11(10):401–404
3. Thornton B, Basu C (2015) Rapid and simple method of qPCR primer design [J]. Methods Mol Biol 1275:173–179

Chapter 7

Detection of miRNA in Cultured Cells or Xenograft Tissues of Breast Cancer

Martin Brown and Meiyun Fan

Abstract

MicroRNA (miRNA) analysis has evolved over the past two decades to become a highly specialized field with broad-reaching applications across a multitude of diseases and cellular processes. The choice of an applicable approach for miRNA quantification will depend on a variety of factors such as cost, time constraints, and throughput. Here, we describe the methods of total RNA isolation, AGO2-bound RNA isolation, miRNA polyadenylation, miRNA-cDNA synthesis, and quantitative real-time polymerase chain reaction for the detection of known miRNAs in cultured cells or xenograft tissues of breast cancer.

Key words AGO2 immunoprecipitation, RNA isolation, miRNA detection, qPCR, Breast cancer

1 Introduction

MicroRNAs (miRNAs), a class of noncoding, small RNA molecules, play a significant role in basic cellular functions such as apoptosis, proliferation, migration, and differentiation [1]. Since the first discovery of the small RNAs produced by the *lin-4* gene in *C. elegans* in 1993 [2], miRNA quantification has become an essential procedure for research aimed to gain in-depth understanding of cancer cell biology [3, 4]. There are several well-established methods for miRNA quantification, including small RNA sequencing, miRNA expression microarray, and reverse transcription real-time quantitative polymerase chain reaction (RT-qPCR) [5–7]. Due to the high sensitivity and relatively low-cost, RT-qPCR remains the method of choice for miRNA quantitation in a variety of biological specimens and the current gold standard for validation of data generated by high-throughput technologies [7]. miRNAs regulate mRNA stability and translation through the action of the RNA-induced silencing complex (RISC) [8]. Based on the endoribonuclease activities and expression patterns of the four human argonaute proteins (AGO1-4), the main components of RISC,

Jian Cao (ed.), *Breast Cancer: Methods and Protocols*, Methods in Molecular Biology, vol. 1406,
DOI 10.1007/978-1-4939-3444-7_7, © Springer Science+Business Media New York 2016

AGO2 is believed to play a key role in miRNA function in mammary gland epithelial cells [9–11]. Consequently, AGO2 immunoprecipitation (AGO2-IP), followed by RNA isolation and miRNA RT-qPCR, represents a direct and feasible approach to quantify functional miRNAs in cells [12–17]. Here, we describe a qPCR-based method for quantitation of total cellular miRNAs and AGO2-bound miRNAs in breast cancer cells.

2   Materials

2.1   Reagents and Equipment for AGO2 Immunoprecipitation

1. Phosphate-buffered saline (PBS): add approximately 900 mL of Milli-Q water (water purified by EMD Millipore Milli-Q Integral Systems) to a 1.5 L container. Add 0.2 g KCl, 0.2 g KH_2PO_4, 8 g NaCl, and 1.45 g $Na_2HPO_4 \cdot 7H_2O$ into the water. Mix the solution on a magnetic stirrer, adjust pH to 7.6 by using 1 N NaOH, adjust final volume to 1000 mL using Milli-Q water, and store at 4 °C.

2. Hypotonic buffer: containing 10 mM Tris (pH 7.5), 10 mM KCl, 2 mM $MgCl_2$, and 1 mM DTT. 100× concentrated stock solutions for each component are made using DNase-/RNase-free H_2O and molecular biology grade chemicals and stored at –20 °C as small aliquots. Immediately before use, complete hypotonic buffer is made by mixing the stock solutions and supplemented with 100 U/mL RNaseOUT (Life Technologies, Carlsbad, CA) and protease inhibitor cocktail (Santa Cruz Biotechnology, Dallas, TX).

3. Wash buffer: complete hypotonic buffer supplemented with 150 mM NaCl and 0.5 % NP-40.

4. High-salt buffer: complete hypotonic buffer supplemented with 400 mM NaCl and 0.5 % NP-40.

5. Antibodies: mouse antihuman AGO2 (Clone 2E12-1C9, Abnova, Taipei City, Taiwan), control mouse IgG kappa chain (Santa Cruz Biotechnology), and anti-mouse IgG-coated magnetic beads (MagnaBind Magnetic Beads, Thermo Scientific, Wilmington, DE).

6. Dounce homogenizer (2 mL) with B pestle.

7. 1.5 mL low-retention microcentrifuge tubes (Thermo Scientific).

8. Orbital shaker in 4 °C refrigerator or cold room.

9. Temperature-regulated centrifuge compatible with 1.5 mL microcentrifuge tubes.

10. Magnetic separation stands for 1.5 mL microcentrifuge tubes.

2.2   Reagents and Equipment for RNA Isolation

1. TRIzol (Life Technologies) (*see* **Note 1**).

2. Molecular biology grade chloroform (Thermo Scientific). Store at room temperature.

3. Molecular grade isopropanol (Thermo Scientific). Store at room temperature.

4. Molecular biology grade absolute ethanol (200 proof, Thermo Scientific). Store at room temperature. To prepare 70 % ethanol, add 15 mL RNase/DNase H_2O into 35 mL ethanol in a 50 mL RNase-/DNase-free tube.

5. GlycoBlue™ Coprecipitant (15 mg/mL, Life Technologies). Store at –20 °C.

6. UV–Vis spectrophotometer such as NanoDrop 2000 (Thermo Scientific).

7. Homogenizer (PowerGen Model 125, Thermo Scientific) for tumor tissues.

8. RNase-/DNase-free microcentrifuge tubes (1.5–2 mL).

2.3 Reagents and Equipment for Polyadenylation and cDNA Synthesis

1. The following buffers used for polyadenylation are provided by NCode miRNA First-Strand Kit (Life Technologies): 5× miRNA reaction buffer, 25 mM $MnCl_2$, 10 mM ATP, Poly A Polymerase, annealing buffer, SuperScript III RT/RNaseOUT enzyme mix, 2× first-strand reaction buffer, DEPC-treated water, and universal RT and qPCR primers. Store all reagents at –20 °C.

2. Floating foam tube rack for 0.2 mL tubes (VWR, Radnor, PA, USA).

3. Benchtop centrifuge compatible with 0.2 mL tubes.

4. Two water baths: set at 37 and 65 °C, respectively.

5. RNase-/DNase-free PCR tubes (0.2 mL).

6. Thermal cycler for PCR.

2.4 Reagents and Equipment for qPCR

1. miRNA-specific primer (*see* **Note 2**).

2. Maxima 2× SYBR Green RT-PCR reaction mix (Thermo Scientific).

3. Microseal "B" Seals (Bio-Rad, Hercules, CA).

4. 96-well PCR plates.

5. qPCR system (e.g., CFX96 Real-Time System, Biorad).

6. Centrifuge compatible with 96-well PCR plates.

3 Methods

3.1 Total RNA Isolation from Cultured Cells

1. Grow cells in a 35-mm or 60-mm dish to reach a density of ~80 % confluence. Remove medium and wash cell with ice-cold PBS once. In a fume hood, add 1 mL TRIzol reagent into the dish and lyse cells by repeated pipetting. Transfer cell lysate into a 1.5 mL centrifuge tube.

2. In a fume hood, add 0.2 mL chloroform into the TRIzol lysate and mix thoroughly by inverting the tubes ten times. Incubate the tubes for 3 min at room temperature. Centrifuge the sample at $12,000 \times g$ for 15 min at 4 °C.

3. In a fume hood, carefully pipette out the aqueous phase (colorless top layer) into a new 1.5 mL RNase/DNase microcentrifuge tube. Avoid drawing any of the interphase or organic layer into the pipette tip when transferring the aqueous phase (*see* **Note 3**).

4. Add 1 μL GlycoBlue Coprecipitant (to aid in pellet visualization) and 0.5 mL of 100 % isopropanol to the aqueous phase; mix by Vortex and incubate at room temperature for 10 min. Centrifuge at $12,000 \times g$ for 10 min at 4 °C.

5. Remove supernatant and wash RNA pellet with 1 mL of 75 % ethanol. Vortex sample to displace pellet from tube bottom and centrifuge at $7500 \times g$ for 5 min.

6. Pipette out ethanol. Centrifuge the tube using a benchtop centrifuge at $2500 \times g$ for 1 min and remove the residue ethanol with a pipette loaded with a gel-loading tip (*see* **Note 4**). Leave the tube uncapped for 2–3 min to allow for residual ethanol evaporation.

7. Suspend purified RNA in 30 μL RNase-/DNase-free water. Place RNA samples at 4 °C for 30 min to allow for complete dissolution.

8. Measure the absorbance of RNA at 260 and 280 nm by using NanoDrop 2000 to determine RNA concentration.

9. Adjust the RNA concentration to 200 ng/μL using RNase-/DNase-free water. Store RNA samples at −80 °C.

3.2 Total RNA Isolation from Xenograft Tumor Tissues

1. Put a fresh tissue fragment (~50 mg) into a 2 mL tube filled with 1 mL TRIzol reagent and use a homogenizer to disrupt the tissue (using PowerGen Model 125 homogenizer: output set at 5, homogenizing 30 s/resting 30 s, repeat three times). Homogenize tumor tissue in a fume hood to limit the exposure to TRIzol reagent.

2. Follow **steps 2–9** of Subheading 3.1 to purify RNA from tumor tissue homogenate.

3.3 Isolation of RNA Associated with AGO2-Based RISC from Cultured Cells

1. Grow cells in three 150-mm dishes to reach a confluence of ~80 %. Discard medium and wash cells with ice-cold PBS once. Collect cells by scrapping cells in ice-cold PBS (3 mL per dish). Pool and transfer cells from the three 150-mm dishes (~3×10^7 cells) into a 15 mL tube. Centrifuge the tube at $1500 \times g$ at 4 °C for 3 min.

2. Suspend cell pellet in 3 mL ice-cold complete hypotonic buffer and incubate on ice for 15 min. Transfer cell suspension into

an ice-chilled Dounce homogenizer. Disrupt the cells to release cytosol by using a "B" pestle with 15 strokes (down plus up equals one stroke). Transfer the disrupted cells into a 15 mL tube and centrifuge at 12,000 x g and 4 °C for 10 min.

3. Pre-clean the cytosol fraction (supernatant from the homogenized cells) with control mouse IgG kappa and anti-mouse IgG-coated magnetic beads to eliminate nonspecific binding: in a 1.5 mL low-retention tube (see **Note 5**), mix 1 mL cytosol fraction with 5 μg control IgG kappa and 50 μL anti-mouse IgG-coated magnetic beads and incubate on an orbital shaker at 4 °C for 1 h. Put tubes in a magnetic separation stands to collect pre-cleaned cytosol fraction.

4. Mix pre-cleaned cytosol fraction (1 mL) with 5 μg anti-AGO2 antibody and 50 μL anti-mouse IgG-coated magnetic beads. After overnight incubation at 4 °C on an orbital shaker, AGO2-IP (immunoprecipitated) beads were washed twice with 1 mL ice-cold wash buffer and once with 1 mL high-salt buffer (hypotonic buffer supplemented with 400 mM NaCl and 0.5 % NP-40).

5. Suspend AGO2-IP beads in 1 mL TRIzol and follow **steps 2–6** of Subheading 3.1 to purify RNA associated with AGO2-based RISC.

6. Dissolve RNA pellet in 10 μL DNase-/RNase-free water.

3.4 Polyadenylation of miRNA

1. Prepare diluted ATP (see **Note 6**): adding 10 μL of 10 mM ATP in 10 μL 1 mM Tris (pH 8.0).

2. Prepare master mix for polyadenylation reaction in a 0.2 mL PCR tube. To prepare polyadenylation mix for n RNA samples, multiply the volumes for the following reagents by ($n+2$):

   - 2 μL miRNA reaction buffer.
   - 1 μL 25 mM MnCl2.
   - 0.4 μL diluted ATP.
   - 0.2 μL poly A polymerase.
   - 1.4 μL DEPC-treated water.

3. Aliquot 5 μL polyadenylation mix into 0.2 mL PCR tubers. Add 5 μL (1 μg) RNA into each tube. For RNA samples with concentration <200 ng/μL see **Note 6** for modified protocol.

4. Vortex the reaction mix and centrifuge briefly to collect contents to the bottom of the tubes.

5. Place the tubes in a foam flotation holder and incubate in a water bath at 37 °C for 15 min. Ensure that all the contents at the bottom of the tubes are immersed in water, but the lids of the tubes are above water surface to avoid contamination from the bath water.

3.5 cDNA Synthesis

1. In a 0.2 mL PCR tube, combine 4 μL of the polyadenylated-RNA from Subheading 3.4 with 1 μL annealing buffer and 3 μL 25 μM Universal RT Primer.

2. Cap and vortex samples.

3. Centrifuge briefly to collect contents to the bottom of the tubes.

4. Incubate the samples in a water bath at 65 °C for 5 min.

5. Place tubes on ice for 1 min.

6. Centrifuge briefly to collect contents to the bottom of the tubes.

7. Add 10 μL of 2× first-strand reaction mix and 2 μL of SuperScript III RT/RNaseOUT enzyme mix into each tube.

8. Centrifuge the tubes briefly to collect contents to the bottom of the tubes.

9. Place tubes in a thermal cycler with a program: 50 °C for 50 min followed by 85 °C for 5 min.

3.6 miRNA qPCR

1. Once the cDNA synthesis is complete, centrifuge the tube briefly and dilute the cDNA 1:10 using RNase-/DNase-free water by adding 180 μL DNase/RNase water into the 20 μL cDNA reaction mix.

2. Prepare qPCR master mix for each miRNA to be examined (*see* **Note 7**) in a 1.5 mL RNase/DNase tube. Determine the number of qPCR reactions required for each miRNA: $n = 3$ replicates × number of RNA samples used for detection of a specific miRNA. To make qPCR master mix for a specific miRNA, multiply the volumes for the following reagents by $(n + 2)$:

   • 6 μL 2× SYBR Green RT-PCR reaction mix.

   • 0.5 μL universal qPCR primer solution (5 μM).

   • 0.5 μL miRNA-specific primer solution (5 μM).

3. Aliquot 7 μL qPCR mix into each well of a 96-well PCR plate. Add 5 μL diluted miRNA-cDNA into each well.

4. Seal the PCR plate with an optically clear adhesive seal (e.g., Microseal "B" Seals).

5. Centrifuge the plate at 625 x g for 3 min.

6. Load the plate into a qPCR system and run the following protocol when Maxima 2× SYBR Green RT-PCR reaction mix is used:

   Step 1: 50 °C for 2 min.

   Step 2: 95 °C for 10 min.

   Step 3: 95° for 15 s.

Step 4: 60 °C for 1 min + plate read.

Repeat **steps 3** and **4** an additional 39 times.

Step 5: 95 °C for 10 s.

Step 6: melt curve at 65–95 °C (0.5 °C increment/30 s).

---

4   Notes

1. TRIzol is toxic and corrosive; use only in a fume hood with appropriate personal protective equipment (PPE). Store at room temperature in a protective case to prevent accidental spill.

2. miRNA-specific primer is a DNA oligo that is identical to the entire mature miRNA sequence. For example, the primer sequence is 5′-CAAAGTGCTTACAGTGCAGGTAG-3′ for hsa-miR-17-5p [caaagugcuuacagugcagguag]. For GC-rich miRNA sequences, it may be necessary to design primers that are truncated by 3–4 bases on the 3′ end to have a melting temperature (Tm) of 60 ± 5 °C. Mature miRBase: the microRNA database.

3. After phase separation, RNA is exclusively in the upper aqueous phase (~50 % of the total volume), and DNA and protein are in the interphase and phenol–chloroform layer. Avoiding contamination of aqueous phase is critical for the quality of the isolated RNA.

4. This step aids to preserve RNA solubility by shorting the air-dry time required for residue ethanol evaporation.

5. The low-retention microcentrifuge tubes improve sample recovery and immunoprecipitation specificity by significantly reducing surface binding.

6. The concentration of RNA purified from AGO2-IP beads is usually in the range of 30–50 ng/μL. Use 5 μL RNA solution (equivalent to 150–250 ng RNA) for cDNA synthesis. Based on the quantity of total RNA, dilute 10 mM ATP in 1 mM Tris (pH 8.0) according to the following formula: ATP dilution factor = 2000/X ng RNA for a 10 μL polyadenylation reaction. For example, if you use 200 ng RNA for a 10 μL polyadenylation reaction, the ATP dilution factor is 2000/100 ng = 20. Dilute the ATP 1:20 by adding 1 μL of 10 mM ATP to 19 μL of 1 mM Tris, pH 8.0.

7. Small nuclear RNA U6 and at least two miRNAs abundantly expressed in breast cancer cells (e.g., miR-21 and let-7 family members) [18] are usually included as internal control for miRNA qPCR for breast cancer cell or tumor tissues.

References

1. Bartel DP (2004) MicroRNAs: genomics, biogenesis, mechanism, and function. Cell 116(2):281–297

2. Lee RC, Feinbaum RL, Ambros V (1993) The *C. elegans* heterochronic gene lin-4 encodes small RNAs with antisense complementarity to lin-14. Cell 75(5):843–854

3. Hayes J, Peruzzi PP, Lawler S (2014) MicroRNAs in cancer: biomarkers, functions and therapy. Trends Mol Med 20(8):460–469. doi:10.1016/j.molmed.2014.06.005

4. Jansson MD, Lund AH (2012) MicroRNA and cancer. Mol Oncol 6(6):590–610. doi:10.1016/j.molonc.2012.09.006

5. Pritchard CC, Cheng HH, Tewari M (2012) MicroRNA profiling: approaches and considerations. Nat Rev Genet 13(5):358–369. doi:10.1038/nrg3198

6. Raabe CA, Tang TH, Brosius J, Rozhdestvensky TS (2014) Biases in small RNA deep sequencing data. Nucleic Acids Res 42(3):1414–1426. doi:10.1093/nar/gkt1021

7. Git A, Dvinge H, Salmon-Divon M, Osborne M, Kutter C, Hadfield J, Bertone P, Caldas C (2010) Systematic comparison of microarray profiling, real-time PCR, and next-generation sequencing technologies for measuring differential microRNA expression. RNA 16(5):991–1006. doi:10.1261/rna.1947110

8. Djuranovic S, Nahvi A, Green R (2011) A parsimonious model for gene regulation by miRNAs. Science 331(6017):550–553. doi:10.1126/science.1191138

9. Liu X, Yu X, Zack DJ, Zhu H, Qian J (2008) TiGER: a database for tissue-specific gene expression and regulation. BMC Bioinformatics 9:271. doi:10.1186/1471-2105-9-271

10. Azuma-Mukai A, Oguri H, Mituyama T, Qian ZR, Asai K, Siomi H, Siomi MC (2008) Characterization of endogenous human Argonautes and their miRNA partners in RNA silencing. Proc Natl Acad Sci U S A 105(23):7964–7969. doi:10.1073/pnas.0800334105

11. Valdmanis PN, Gu S, Schuermann N, Sethupathy P, Grimm D, Kay MA (2012) Expression determinants of mammalian argonaute proteins in mediating gene silencing. Nucleic Acids Res 40(8):3704–3713. doi:10.1093/nar/gkr1274

12. Hassan MQ, Gordon JA, Lian JB, van Wijnen AJ, Stein JL, Stein GS (2010) Ribonucleoprotein immunoprecipitation (RNP-IP): a direct in vivo analysis of microRNA-targets. J Cell Biochem 110(4):817–822. doi:10.1002/jcb.22562

13. Burroughs AM, Ando Y, de Hoon MJ, Tomaru Y, Suzuki H, Hayashizaki Y, Daub CO (2011) Deep-sequencing of human Argonaute-associated small RNAs provides insight into miRNA sorting and reveals Argonaute association with RNA fragments of diverse origin. RNA Biol 8(1):158–177

14. Easow G, Teleman AA, Cohen SM (2007) Isolation of microRNA targets by miRNP immunopurification. RNA 13(8):1198–1204. doi:10.1261/rna.563707

15. Tan LP, Seinen E, Duns G, de Jong D, Sibon OC, Poppema S, Kroesen BJ, Kok K, van den Berg A (2009) A high throughput experimental approach to identify miRNA targets in human cells. Nucleic Acids Res 37(20), e137. doi:10.1093/nar/gkp715

16. Thomson DW, Bracken CP, Goodall GJ (2011) Experimental strategies for microRNA target identification. Nucleic Acids Res 39(16):6845–6853. doi:10.1093/nar/gkr330

17. Hafner M, Landthaler M, Burger L, Khorshid M, Hausser J, Berninger P, Rothballer A, Ascano M Jr, Jungkamp AC, Munschauer M, Ulrich A, Wardle GS, Dewell S, Zavolan M, Tuschl T (2010) Transcriptome-wide identification of RNA-binding protein and microRNA target sites by PAR-CLIP. Cell 141(1):129–141. doi:10.1016/j.cell.2010.03.009

18. Fan M, Krutilina R, Sun J, Sethuraman A, Yang CH, Wu ZH, Yue J, Pfeffer LM (2013) Comprehensive analysis of microRNA (miRNA) targets in breast cancer cells. J Biol Chem 288(38):27480–27493. doi:10.1074/jbc.M113.491803

Chapter 8

Pyrosequencing Analysis for Breast Cancer DNA Methylome

Cem Kuscu and Canan Kuscu

Abstract

Unraveling DNA methylation profile of tumor is important for the diagnosis and treatment of cancer patients. Because of the heterogeneity of clinical samples, it is very difficult to get methylation profile of only tumor cells. Laser capture Microdissection (LCM) is giving us a chance to isolate the DNA only from the tumor cells without any stroma cell's DNA contamination. Once we capture the breast tumor cells, we can isolate the genomic DNA which is followed by the bisulfite treatment in which unmethylated cytosines of the CG pairs are converted into uracil; however, methylated cytosine does not go into any chemical change during this reaction. Next, bisulfite treated DNA is used in the regular PCR reaction to get a single band PCR amplicon which will be used as a template for the pyrosequencing. Pyrosequencing is a powerful method to make a quantitative methylation analysis for each specific CG pair.

Key words Breast cancer, Laser capture microdissection (LCM), DNA methylation, Bisulfite conversion, Pyrosequencing

1 Introduction

Breast cancer is the most common form of cancer among women and it is the second most deadly cancer after lung cancer [1]. Since tissue samples taken from cancer patients are heterogeneous, and therefore are composed of a mixture of tumor cells and surrounding stroma cells, experimental techniques for the analysis of cancer cells have been difficult. However, improvements have been made over the years, including Laser Capture Microdissection (LCM), which has been used for the analysis of histopathology samples since 1996 [2]. In this technique, laser is accompanied with the objective of the microscope and focuses on the tissue sections. One can draw a border around target cells on the computer image of histology slides which is called "element" to dissect the cells in the small size of tube which will be used for future DNA or RNA isolation. Therefore, LCM improves the quality of data from clinical samples and gives us a trustable output [3].

Jian Cao (ed.), *Breast Cancer: Methods and Protocols*, Methods in Molecular Biology, vol. 1406,
DOI 10.1007/978-1-4939-3444-7_8, © Springer Science+Business Media New York 2016

Along with genetic alterations of individuals, recent findings shed a light on the importance of epigenetic aberrations for cancer progression [4, 5]. Epigenetic mechanisms are basically divided into two main groups called DNA methylation and histone modifications. Some studies include miRNA as part of the epigenetic mechanism. The methylation of the fifth carbon on the cytosine (5mC) has been known as the main mechanism of DNA methylation for gene regulation in eukaryotes, but very recent studies demonstrated that sixth nitrogen on the adenosine (6mA) can also be methylated and change the gene expression in eukaryotes as well [6, 7]. For the rest of the chapter, the term of DNA methylation is used only for the methylated cytosine (5mC). In normal cells, most of the DNA methylation takes place in repetitive genomic regions, such as LINES (long interspersed transposable elements) and SINES (short interspersed transposable elements) to maintain genomic integrity [8]. Besides random and diverse methylation of cytosines in these regions, DNA methylation is usually concentrated in CpG islands (CGI), which are described with the following formula; minimum 200-bp stretch of DNA with a minimum $C + G$ content of 50 % and an Obs_{CpG}/Exp_{CpG} in excess of 0.6 [9]. Based on this criterion, there are approximately 29,000 CpG islands (CGI) in the human genome. Besides the localization within promoters of genes, some CpG islands exist in the intergenic or intragenic (intron) regions [10]. Half of all human promoters have been reported to have CGIs, but these CGIs within promoters are generally not methylated. Increased level of methylation in the CpG island around promoters inhibits transcription either directly by blocking the access of specific transcription factors to the promoter of the genes or indirectly by recruiting methyl-CpG-binding domain (MBD) proteins which might recruit histone-modifying and chromatin-remodeling complexes that cause the condensation of the chromatin [11]. In addition to the basic role of DNA methylation in the development of organisms and cellular differentiation via altering gene expression profiles, aberrant DNA methylation patterns have also been linked to nearly all types of cancer [12]. The role of methylation in cancer progression is primarily considered as a molecular instrument for hypermethylation of promoters in order to silence tumor suppressor genes, and majority of studies related with this topic focus in this direction [13, 14]. On the other hand, hypomethylation of oncogenes or growth-related genes is another mechanism related with cancer progression. Almost 30 years ago, Feinberg and Vogelstein demonstrated for the first time that hypomethylation in the promoters of oncogenes, such as c-Ha-ras and c-Ki-ras, induced the formation of tumors in colon and lung tissues [15]. Therefore, monitoring the alteration of DNA methylation level on specific loci, especially the promoter region of tumor suppressor and oncogenes, is very critical for cancer patients. Frommer and his colleagues developed one of the most powerful techniques in which unmethylated cytosines, not methylated ones, are converted into uracil after bisulfite treatment. Finally, we can identify the status of the cytosine after the Sanger sequencing of the target region [16, 17].

In 1996, another sequencing technique called pyrosequencing was developed by Mostafa Ronaghi and Pal Nyrén [18, 19]. In their new sequencing approach, they immobilize one strand of DNA with the help of biotinylated primer and then they add the pyrosequencing primers into the solution. Later, they add and remove A, G, C, and T solution for each nucleotide synthesis. In each step, pyrophosphate (PPi) is released because of the addition of one of dNTPs to the growing strand with the help of DNA polymerase. ATP sulfurylase catalyzes the reaction between this free PPi and APS (adenosine 5′ phosphosulfate) after each step to produce an ATP which is used as a substrate in the next step. Luciferin is converted into oxyluciferin in the presence of ATP, producing light. Then all free single nucleotides and ATP are degraded by another enzyme called apyrase. Amount of light produced at each step is used to quantify the data and predict the percentage of nucleotide for some clinical analysis. Main applications of pyrosequencing are quantitative analysis of sequence variants for allele frequency in the population, analysis of mutations, bacterial/fungal typing, and analysis of DNA methylation ratio on specific CG pairs. The main disadvantage of pyrosequencing is the short read outcome that is usually around 50–60 bp.

2   Materials

2.1   Cell Cultures or Tissues

1. Breast cancer MCF-7 and MDA-MB-231 cell lines were purchased from ATCC (Manassas, VA) and were cultured in Dulbecco's Modified Eagle's Medium (DMEM) (Invitrogen) plus 10 % FBS. Breast cancer invasive and in situ samples were obtained from the Pathology Department of Stony Brook University.

2.2   FFPE Tissue Preparation

1. Breast cancer specimen (right after surgery, standard sample size: $0.5 \times 1 \times 1$ cm).
2. 10 % neutral formalin.
3. Histology cassette (Tissue-Tek® Uni-Cassettes, Sakura).
4. 70, 80, and 95 % EtOH.
5. Xylene.
6. Paraffin.
7. Hot water bath (60 °C).
8. Cooling station (<–10 °C).

2.3 Hematoxylin and Eosin Staining

1. Microtome (Model 48577-60).
2. SuperFrost Plus slides.
3. 50 °C water bath.
4. 65 °C incubator.
5. Xylene
6. 70, 80, 95, and 100 % EtOH.
7. Hematoxylin solution (*see* **Note 1**).
8. Acid alcohol: 0.1 % HCl, 50 % EtOH in distilled water.
9. Scott's tap water: 2 g sodium bicarbonate and 10 g $MgSO_4$ in 1 l of distilled water.
10. Eosin solution.
11. Glass or plastic jars.
12. Mounting media.
13. Forceps and brushes.

2.4 Laser Capture Microdissection

1. Leica laser microscope.
2. PALM RoboSoftware.
3. 0.5 ml eppendorf tube.
4. 70 % EtOH.
5. Microcentrifuge.

2.5 Genomic DNA Isolation

1. Qiagen DNeasy Tissue Kit (Cat # 69504).
2. Tabletop centrifuge (up to 13 k rpm).
3. Xylene.
4. RNase.

2.6 Identification of CpG Island

1. MethPrimer: http://www.urogene.org/cgi-bin/methprimer/methprimer.cgi.
2. Bioinformatics.org: http://www.bioinformatics.org/sms2/cpg_islands.html.
3. CpG Island Searcher: http://www.uscnorris.com/cpgislands2/cpg.aspx.
4. UCSC Genome Bioinformatics: https://genome.ucsc.edu/index.html.

2.7 Primer Design and CpG Assay Design

1. MethPrimer [20]: http://www.urogene.org/methprimer/index.html.
2. MethMarker: http://methmarker.mpi-inf.mpg.de/.
3. RepeatMasker: http://www.repeatmasker.org/.
4. Pyro Q CpG (Biotage) and PSQ.

2.8 Bisulfite Treatment and Bisulfite-Specific PCR

1. EZ DNA Methylation-Gold™ Kit (Zymo Research).
2. PCR machine.
3. High fidelity DNA polymerase or HotStarTaq DNA Polymerase (store at –20 °C).
4. 5′-biotinylated and HPLC purified forward primers (10 nM, Sigma or Eurofins).
5. Unlabelled reverse PCR primer (desalted, 10 nM).
6. QIAquick PCR Purification Kit (Qiagen, Cat. No. 28104).
7. Human non-methylated control DNA (D5014-1, Zymo Research).
8. HpaII restriction endonuclease (NEB).

2.9 Gel Electrophoresis

1. Agarose.
2. Gel electrophoresis system.
3. TAE buffer (50× stock) : 242 g Tris base, 57.1 ml glacial acetic acid, and 100 ml of 500 mM EDTA (pH 8.0) solution in 1 l of distilled water.
4. 6× loading dye: 3 ml glycerol (30 %), 25 mg bromophenol blue (0.25 %) in 10 ml distilled water.
5. UV gel analyzer.
6. MinElute Gel Extraction Kit (if necessary) (Qiagen).

2.10 Preparation of Pyrosequencing Template

1. Heat block up to 80 °C.
2. Microplate shaker.
3. Troughs and 96-well vacuum prep.
4. 96-well pyrosequencing plate.
5. Streptavidin Sepharose High Performance Beads (keep at 4 °C).
6. Annealing buffer: 2.42 g Tris and 0.43 g magnesium acetate-tetrahydrate in 1 l distilled water (pH: 7.6).
7. 2× Binding buffer: 1.21 g Tris, 117 g NaCl, 0.292 g EDTA, and 1 ml Tween 20 in 1 l distilled water (pH:7.6).
8. 70 % EtOH.
9. Denaturation solution: 0.2 M NaOH.
10. Washing buffer: 1.21 g Tris in 1 l distilled water (pH: 7.6).
11. Unlabeled internal pyrosequencing primer.
12. Adhesive sealing film for 96-well plate.

2.11 Pyrosequencing

1. Pyromark Q24 or equivalent instrument.
2. Pyromark Q24 or equivalent cartridge.

2.12 Run Sample and Data Analysis

Pyro Q CpG Software.

3   Method

3.1   Cell Cultures and Tissues

Cell lines or tissues both can be used for the pyrosequencing analysis of DNA methylome. For cell lines, adherent cells need to be harvested with trypsin (0.05 %) for 2–5 min at 37 °C incubation. Then, cells will be washed with PBS and pellet can be used for the DNA isolation in Subheading 3.5. If you have paraffin embedded tissue, you can continue to Subheading 3.3 otherwise breast tissues should be prepared immediately after the surgery under the supervision of pathologist.

3.2   FFPE Tissue Preparation

1. Dissect the breast tumors with these dimensions; 0.5 cm (depth) × 1 cm × 1 cm and place inside a histology cassette between two small sponges. Put the cassette immediately to 10 % formalin solution and incubate for 1–2 days at room temperature (*see* **Note 2**).

2. Use tap water to wash the fixed tissue for half hour. Dehydrate the tissues by 30 min incubation of 70, 80, 95 % of EtOH, followed by two times of 100 % EtOH incubation, 1 h each.

3. Wash the tissue with xylene twice, 1 h each. At the same time, start to prepare paraffin by heating it to 55–60 °C.

4. Pour a little hot paraffin into the mold, then embed the fixed tissue on top of hot paraffin. Orient the specimen if it is necessary, otherwise fill the plastic cover with hot paraffin. Once mold is filled with hot paraffin, move the mold onto the cooling station (<–10 °C).

5. Take the paraffin block from mold once the wax is hard.

3.3   Hematoxylin and Eosin Staining

1. Set up water bath to the 50 °C. Change the blade of microtome with a fresh one and set up the thickness to 5 μm. FFPE sections were cut and collected from block face and transferred onto a prewarmed water bath with the help of brush (*see* **Notes 3** and **4**). Let paraffin sections incubate on the surface of warm water for 5 min, then you can transfer tissue sections on the surface of SuperFrost Plus slides gently. Mark your samples with pencil; do not use pen (*see* **Note 5**).

2. Transfer all your slides to the 65 °C incubator or small oven for 2 h. This treatment melts most of the paraffin around your tissues.

3. After drying the sections on a glass slide, the sections are merged into xylene solution for 10 min, twice. The sections then are rehydrated in serial ethanol solution from 100 % EtOH to distilled water in the following order: 100, 95, 70, 50 % EtOH and water, 10 min each.

4. The slides are stained with a nuclear dye (Hematoxylin Gill-1 or Mayer solution) for 5 min followed by rinsing with distilled

water. Then, the slides are incubated in acid alcohol solution for 3 s. After washing the slides in tap water, they are treated with eosin (counter stain) for 30 s to 1 min.

5. The slides then are dehydrated in several ethanol solutions in the following order: 70, 95, and 100 % EtOH, 10 min each. Finally, the slides are treated with xylene for 5 min, twice. The first section of every six cut should be covered with coverslips and used as a reference slide for the others.

6. The slides can be stored at room temperature.

3.4 Laser Capture Microdissection

1. First tumor cells should be identified on the reference slides for each H and E (hematoxylin–eosin)-stained sections. In this step, it is recommended that you should get help from a pathologist if you do not have enough experience to differentiate the tumor cells (in situ or invasive) from surrounding stroma cells.

2. By using the reference slide you can identify the location of tumor cells and mark their location on the bottom of the slide with a pen. Then, place the marked slide on the microscope stage with the help of slide clips. Please see the summary of the setup in Fig. 1.

3. Put 40 μl of 70 % EtOH to the 0.5 ml eppendorf tube's cap. Invert the tube carefully and place it on the tube holder of the microscope on top of the slide. Distance between the slide and the cap should be minimized (*see* **Note 6**).

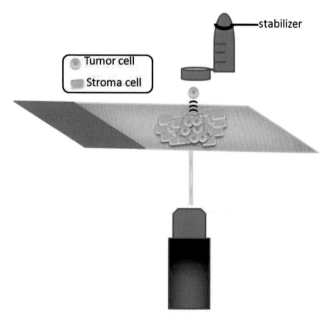

Fig. 1 Positions of the tissue slide, collection tube, and objectives in the microscope

4. Open the PALM ROBO or equivalent application on the computer. First, set up 10× objective on the microscope and observe the image on the screen. By taking advantage of previous mark on the bottom of the slide, move the stage on that position. Optimize the "focus" and "light" on the program and try to see image on the computer. The image quality on the computer might be very low since there is no mounting media and coverslip on the slide. If you see very blurry image, change the objective into 5× and adjust the distance between the slide and eppendorf's cab.

5. Draw a line around the tumor cells on the computer image to catapult the individual points out of your sample. These defined areas are called "elements". You can use graphics bar of PALM RoboSoftware to define different sizes and shapes of elements. You can save your elements into your computer and calculate the catapulted area for your quantitative analysis if you need it (*see* **Note 7**).

6. Set up the speed of the laser to 2000 μm/s for 5 μm section. If it is larger than 5 μm, reduce the speed. Two other parameters need to be adjusted are "UV-energy" and "UV-focus". Initially set up the "UV-energy" to 82 and the "UV-focus" to 76. After the first couple of shots, you need to adjust these values according to the softness of the tissue. If cells are not dissected from the surface, you can increase the energy and focus. The specimens were then subjected to the dissection by hitting the laser in the designated area on the glass slide and cells were collected in the inverted microcentrifuge cap. At the end of dissection, 200 μl of ethanol was added into the microcentrifuge tube. Tubes were centrifuged at 13,000 rpm (17,000–18,000×*g*) for 15 min. After centrifugation, most of the ethanol was removed from tube and cells were used to extract the DNA (*see* **Note 8**).

3.5 Genomic DNA Isolation

1. Use the whole pellet obtained from **step 6** of Subheading 3.4 for DNA isolation. Perform the DNA isolation according to the manufacturer's protocol (Qiagen DNeasy Tissue Kit). Since you have already treated the tissue with xylene at the end of H and E staining, it is not necessary to treat the pellet with xylene. If your pellet size is large and if it seems to contain wax, treat your pellet with xylene according to the manufacturer's protocol. Since total volume of DNA solution for the bisulfite treatment is around 20 μl, elute your DNA in 20–30 μl of water rather than 100 μl. Or elute the DNA in 100 μl water and concentrate the DNA with vacuum dryer, if the latter is available.

2. Measure the concentration of your DNA with NanoDrop, and aliquot 200–500 ng DNA for the next step. If you have less than 200 ng DNA, continue with your DNA solution.

However, if the total amount of your DNA is less than 500 pg, you need to prepare more slides and perform more laser capturing.

3.6   Identification of CpG Island

1. Based on the bioinformatics tool mentioned in the Subheading 2.6, find your potential CpG island around your target genes. By blatting this region in UCSC genome browser, you can confirm the presence of the CpG island and download the CpG island with 100 bp flanking regions that can be used for the primer design in the next step.

3.7   Primer Design and CpG Assay Design

1. Primer design for bisulfite treated DNA is the critical step for the DNA methylation experiment since all cytosine in the genome except the cytosine of CpG pairs converted into uracil, however the status of cytosine in the CpG pairs after the bisulfite treatment depends on the methylation profile. Therefore, bisulfite sequencing primers should align to the CpG free region. Therefore, it is useful to include some flanking region around your CpG island before starting primer design.

2. By using the "repetitive masker" from Subheading 2.7, you can eliminate repetitive region for your primer design.

3. Some important criteria for pyrosequencing PCR primers;

    (a) They should not form dimers or hairpin structures.

    (b) Length of primers should be between 20 and 30 bp with an annealing temperature of 50–60 °C.

    (c) Product size should be between 100 and 300 bp, 200 bp is the optimum length (It is very difficult to find CpG-free regions for some parts of genome in which CpG intensity is very high; you might increase the size of your PCR fragment up to 500 bp in these cases).

    (d) One of the forward or reverse primers should be biotinylated at their 5′ end.

    (e) There is no large "T" stretches in the target sequence.

4. Biotage PSQ program is one of the programs that can help you to design the bisulfite-specific PCR primers and sequencing primers. First you need to convert all "CG" pairs into "YG" pairs in your target sequence, and then convert all "C" into "T" (Fig. 2). Import this modified DNA sequence into sequence editor. On the sequence tab, you can set your region where you want to design your PCR primer and sequence primer. Hit run button to start analysis. The color of the primer on the right tab demonstrates the quality of the primer sets; Blue primers are of high quality, yellow primers are of medium quality, orange primers are of low quality, and red primers are

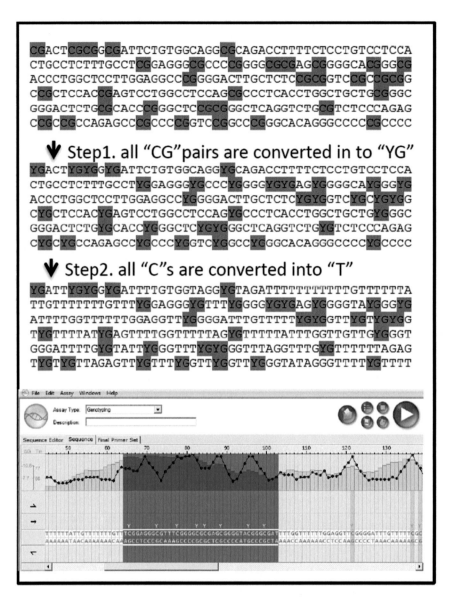

Fig. 2 Preparation of your target sequence for primer design. *Top panel*: original sequence of your target. *Middle panel*: All "CG" pairs converted into "YG" and *shaded with gray*. *Bottom panel*: All other remaining C's in the original sequence are converted into T's. Lowest section demonstrates that how the modified sequence looks like once you import your modified sequence into the Pyro Q CpG program. *Blue shaded area* was chosen for the first assay

of very low quality primer with some severe problem. Usually scores for the primers are in the range of 50–70 out of 100. Warnings for the primers should be taken into consideration before ordering, and they are sometimes eliminated by changing the target site or with a few base pair changes in the position of the primers. You can order blue or yellow primers for your assay and do not forget to add biotin to one of the forward or reverse primers. If your amplicon size is larger than 100 bp, you can design several sequencing primers since each sequencing primer only reads up to 50 bp in the pyrosequencing run.

3.8 Bisulfite Treatment and Bisulfite-Specific PCR

1. Treat up to 500 ng isolated DNA from the dissected cells and unmetylated control DNA by using EZ DNA methylation Gold kit (Zymo Research) according to the manufacturer's protocol. Suggested treatment on the thermal cycler is:

   98 °C for 10 min, 64 °C for 2.5 h.

   Alternative incubation for bisulfate conversion:

   98 °C for 10 min, 53 °C for 30 min, (53 °C for 6 min + 37 °C for 30 min) for eight cycles,

2. DNA can be eluted in 20 µl water or elution buffer and 2 µl of the elute should be used in the BSP PCR reaction (*see* **Note 9**). High fidelity DNA polymerase or HotStarTaq DNA Polymerase should be used for the amplification of the target region. PCR was performed with the following conditions; 95 °C for 5 min, (94 °C for 30 s, 56 °C for 30s, and 72 °C for 30s) for at least 50 cycles, 72 °C for 7 min (*see* **Note 10**). You can also try SYBR Green 2× mixture if you have some difficulties to get a single band PCR amplicon.

3.9 Gel Electrophoresis

1. Run your PCR reaction on 2 % agarose gel prepared with 1× TAE buffer. Run 4 µl of the total PCR reaction (50 µl) on the gel. If you have a single band for your target region, you can use the rest of PCR reaction for the pyrosequencing reaction.

2. If you have some nonspecific bands in your gel results, you need to optimize your PCR conditions to get a single band. If you do not eliminate the unspecific bands, load whole PCR reaction into the gel and extract the desired band from the gel by using minElute Gel Extraction kit (Qiagen).

3.10 Preparation of Pyrosequencing Template

1. Bring all buffers and components at room temperature and incubate for 10 min.

2. In each well of 96-well PSQ plate, add 0.3 µM–0.4 µM sequencing primer in the presence of 40 µl of annealing buffer.

3. In 96-well PCR plate, add 36 µl of PCR reaction, 4 µl of Sepharose bead (undiluted), and 40 µl of 2×-binding buffers (*see* **Note 11**). Prepare two wells as a control (no DNA) for each internal sequencing primer.

4. Seal the plates and place it on microplate-shaker for 10 min at room temperature (1400 rpm or 250–300 × g).

5. Wash the vacuum tool with distilled water twice (fill the trough with water and place the vacuum tool inside, turn on the vacuum and wait for 30 s).

6. Remove PCR plate from shaker and immediately place filter tips of vacuum tool carefully into the PCR plate.

7. After capturing the DNA and beads on the filter tip of vacuum, you need to do several washes with the following solutions: 70 % EtOH, denaturation solution, and washing buffer.

8. Fill the trough with 70 % EtOH and place the vacuum tool into the trough. Turn on the vacuum and let it flush through the filters for 5 s.

9. Repeat **step 8** with denaturation buffer and washing buffer, 5 s each.

10. Hold the vacuum tool perpendicular and let it dry for 5 s, and return it to the horizontal position.

11. Close the vacuum pump (*see* **Note 12**) and place the filter tips which have only biotinylated DNA strand into the PSQ plate (prepared at **step 2**). Shake the vacuum gently few times and leave it for a few seconds inside the PSQ plate. This step releases the bound DNA strand from filter into the solution which has the sequencing primers and annealing buffer.

12. Annealing of primers to the DNA: incubate PSQ 96-well plate on 80 °C heat block for 2 min (not more than 2 min) and place it back on a cold surface.

3.11 Pyrosequencing

1. Put your PSQ plate in your instrument and insert the right cartridge gently.

2. Fill the cartridge by using the right concentration and volume of solution E (enzyme mixture), solution S (substrate mixture), and the four nucleotides dATP-a-S (A), dCTP (C), dGTP (G), and dTTP (T) according to the manufacturer's protocol (*see* **Note 13**) .

3.12 Run Sample and Data Analysis

1. For each pyrosequencing primer you are using for individual assay, you should enter the input sequence as a "sequence to analyze". Program automatically produces a dispensation order for your fragment.

2. For each sequencing primer, choice the well on the computer with the appropriate assay.

3. Click run button on the computer. Enzymes, substrate, and four nucleotides should be dispensed according to the predetermined order.

4. After the pyrosequence run, each well on the plate is shown with color with the following code:
   Blue wells = pass; Yellow or Orange wells = might be corrected with manual edit; Red wells = failed

5. Do not forget to check negative control wells in which you have only primer. You should not see any signal in those wells.

6. Pyro Q CpG Software will demonstrate two graphs at the end, one is for theoretical histogram and the second one is actual pyrogram showing the intensity of peaks. On both graphs, *x*-axes show the dispensation order of your target sequence. While *y*-axis shows number of nucleotides in the histogram, it shows the relative peak intensity in the pyrogram.

7. Since cytosine in the CG pair has two possible outcomes after bisulfite sequencing, it is represented as a gray bar with two nucleotides (T and C). "G" is not represented immediately after gray shaded "TC" because the dispensing order sometimes puts random nucleotides before "G".

   On top the gray boxes, calculated percentage of methylation is shown for each cytosine of CG pairs (Fig. 3).

8. You can trust the percentage of DNA methylation if the values are in the range of 5–95 % because of the 5 % deviation rate between the biological replicates.

4 Notes

1. There are three types of Hematoxylin Gill solutions: 1, 2, and 3. Their names are based on the concentration of the dye; Gill-1 has 2 g dye per 1 ml, Gill-2 has 4 g, and Gill-3 has 6 g per liter. Depending on the intensity of nuclear stain, you can choose one of these solutions. Protect these solutions from dye and store at room temperature. There is also Hematoxylin Mayer solution in which dye concentration is the half of the Gill-1 solution.

2. There is no true universal fixative, but 10 % neutral buffered formalin is the most commonly used fixative for the specimen preparation. Since it is carcinogenic and can cause eye, skin, and respiratory tract irritation, it should be prepared inside the hood.

3. Do not touch the sections and wax samples with your finger. Instead, use forceps or brush to pick up specimen from block face.

4. If you have difficulties to get one single smooth layer of section, cooling the surface of paraffin block on the ice might help. You can also try moisturizing the surface of the wax paraffin.

5. Pencil mark is permanent during the xylene and alcohol treatment. Numbering of each slide is important since we use first section as a reference slide for the next five or six sections. We cover the first section with glass coverslip in the presence of mounting media and use it for localization of the tumor cells during the laser capture microdissection. Do NOT cover the other slides.

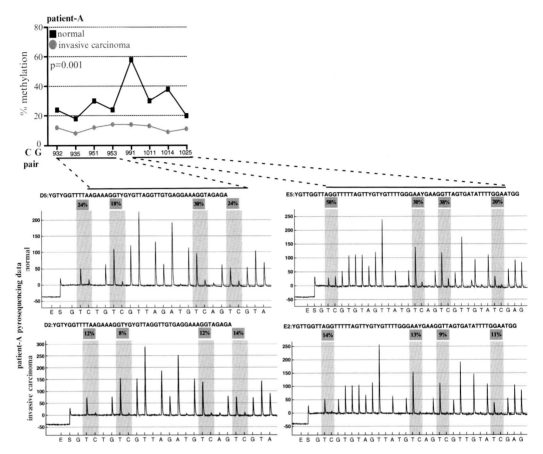

Fig. 3 Analysis of Pyrosequence Run. KIAA1199 (CEMIP) promoter region was amplified from bisulfite treated genomic DNA with two different primer pairs shown by two *solid lines* on the *y*-axis of the top graph. Upper pyrograms demonstrate the methylation profile of normal (benign) tissue; bottom pyrograms show the methylation profile of LCM dissected cancer cells. Whole methylation percentages are summarized on the top graph. Breast cancer patients have higher expression of KIAA1199 with respect to the normal (benign) tissue because of the hypomethylation on its promoter region

6. To collect the dissected cells from surface of the slide into the cap, the cap should be in close proximity to the surface. Be aware of the possible risk that the slide might get wet if the meniscus of the solution gets in touch with the surface of the slide.

7. One cell has approximately 57 μm² area on the surface of the slide. One single human cell with 2N karyotype has 6 pg DNA. Since the lower limit of Zymo research kit for bisulfite conversion is 500 pg, you need to isolate at least $500/6 = 84$ cells. Quality of DNA in FFPE tissue is very low, and therefore we multiply this amount with 10, so $84 \times 10 \times 57$ μm² $= 48,000$ μm² area should be selected as a minimum total "element" on one slide.

8. After the centrifugation, the solution might turn pinkish because of the hematoxylin and eosin staining of the cells. This color change could be a sign of presence of cell in the cap.

9. Treat 500 ng of unmethylated control DNA with EZ DNA methylation Gold kit. Use the MSP-control primer with bisulfite treated control DNA for the PCR reaction, and digest the PCR product with HpaII. If your bisulfate conversion works well, HpaII should not digest the PCR product.

10. Pyro-PCR should run at least 50 cycles to extinguish the extra nucleotides and primers in the PCR solution.

11. 20 μl PCR reaction from genomic DNA is usually enough for pyrosequencing reaction; however, you should increase the amount to 36–40 μl to increase the quality of signal because of the FFPE treated and laser captured DNA.

12. Closing the vacuum pump is very critical at this step; if you leave the pump on, you would suck up the primer solutions from the PSQ plate.

13. Since dATP can react with luciferin to produce photons, you should use modified dATP, called "dATPαS." It is not a substrate of luciferase and usage of the modified dATP reduces the noise ratio during the reaction.

References

1. Jemal A, Siegel R, Xu J, Ward E (2010) Cancer statistics, 2010. CA Cancer J Clin 60(5): 277–300

2. Emmert-Buck MR, Bonner RF, Smith PD, Chuaqui RF, Zhuang Z, Goldstein SR, Weiss RA, Liotta LA (1996) Laser capture microdissection. Science 274(5289):998–1001

3. Espina V, Heiby M, Pierobon M, Liotta LA (2007) Laser capture microdissection technology. Expert Rev Mol Diagn 7(5):647–657

4. Sharma S, Kelly TK, Jones PA (2010) Epigenetics in cancer. Carcinogenesis 31(1):27–36

5. Dawson MA, Kouzarides T (2012) Cancer epigenetics: from mechanism to therapy. Cell 150(1):12–27

6. Greer EL, Blanco MA, Gu L, Sendinc E, Liu J, Aristizabal-Corrales D, Hsu CH, Aravind L, He C, Shi Y (2015) DNA methylation on N(6)-Adenine in C. elegans. Cell 161(4):868–878

7. Zhang G, Huang H, Liu D, Cheng Y, Liu X, Zhang W, Yin R, Zhang D, Zhang P, Liu J, Li C, Liu B, Luo Y, Zhu Y, Zhang N, He S, He C, Wang H, Chen D (2015) N(6)-methyladenine DNA modification in Drosophila. Cell 161(4):893–906

8. Robertson KD (2005) DNA methylation and human disease. Nat Rev Genet 6(8):597–610

9. Gardiner-Garden M, Frommer M (1987) CpG islands in vertebrate genomes. J Mol Biol 196(2):261–282

10. Saxonov S, Berg P, Brutlag DL (2006) A genome-wide analysis of CpG dinucleotides in the human genome distinguishes two distinct classes of promoters. Proc Natl Acad Sci U S A 103(5):1412–1417

11. Portela A, Esteller M (2010) Epigenetic modifications and human disease. Nat Biotechnol 28(10):1057–1068

12. Jaenisch R, Bird A (2003) Epigenetic regulation of gene expression: how the genome integrates intrinsic and environmental signals. Nat Genet 33(Suppl):245–254. doi:10.1038/ng1089ng1089 [pii]

13. Hayslip J, Montero A (2006) Tumor suppressor gene methylation in follicular lymphoma: a comprehensive review. Mol Cancer 5:44

14. Esteller M (2002) CpG island hypermethylation and tumor suppressor genes: a booming present, a brighter future. Oncogene 21(35):5427–5440

15. Feinberg AP, Vogelstein B (1983) Hypomethylation of ras oncogenes in primary

human cancers. Biochem Biophys Res Commun 111(1):47–54

16. Clark SJ, Harrison J, Paul CL, Frommer M (1994) High sensitivity mapping of methylated cytosines. Nucleic Acids Res 22(15):2990–2997

17. Frommer M, McDonald LE, Millar DS, Collis CM, Watt F, Grigg GW, Molloy PL, Paul CL (1992) A genomic sequencing protocol that yields a positive display of 5-methylcytosine residues in individual DNA strands. Proc Natl Acad Sci U S A 89(5):1827–1831

18. Ronaghi M, Karamohamed S, Pettersson B, Uhlen M, Nyren P (1996) Real-time DNA sequencing using detection of pyrophosphate release. Anal Biochem 242(1):84–89

19. Ronaghi M, Uhlen M, Nyren P (1998) A sequencing method based on real-time pyrophosphate. Science 281(5375):363, 365

20. Li LC, Dahiya R (2002) MethPrimer: designing primers for methylation PCRs. Bioinformatics 18(11):1427–1431

Part III

Isolation of Breast Cancer Cells

Chapter 9

Vita-Assay™ Method of Enrichment and Identification of Circulating Cancer Cells/Circulating Tumor Cells (CTCs)

Shaun Tulley, Qiang Zhao, Huan Dong, Michael L. Pearl, and Wen-Tien Chen

Abstract

The ability to capture, enrich, and propagate circulating cancer cells/circulating tumor cells (CTCs) for downstream analyses such as ex vivo drug-sensitivity testing of short-term cultures of CTCs, single cell sorting of CTCs by fluorescence activated cell sorting (FACS), animal injection tumor and/or metastasis formation studies, next generation sequencing (NGS), gene expression profiling, gene copy number determination, and epigenomic analyses is of high priority and of immense importance to both the basic research and translational/clinical research communities. Vitatex Inc.'s functional cell separation technology, constructed as Vita-Assay™ (AG6W, AN6W, AR6W) culture plates, is based on the preferential adhesion of invasive rare blood cells of tissue origin to a tissue or tumor microenvironment mimic—the so-called cell adhesion matrix (CAM), which has a demonstrated ability to enrich viable CTCs from blood up to one-million fold.

The CAM-scaffold allows for the functional capture and identification of invasive CTCs (iCTCs) including invasive tumor progenitor (TP) cells from cancer-patients' blood. CAM-captured CTCs are capable of ingesting the CAM (CAM+) itself. Green and red fluorescent versions of Vita-Assay™ (AG6W and AR6W) allow for direct visualization of CAM-uptake by cancer cells. Vita-Assay™ CAM-enrichment has allowed for sensitive multiplex flow cytometric and microscopic detection of iCTCs from patients with cancers of the breast, ovary, prostate, pancreas, colorectum, and lung; it has also been successfully utilized for ex vivo drug-sensitivity testing of ovarian-cancer patient CTCs. The CAM enrichment method is equally suitable for the separation of iCTCs and TP cells in ascites and pleural fluid.

Key words Cancer cell capture, Cell invasion, Metastasis detection, Rare-cell enrichment, Circulating cancer cells, Circulating tumor cells, iCTCs, Vita-Assay™, Cell-adhesion matrix, Vitatex Inc

1 Introduction

The development of human cancer is a multistep process driven by genome instability which ultimately results in the acquisition of at least eight major biological capabilities of cancer cells, namely, sustained proliferative signaling, evasion of growth suppressors, an increased resistance to cell death, an unlimited replicative potential (immortality), the ability to induce angiogenesis, evasion of

Jian Cao (ed.), *Breast Cancer: Methods and Protocols*, Methods in Molecular Biology, vol. 1406,
DOI 10.1007/978-1-4939-3444-7_9, © Springer Science+Business Media New York 2016

immune destruction, a reprogramming of energy metabolism, and activation of the multistep process of invasion and metastasis (invasion–metastasis cascade) [1, 2].

Metastasis, or dissemination of primary tumor cells ("seeds") through the blood and lymph systems to distant secondary-sites/organs ("soil") in the body, involves a cascade of sequential and discernible steps including the loss of cellular adhesion; an increase in cell motility; invasion of adjacent tissues; entry into, survival in, and transport through the circulation; arresting in a secondary site; and eventual extravasation and growth in the new tissue/organ [3–7]. Indeed, it is this very process, cancer metastasis, which is responsible for more than 90 % of cancer-associated deaths [5, 8–11]. Despite encompassing such a lethal series of steps in neoplastic disease progression, experimental models indicate that of the millions of cancer cells that escape the primary tumor and are continuously being dispersed through the body, only a small number of these cells (<0.1 %) are able to survive in the circulation (viable CTCs), reach a distant organ, survive in a dormant state in the secondary site, evade the immune system and any systemic therapy a cancer patient is receiving, and grow into an overt metastatic lesion [7, 12–17].

It should therefore come as no surprise the immense interest shown by cancer biologists and clinical oncologists alike and the impetus being put on capturing, enriching, identifying, and studying these extremely rare, viable, metastasis-initiating CTCs/cancer stem/tumor progenitor cells, the key players in the transition from localized to systemic disease [4, 7, 9, 13, 18]. Basic cancer research studies in areas such as the molecular characterization of CTCs as well as clinical applications such as patient CTC enumeration and sequential CTC-monitoring of cancer patient blood during systemic treatment are of high value and importance. The main obstacle in studying CTCs is of course the rarity that they appear in blood at approximately one CTC in a background of a billion red blood cells, and millions (10^6–10^7) of white blood cells and at an average frequency on the order of 10–100 CTC per mL of whole blood [13, 19–22].

The Vita-Assay™ CTC-enrichment platform has been specifically designed to functionally capture and enrich (up to one million-fold) these rare, viable, and invasive CTCs (iCTCs) and to exploit their preferential adherence to the CAM surface attributable to their inherent invasive nature and high avidity for the extracellular matrix (ECM), with the added advantage that CAM-captured CTCs can be identified after subsequent ingestion of fluorescent-CAM (CAM+) [18, 23–26]. Furthermore, since the functional proclivities to degrade and ingest the ECM are major acquired capabilities of invasive and metastatic cells, CAM+ cells represent a unique way to identify and enrich CTCs, and in all likelihood these iCTCs encompass resident TP or cancer stem cells [18, 23–26].

The CAM method allows for the capture of iCTCs indiscriminant of tumor cell type (primary tumor origin), cell size, CTC morphology or expression of particular protein-markers and thus offers a robust, comprehensive opportunity to capture true metastasis-initiating cells. Vita-Assay™ has allowed for sensitive multiplex flow cytometric and microscopic detection of CTCs/iCTCs from patients with cancers of the breast, ovary, prostate, pancreas, colorectum, and lung to date, and has been used to relate CTC enumeration and prognosis, as well as to generate genotypic and phenotypic data from cultured CTCs captured from the blood of prostate, ovarian, and breast cancer patients [18, 23–26]. Since patient's CTCs captured using Vita-Assay™ plates are viable, continued culture of CTCs in the same plate is suitable for ex vivo drug-sensitivity and resistance testing.

In conclusion, the Vita-Assay™ method enables the efficient capture, enrichment, and identification of rare, invasive, and viable CTCs, allowing for an array of downstream analyses/applications of both cancer research (CTC culturing, mutational analysis, molecular characterization, ex vivo drug selection studies, etc.) and clinical investigation (CTC enumeration, prognostic value, and sequential measurement of patient CTC number to monitor treatment response).

2   Materials

2.1   Vita-Assay™

1. Vita-Assay™ plate (Vitatex, Stony Brook, NY, USA): Each 6-well plate is coated with either plain-CAM (Vita-Assay™ AN6W, Product # 102.01 N) or red fluorescent-CAM (Vita-Assay™ AR6W, Product # 102.02R) or green fluorescent-CAM (Product # 102.03G). Each plate is equivalent to six assays and allows for the processing of six distinct patient samples on a single plate (*see* **Note 1**).

2. Each Vita-Assay™ also includes three 100-µL tubes of Cell Releasing CAM Enzyme (*see* **Note 2**). Cell Releasing CAM Enzyme must be stored in a freezer (–20 °C) immediately upon receipt (*see* **Notes 3** and **4**).

2.2   Necessary Materials Not Supplied with Vita-Assay™

1. BD Vacutainer® 10.0 mL sodium heparin tubes are recommended (*see* **Note 5**).

2. Antibodies for Flow Cytometry: Positive identification of iCTCs: Use (1) PE-conjugated TP marker (PE-anti-seprase/PE-anti-CD44 antibody mix (Vitatex). Alternatives are (2) PE-EpCAM (BioLegend), or antibodies against (3) CA125 (ovarian), (4) CA19-9 (pancreatic), (5) HER2 (breast), (6) PSMA (prostate). For exclusion of hematopoietic lineage (HL) or immune cells, use APC-conjugated anti-CD45 (BD).

3. Antibodies for Microscopy: Positive identification of iCTCs: FITC-conjugated TP marker (FITC-anti-seprase/anti CD44 antibody mix (Vitatex) or FITC-anti-Epi mix of ESA (Biomeda), EPCAM (BioLegend), and BerEp4 (Dako). For exclusion of hematopoietic lineage (HL) or immune cells, use mouse anti-human CD45 (BD).

4. Dako color reaction Kit: (1) secondary antibody-biotinylated link, (2) streptavidin-AP, (3) BCIP/NBT substrate system.

5. 7-aminoactinomycin D (7-AAD).

6. Hoechst 33342 or DAPI.

7. 10× BD FACS Lysing solution (BD).

8. A low-speed, swing-out bucket centrifuge for pelleting cells.

9. A temperature-controlled CO_2 incubator.

10. A vacuum aspirator.

11. Sterile 15-mL polystyrene conical centrifuge tubes.

12. Sterile 50-mL polystyrene conical centrifuge tubes.

13. 5-mL polystyrene tube with cell strainer cap.

14. 10× Red cell lysis buffer: 1.54 M NH_4Cl, 100 mM $KHCO_3$, 1 mM EDTA pH 8.0. Either prepare or purchase commercially available RBC Lysis Buffer 10× (BioLegend).

15. Standard complete medium containing 10 % fetal calf serum. Alternatively, complete cell culture (CCC) medium: 1:1 mixture of Dulbecco's modified Eagle's medium (DMEM) and RPMI 1640 supplemented with 10 % calf serum, 5 % Nu-serum, 2 mM L-glutamine, 1× penicillin–streptomycin Solution.

16. Sterile 1× phosphate buffered saline (PBS), pH 7.4.

17. Sterile 1× phosphate buffered saline containing **Ca2+ (Ca2+ PBS)**, pH 7.4.

18. Sterile 0.2 % bovine serum albumin (BSA) in 1× PBS, pH 7.4.

19. Adjustable water bath set to 37 °C.

20. Cytocentrifuge system (such as Statspin cytofuge).

21. Trypsin–EDTA solution.

22. Standard fluorescent microscope system with phase-contrast, blue, green, and red channels.

3   Methods

Carry out procedures at room temperature and under sterile conditions unless otherwise specified.

3.1 Vita-Assay™: Cell Isolation Protocol

1. Blood Collection: Using your institution's recommended procedure for blood standard venipuncture, collect blood into one or more BD Vacutainer® 6.0 mL sodium heparin tubes or 6.0 mL lithium heparin tubes or BD Vacutainer® 10.0 mL sodium heparin tubes. Transport tubes to laboratory (*see* **Notes 6** and **7**).

2. Preparation of the nuclear cell fraction by red blood cell (RBC) lysis: For preparation of nuclear cells from whole blood, red blood cells in the blood samples will be lysed by mixing blood and 1× RBC Lysis Buffer at a ratio of 1:25 at 20–25 °C. For a 2-mL blood sample, mix 2-mL blood and 48-mL 1× RBC Lysis Buffer at 20–25 °C in a sterile 50-mL conical tube.

3. Rotate the 50-mL tube with blood mix on a rotator at low speed (10 rpm, Stovall Low Profile Roller) for 5 min at 20–25 °C.

4. Pellet nuclear cells by centrifugation to $350 \times g$ for 5 min. Carefully remove supernatant.

5. Add 4-mL of a Complete Cell Culture (CCC) medium to each 50-mL conical tube to resuspend the cell pellet, resulting in a total of 4-mL cell suspension in CCC medium containing the cells derived from 2-mL whole blood.

6. Loading cell suspension into wells of Vita-Assay™ plates: Aliquot 2-mL of the cell suspension into each of the 6 wells of the Vita-Assay™ plate coated with CAM. Culture cells in a 5–7 % CO_2 incubator at 37 °C for 1 h to allow adherence of tumor cells into CAM to obtain the enriched CAM-avid cells.

7. Wash away floating cells and culture CAM-avid cells: After 1 h incubation above, remove unattached cells and existing medium and discard into a waste container.

8. Wash wells one time by pipetting 2-mL of CCC medium gently down the side of each well.

9. Move the plate in a horizontal circle six (6) times and discard wash solution.

10. Add 2-mL of CCC medium into each well and culture cells for 18 h to enhance signal of CAM uptake by iCTCs (only AG6W and AR6W will display CAM uptake signal) (*see* **Notes 8** and **9**).

11. After 18 h cell incubation in each well remove any unattached cells along with medium into a waste container.

12. Freshly prepare CAM enzyme working solution by diluting the provided frozen enzyme with 2-mL of 1× phosphate buffered saline containing Ca^{2+} (Ca^{2+} PBS), pH 7.4 for use in **step 14** (*see* **Note 10**).

13. Wash wells three times by pipetting 2-mL of 1× PBS into each well each time, followed by moving the plate in horizontal circle three (3) times and discarding the wash solution.

14. Add 1-mL of the CAM enzyme working solution into each washed Vita-Assay™ well.

15. Place Vita-Assay™ plate in a CO_2 incubator at 37 °C for 10 min to enzymatically dissolve CAM and release tumor cells into suspension.

16. Optional Step: Skip this step when processing cancer-patient blood samples. Only follow this step for experiments in which cancer cell-lines are spiked-in to healthy donor blood (CTC model experiments) and only if CAM-enzyme alone does not detach tumor cells, which are usually two to four times bigger than co-isolating hematopoietic cells. Add 1-mL of the trypsin–EDTA solution into each Vita-Assay™ well, and place in a CO_2 incubator at 37 °C for 5 min to detach remaining adherent cells.

17. Transfer cell suspension into a new 15 mL conical centrifuge tube.

18. Wash the well (sequentially with other wells of the same experimental condition) of Vita-Assay™ plate one (1) time with 3-mL of CCC medium.

19. Transfer the wash into the 15 mL tube with enzymatically released cells (*see* **Note 11**).

20. Concentrate cells by centrifuging the 15 mL conical tube at $350 \times g$ for 5 min.

21. After centrifugation, remove supernatant by gentle aspiration, retaining the last 100 μL containing the enriched cell fraction (*see* **Note 12**). The 100 μL cell suspension can be processed for cellular analyses including, but not limited to, enumeration by automated flow cytometry (Subheading 3.2 that follows below); validation of cell identity by microscopy (Subheading 3.3 below); and CTC cell culture (Subheading 3.4 below).

3.2 CTC Enumeration Using Automated Flow Cytometry

1. From **step 21** of Subheading 3.1 above, loosen and resuspend the cell pellet by pipetting up and down five times.

2. Fixation: Fix cells by adding 1.0 mL of 1× BD FACS Lysing solution to the 100 μL of cell suspension and incubating at 20–25 °C for 10 min.

3. After incubation in fixative, add 3-mL of PBS containing 0.2 % BSA, mix and concentrate cells by centrifuging the 15 mL conical tube at $350 \times g$ for 5 min.

4. After centrifugation remove supernatant by gentle aspiration, retaining the last 80 μL containing the enriched fixed-cells. The fixed cell suspension could be stored at 2–8 °C at this point.

5. Antibody and nucleic acid dye staining: Add 6 μL PE-anti-seprase/CD44 antibody cocktail (or 6 μL PE-anti-Epi), 10 μL APC-anti-HL antibody and 10 μL of 50 μg/mL stock of 7AAD to the 80 μL cell suspension and stain cells at 20–25 °C in the dark, for 30 min.

6. Washing of stained cells: Add 3-mL of PBS containing 0.2 % BSA.

7. Collect cells by centrifugation at $350 \times g$ for 5 min.

8. Remove supernatant and save the last 500-μL containing fixed, stained cell suspension in sterile PBS containing 0.2 % BSA.

9. Preparation for flow cytometric cell counting: Particulates in fixed, stained cells must be filtered away using a polystyrene tube with cell strainer cap. Collect cells by centrifugation at $350 \times g$ for 5 min.

10. Count cells with a multiplex flow cytometer, i.e., BD FACSCalibur™ (*see* **Note 13**).

11. Reporting: Flow cytometry plots and/or tables of enumeration of CTCs and/or iCTCs and immune cells per 1-mL of blood are shown (Fig. 1).

3.3 Validation of CTC Identity (on Vita-Assay™ AR6W) by Microscopy

The number of CTCs or iCTCs ascertained by flow cytometry can be validated using microscopy for cellular morphology of CTCs as compared to immune cells, observed as positive for CAM uptake (CAM+) and Epi or TP expression, and negative for hematopoietic lineage (HL) markers, as described in Protocol below. The microscopy step offers a clear morphological discrimination of tumor cells from hematopoietic (immune) cells.

1. Microscopy—Validation of iCTCs, CAM-avid CTCs, and immune cells: From **step 21** of Subheading 3.1, Loosen the cell pellet by pipetting up and down five times.

2. Fixation: Fixation can be done by adding 100 μL of 2 % para-formaldehyde in PBS, pH 7.3, to the 100 μL cell suspension, and incubate at 20–25 °C for 5 min.

3. Add 3-mL of PBS containing 0.2 % BSA, mix and concentrate cells by centrifuging the 15 mL conical tube at $350 \times g$ for 5 min, and remove supernatant by gentle aspiration, retaining the last 100 μL containing the enriched cells. The fixed cell suspension could be stored at 2–8 °C at this point.

4. Antibody and nucleic acid dye staining: Add 8 μL anti-HL antibody mix to the 100 μL cell suspension and incubate for 20 min, followed by staining using Dako kit: secondary-antibody biotin-conjugate, Streptavidin-AP enzyme, and BCIP/NBT substrate according to manufacturer's specifications (positive cells will stain blue/purple). Wash 1× in PBS containing 0.2 % BSA, and add 8 μL of FITC-TP mix or FITC-Epi mix, and add 1 μL 300 μM DAPI (or equivalent Hoechst) and stain cells at 20–25 °C in the dark, for 30 min.

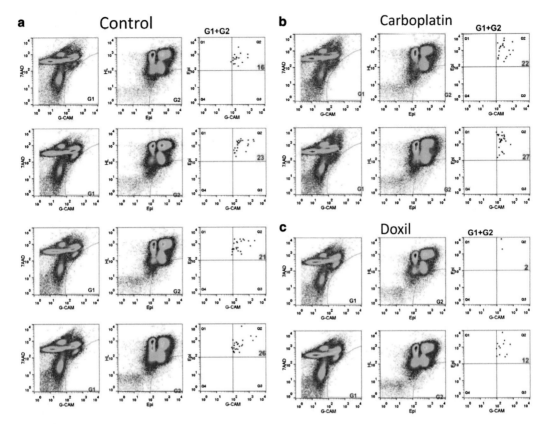

Fig. 1 Flow cytometric analysis and enumeration of iCTC and immune cells captured on Vita-Assay™ AG6W from 0.67 mL blood of a recurrent ovarian cancer patient (SB448-1). Cells were cultured (ex vivo) for up to 72 h in the absence (control) (**a**), or presence of chemotherapeutic drugs demonstrating drug-resistance (approximately equal counts of iCTCs) with (**b**) Carboplatin or indicating drug-sensitivity (lowered iCTC counts) with (**c**) Doxil (Doxorubicin). Gates were made on G-CAM+ (CAM uptake)/7AAD+ fixed cells (G1) to exclude platelets or non-cellular particles (*left panel*, events on 10^0 7AAD) that also exclude HL$^{+/dim}$ events (G2). Events that overlap between G1 and G2 are identified and enumerated as iCTC counts shown in *red* on the *right panel* of the plot for each blood sample. Immune cells were shown as 7AAD+ HL+ clusters in *left* two panels. Note that none of the experiments exhibited artifacts that were not possible to analyze; accordingly, the assay failure rate for the iCTC flow cytometry detection is estimated as 0 %. When cross samples were compared, iCTC and immune cell counts were standardized to per 1-mL blood counts

5. Washing of stained cells: Add 3-mL of sterile PBS containing 0.2 % BSA. Collect cells by centrifugation at $350 \times g$ for 5 min. Remove supernatant and save the last 200-µL containing fixed, stained cell suspension in sterile PBS containing 0.2 % BSA.

6. Preparation for microscopy: To concentrate isolated cells to 7-mm diameter area on a microscopic glass slide for microscopic analysis, cytospin preparation should be performed using devices, i.e., StatSpin cytofuge and Filter Concentrators.

7. After concentrating cells on glass slide, place a drop of mounting medium on the cell sample on the microscopic slide.

iCTCs: CAM⁺ TP⁺ NA⁺ HL⁻
Immune cells: CAM⁻ TP⁻ NA⁺ HL⁺

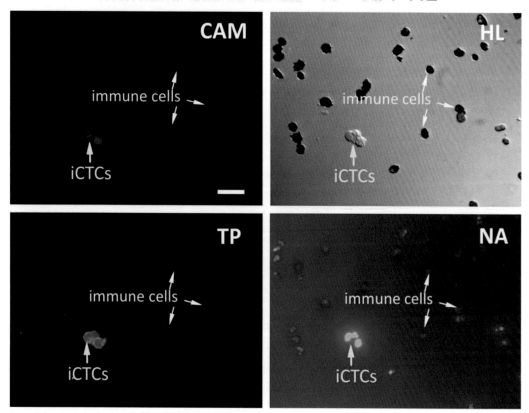

Fig. 2 Validation of CAM+ flow cytometric isolation of iCTCs by microscopic imaging. Phase contrast microscopy and fluorescence microscopy were used to image cells from 1 mL of blood of an ovarian cancer patient (SB423-2) which were captured and identified using Vita-Assay™ AR6W. In this field three iCTCs were identified and observed to be CAM+ (invaded into and ingested *red-fluorescent* CAM), TP+ (CD44+/seprase+) and NA+ (nucleic acid using Hoechst 33342). Take note of the numerous hematopoietic lineage cells (HL+) stained with *blue* color substrate-conjugated secondary antibody against primary CD45 antibody, and were unable to ingest CAM (CAM negative), and were also tumor/tumor progenitor marker negative (TP negative). Bar in CAM panel =40 μm

Examine and record images of cell types (CAM+, TP+, NA+, CD45–) using a fluorescence microscope equipped with multiple filters (Fig. 2).

3.4 CTC Cell Culture for Downstream Applications

CTCs in blood and captured in the Vita-Assay™ plate can be propagated in culture in the same plate in CCC media for a period of at least 14 days without loss of cellular viability, whereas the majority of co-isolating hematopoietic cells are lost within 7 days (Fig. 3). The culture of viable CAM-avid tumor and immune cells for a period of time allows for sufficient expansion of CTCs for use in a

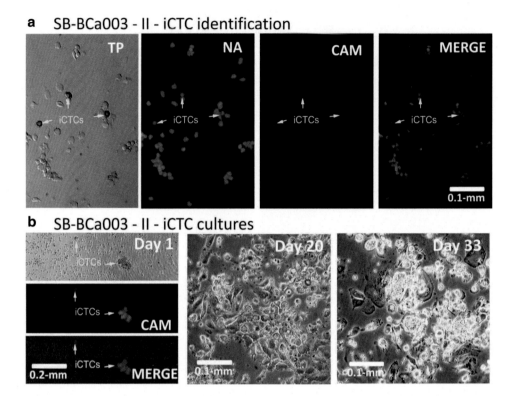

Fig. 3 (**a**) Fluorescence microscopic images of cells captured on Vita-Assay™ AR6W and derived from 1 mL blood of a patient with stage II breast cancer (SB-BCa003-II). iCTCs were identified as TP (anti-CD44/anti-seprase) positive and were stained with *red* color substrate-conjugated secondary antibodies against primary TP antibodies that targeted iCTCs, in addition to being nucleic acid positive (NA) as determined with Hoechst 33342 dye+, and CAM+ having ingested red-fluorescent CAM. Single *yellow arrows* depict iCTCs. (**b**) Vita-Assay™ AR6W CAM-enriched cells were cultured in CCC media on the CAM scaffold for 1–33 days. Live iCTCs were photographed under phase contrast microscopy and fluorescence microscopy (red-fluorescent CAM, to reveal CAM uptake/labeling of tumor cells). iCTCs propagated as time increased. On day 1 (**b**, *left panels*), tumor cells (iCTCs) were seen to associate with and ingest/uptake CAM and morphologically presented as solitary cells and in a cluster (*yellow arrows*) larger than hematologic cells (smaller cells seen under phase). By day 20 (**b**, *middle panel*), iCTCs were observed to have propagated and expanded in culture, whereas co-isolating hematologic cells decreased in number and were not apparent in the field. By day 33 (**b**, *right panel*), tumor cells grew in clusters with large epithelioid cells

variety of downstream applications, for example, to harvest cells and isolate DNA, RNA, or protein for CTC mutational and/or molecular analyses. In addition, viable CAM-avid tumor cells can be applied in ex vivo drug sensitivity testing and mouse injection/ tumor formation studies that are intended for improved therapeutic intervention. Simply follow **steps 1–10** of Subheading 3.1. Cells from that point can be cultured on CAM in CCC media. Below is a brief overview of the procedure if testing out an inhibitor or drug on CTCs in culture.

1. Blood Collection/Shipping: For culture and experimental testing of viable CTCs a larger volume (for example 10–20 mL

or if possible more) of blood is preferred. Venous blood will be drawn from a cancer patient using three BD Vacutainer® 10.0 mL green capped, sodium heparin tubes. Blood samples can be collected in the clinic and shipped using overnight express shipping from distant sites at refrigerated temperature (2–8 °C) to lab.

2. Red Blood Cell Lysis: Prepare $1–10 \times 10^6$ nuclear cells from 1-mL blood using red blood cell lysis buffer.

3. Suspend cells in 2-mL CCC medium and seed cells for 1 h in 1 well of Vita-Assay™ plate.

4. CTC Enrichment and Experimental Testing: After the 1 h incubation, remove floating cells and replace fresh medium ± inhibitor/drug/or any combination thereof in each well of Vita-Assay™ plates. It is highly recommended to conduct such studies in duplicate if not triplicate (*see* **Note 14**).

5. Flow Cytometry—Cell Identification and Enumeration: After at least 2 days in culture, rotate and wash the cellular layer 3× with PBS.

6. Perform enzymatic digestion of CAM using CAM-enzyme.

7. Collect CAM-bound cells, fix cells, and stain cells with APC-anti-hematopoietic lineage (HL) antibody markers, PE-anti-TP antibody markers, and nucleic acid (NA) dye 7AAD for enumeration of iCTCs and immune cells by flow cytometry (*see* **Note 15**).

4 Notes

1. One Vita-Assay™ plate can be used for six assays or six rare cell separations. $1–15 \times 10^6$ nuclear cells from 1–3 mL of whole blood (after red blood cell lysis) should be loaded into 1 well of the 6-well plate.

2. CAM-enzyme is used for enzymatic digestion of CAM and release of CTCs/endothelial progenitor cells from CAM.

3. Immediately prior to use, dilute the frozen cell releasing CAM enzyme by adding 2-mL of 1× phosphate buffered saline containing Ca^{2+} (Ca^{2+} PBS), pH 7.4, to make CAM enzyme working solution.

4. Any unused CAM-enzyme working solution can be stored at 2–8 °C for up to 5 days.

5. Other suitable recommended tubes are BD Vacutainer® 6.0 mL sodium heparin tubes (BD catalog # 367878) or 6.0 mL lithium heparin tubes (BD catalog # 367886).

6. Blood in collection tubes may be stored for up to 48 h at 2–8 °C prior to cell enrichment steps.

7. All pipetting steps from this point should be conducted in a laminar flow hood using sterile technique.

8. The 6 wells can be used for a simple experiment (by combining all cells collected) or six (6) experimental points.

9. CAM-avid cells can continue to be cultured by adding 2-mL CCC medium into each well and transferring plate to incubator.

10. Any unused CAM enzyme working solution can be stored at 2–8 °C for up to 5 days.

11. In case of combining all cells collected, the 6 wells can be sequentially washed each time with the same 3-mL of CCC medium and transferring the wash at the end of each round into the 15 mL tube.

12. Use careful aspiration of the supernatant (by pipetting) to prevent any cell loss.

13. Gating/CTC enumeration can be done directly on BD FACSCalibur™. Our lab typically saves the raw flow cytometry data and analyzes it using FlowJo software.

14. For example, a total of 18 wells can be used to culture cells—among 18 wells, 4 wells without any drug or inhibitor (control); cells in 14 remaining wells can be treated with specific drugs, inhibitors, or any other experimental test the end user might want to try.

15. Refer to Fig. 1 as it demonstrates iCTC counts from 3-day cultures of cells treated with or without two cytotoxic drugs, demonstrating the ability of the Vita-Assay™ to be used in drug sensitivity testing of CTCs.

Acknowledgement

This study was supported by Small Business Innovative Research (SBIR) grant R44CA140047 from the NCI awarded to Vitatex Inc. that holds a subcontract with Stony Brook Medicine.

References

1. Hanahan D, Weinberg RA (2011) Hallmarks of cancer: the next generation. Cell 144(5):646–674. doi:10.1016/j.cell.2011.02.013, S0092-8674(11)00127-9 [pii]

2. Hanahan D, Weinberg RA (2000) The hallmarks of cancer. Cell 100(1):57–70, S0092-8674(00)81683-9 [pii]

3. Gupta GP, Massague J (2006) Cancer metastasis: building a framework. Cell 127(4): 679–695. doi:10.1016/j.cell.2006.11.001, S0092-8674(06)01414-0 [pii]

4. Oskarsson T, Batlle E, Massague J (2014) Metastatic stem cells: sources, niches, and vital pathways. Cell Stem Cell 14(3):306–321. doi:10.1016/j.stem.2014.02.002, S1934-5909(14)00053-8 [pii]

5. Spano D, Heck C, De AP, Christofori G, Zollo M (2012) Molecular networks that regulate

cancer metastasis. Semin Cancer Biol 22(3):234–249. doi:10.1016/j.semcancer.2012.03.006, S1044-579X(12)00055-7 [pii]

6. Fidler IJ (2003) The pathogenesis of cancer metastasis: the 'seed and soil' hypothesis revisited. Nat Rev Cancer 3(6):453–458. doi:10.1038/nrc1098, nrc1098 [pii]

7. Chambers AF, Groom AC, MacDonald IC (2002) Dissemination and growth of cancer cells in metastatic sites. Nat Rev Cancer 2(8):563–572. doi:10.1038/nrc865, nrc865 [pii]

8. Mehlen P, Puisieux A (2006) Metastasis: a question of life or death. Nat Rev Cancer 6(6):449–458. doi:10.1038/nrc1886, nrc1886 [pii]

9. Monteiro J, Fodde R (2010) Cancer stemness and metastasis: therapeutic consequences and perspectives. Eur J Cancer 46(7):1198–1203. doi:10.1016/j.ejca.2010.02.030, S0959-8049(10)00157-7 [pii]

10. Nguyen DX, Bos PD, Massague J (2009) Metastasis: from dissemination to organ-specific colonization. Nat Rev Cancer 9(4):274–284. doi:10.1038/nrc2622, nrc2622 [pii]

11. Pantel K, Brakenhoff RH (2004) Dissecting the metastatic cascade. Nat Rev Cancer 4(6):448–456. doi:10.1038/nrc1370, nrc1370 [pii]

12. Kang Y, Pantel K (2013) Tumor cell dissemination: emerging biological insights from animal models and cancer patients. Cancer Cell 23(5):573–581.doi:10.1016/j.ccr.2013.04.017, S1535-6108(13)00182-7 [pii]

13. Joosse SA, Gorges TM, Pantel K (2014) Biology, detection, and clinical implications of circulating tumor cells. EMBO Mol Med. doi:10.15252/emmm.201303698. emmm.201303698 [pii]

14. Fidler IJ (1970) Metastasis: quantitative analysis of distribution and fate of tumor emboli labeled with 125 I-5-iodo-2′-deoxyuridine. J Natl Cancer Inst 45(4):773–782

15. Bockhorn M, Jain RK, Munn LL (2007) Active versus passive mechanisms in metastasis: do cancer cells crawl into vessels, or are they pushed? Lancet Oncol 8(5):444–448. doi:10.1016/S1470-2045(07)70140-7, S1470-2045(07)70140-7 [pii]

16. Butler TP, Gullino PM (1975) Quantitation of cell shedding into efferent blood of mammary adenocarcinoma. Cancer Res 35(3):512–516

17. Zhe X, Cher ML, Bonfil RD (2011) Circulating tumor cells: finding the needle in the haystack. Am J Cancer Res 1(6):740–751

18. Lu J, Fan T, Zhao Q, Zeng W, Zaslavsky E, Chen JJ, Frohman MA, Golightly MG, Madajewicz S, Chen WT (2010) Isolation of circulating epithelial and tumor progenitor cells with an invasive phenotype from breast cancer patients. Int J Cancer 126(3):669–683. doi:10.1002/ijc.24814

19. Alix-Panabieres C, Schwarzenbach H, Pantel K (2012) Circulating tumor cells and circulating tumor DNA. Annu Rev Med 63:199–215. doi:10.1146/annurev-med-062310-094219

20. Alix-Panabieres C (2012) EPISPOT assay: detection of viable DTCs/CTCs in solid tumor patients. Recent Results Cancer Res 195:69–76. doi:10.1007/978-3-642-28160-0_6

21. Miller MC, Doyle GV, Terstappen LW (2010) Significance of circulating tumor cells detected by the cell search system in patients with metastatic breast colorectal and prostate cancer. J Oncol 2010:617421. doi:10.1155/2010/617421

22. Ghossein RA, Bhattacharya S, Rosai J (1999) Molecular detection of micrometastases and circulating tumor cells in solid tumors. Clin Cancer Res 5(8):1950–1960

23. Pearl ML, Zhao Q, Yang J, Dong H, Tulley S, Zhang Q, Golightly M, Zucker S, Chen WT (2014) Prognostic analysis of invasive circulating tumor cells (iCTCs) in epithelial ovarian cancer. Gynecol Oncol 134(3):581–590. doi:10.1016/j.ygyno.2014.06.013, S0090-8258(14)01053-1 [pii]

24. Friedlander TW, Ngo VT, Dong H, Premasekharan G, Weinberg V, Doty S, Zhao Q, Gilbert EG, Ryan CJ, Chen WT, Paris PL (2014) Detection and characterization of invasive circulating tumor cells derived from men with metastatic castration-resistant prostate cancer. Int J Cancer 134(10):2284–2293. doi:10.1002/ijc.28561

25. Paris PL, Kobayashi Y, Zhao Q, Zeng W, Sridharan S, Fan T, Adler HL, Yera ER, Zarrabi MH, Zucker S, Simko J, Chen WT, Rosenberg J (2009) Functional phenotyping and genotyping of circulating tumor cells from patients with castration resistant prostate cancer. Cancer Lett 277(2):164–173. doi:10.1016/j.canlet.2008.12.007, S0304-3835(08)00940-3 [pii]

26. Fan T, Zhao Q, Chen JJ, Chen WT, Pearl ML (2009) Clinical significance of circulating tumor cells detected by an invasion assay in peripheral blood of patients with ovarian cancer. Gynecol Oncol 112(1):185–191. doi:10.1016/j.ygyno.2008.09.021, S0090-8258(08)00732-4 [pii]

Chapter 10

Breast Cancer Stem Cell Isolation

Xuanmao Jiao, Albert A. Rizvanov, Massimo Cristofanilli, Regina R. Miftakhova, and Richard G. Pestell

Abstract

Cells within the tumor are highly heterogeneous. Only a small portion of the cells within the tumor is capable to generate a new tumor. These cells are called cancer stem cells. Theoretically, cancer stem cells are originally from normal stem cells or early progenitor cells which accumulate the random mutations and undergo an altered version of the normal differentiation process. The cancer stem cell drives tumor progression and its recurrence. Thus, the technique to identify and purify the cancer stem cell is the key in any cancer stem cell research. In this protocol, we provide the basic technology of identification and purification of breast cancer stem cells as well as further functional assays to help the researchers achieve their research goals.

Key words Breast cancer stem cells, Circulating tumor cells, Mammosphere formation, Fluorescence-activated cell sorting (FACS), Transplantation assays

1 Introduction

In 2003, Al-Hajj and colleagues reported for the first time that a small number of $CD44^+CD24^{-/low}Lineage^-$ primary tumor cells are able to initiate tumor formation in a mammary pad of NOD/SCID mice [1]. Several studies further demonstrated that $CD44^+CD24^-/low$ phenotype cannot be ubiquitously used to identify SC in all breast cancer subtypes [2]. Later, aldehyde dehydrogenase 1 (ALDH1) activity was shown to be a better predictive marker of chemoresistant BCSC as compared to CD44/CD24 combination [3]. However, neither ALD^{Hhigh} nor $CD44^+CD24^{-/low}$ cells show 100 % of sphere formation ability in vitro. Sorted population of $CD44^+CD24^{-/low}$ cells which were epithelial specific antigen (ESA) positive or ALDH1 positive had higher tumor generation capacity in vivo than $CD44^+CD24^{-/low}$ cells alone [4], suggesting that a combination of three and more markers is required to identify BCSC in all breast cancer subtypes.

Xuanmao Jiao and Albert A. Rizvanov contributed equally.

Jian Cao (ed.), *Breast Cancer: Methods and Protocols*, Methods in Molecular Biology, vol. 1406,
DOI 10.1007/978-1-4939-3444-7_10, © Springer Science+Business Media New York 2016

1.1 BCSC Markers

CD44 is an abundantly expressed transmembrane glycoprotein. CD44 has high specificity for hyaluronic acid, but also can interact with extracellular matrix (ECM) proteins (osteopontin, collagen, laminin) and non-ECM ligands (matrix metalloproteinases) [5]. Owing to numerous posttranscriptional and posttranslational modifications, specific isoforms of CD44 control a wide variety of cell functions, including cell adhesion, migration, homing, and transmission of growth signals [6]. The siafucosylated isoform of CD44 known as HCELL (hematopoietic cell E-selectin/L-selectin ligand) ensures high cell tropism towards bone tissue [7]. Recently it has been shown that ~70 % of patients with early bone metastasis have tumors enriched by CD44+CD24$^{-/low}$ cells [8]. Various isoforms of CD44 are overexpressed in almost every cancer type [9].

Similarly to CD44, CD24 is involved in cell adhesion and metastasis [10] and highly expressed in various cancer types [11]. In contrast to differentiated cells, cancer progenitor and stem cells possess low or absent expression of CD24 [12].

Aldehyde dehydrogenase 1 (ALDH1) family proteins are required for aldehyde detoxification and retinal oxidation. All-trans-retinoic acid serves as a ligand of retinoic receptors (RAR, RXR) and nuclear receptor family transcriptional factors, which regulate cell growth and differentiation [13]. Transplantation of 300 patient-derived ALDH1-positive primary breast cancer cells resulted in 100 % tumor formation in the murine mammary fat pad, while in contrast, transplantation of 300 ALDH1-negative cells yielded zero mouse tumors [14]. ALDH1 expression negatively correlates with breast cancer patient survival [4]. Furthermore, gene expression data of 3455 breast cancer patients indicate that only ALDH1A1 activity level, one of the six ALDH1 family members, correlates with poorer overall survival of BC patients [15].

Epithelial cell adhesion molecule (EpCAM), also named as epithelial-specific antigen (ESA) or CD326, is a transmembrane protein required for breast cancer cell proliferation, migration, and invasion [16] [17]. Remarkably, EpCAM was the very first target of monoclonal antibody-based cancer therapy [18]. In early experiments, the tumor formation ability of Lin$^-$ESA$^+$CD44$^+$CD24$^{-/low}$ cells was compared with ESA$^-$CD44$^+$CD24$^{-/low}$ cells [1]. Only ESA$^+$ cells were able to generate tumors in NOD/SCID mice. In addition, only ESA$^+$CD44$^+$CD24$^{-/low}$ cells had the capacity to regenerate tumors in a serial transplantation experiments.

1.2 Mammosphere Formation Assay

The first sphere formation assay was developed by Reynolds et al. in 1992 for neural stem cells [19]. A decade later the protocol was modified to assess the activity of normal breast cancer stem cells [20]. The mammosphere formation assay is a powerful and informative technique, but it is important to remember that quiescent stem cells may not form spheres in a short assay time [21]. Furthermore, not only BCSC but also some progenitor cells can form spheres [22]. This may lead to an overestimation of stem cell percentage based on a sphere count. However, it is believed that only BCSC can regenerate mammospheres during repetitive mammosphere passage.

1.3 In Vivo Transplantation

The efficiency of tumor formation in xenograft animal models correlates with the number of SC [14]. Limiting diluting transplantation studies were widely used to identify the percent of BCSC at the time when our knowledge on BCSC markers was limited. This time-consuming technique is less frequently used and xenotransplantations of a small number or single BCSC, which was sorted based on SC markers, have become the more commonly used approach.

1.4 Origin of BCSC

Two cell types can be distinguished in mammary glands: luminal epithelial cells and basal myoepithelial cells. Myoepithelial cells in turn can be subdivided into cells with basal epithelial and mesenchymal phenotypes. Breast cancer $CD44^+/CD24^-$ cells demonstrate undifferentiated basal mesenchymal cell properties, whereas more differentiated $CD44^+CD24^+$ cells exhibit basal epithelial cell features [2]. At the same time, accumulating evidence suggest that metastasis initiating cells (MIC) originate from primary human breast cancer cells of luminal origin [23]. These confusing conclusions raised the question if the MIC population belonged to the BCSC population. Recent studies demonstrate that 66.7 % of the patients with metastatic breast cancer have CK^+ circulating tumor cells in their peripheral blood. 80 % of the patients with CK^+ cells have $CD44^+CD24^{-/low}$ population [24].

1.5 BCSC, Therapy and Recurrence

Breast cancer recurrence arises in ~40 % of patients. In half of the clinical cases, cancer reappears later than 5 years. These observations led to the discovery of two main features of BCSC: First, a population of BCSC was shown to be resistant to chemotherapy, radiotherapy, and endocrine therapy [25]. And second, dormant BCSC, which can reactivate their tumorigenic potency in response to specific signals [26]. In this regard, inhibition of bone morphogenetic protein (BMP) leads to an activation of dormant breast cancer cells in the lungs [27].

2 Materials

2.1 Reagents and Culture Dishes

Collagenase.

Hyaluronidase.

Dispase.

DNase I.

EGF.

Insulin.

Cholera toxin.

Hydrocortisone.

B-27 supplement (50×),

Human FGF basic.

Heparin sodium salt.

BD Matrigel Matrix, Growth Factor Reduced.

DMEM/F12 50/50 medium.

0.25 % trypsin–2.1 mM EDTA 1×.

Trypan blue stain 0.4 %.

ALDEFLUOR kit.

Purified rat anti-mouse CD16/CD32 (2.4G2).

Normal mouse IgG.

Normal rat IgG.

APC-rat anti-mouse CD31 (MEC 13.3).

APC-rat anti-mouse CD45 (30-F11).

APC-rat anti-mouse Ter-119.

APC-anti-human lineage cocktail (CD3, CD14, CD16, CD19, CD20, CD56),

PE-rat anti-mouse CD24 (M1/69).

PE-mouse anti-human CD24.

PE/Cy5-rat anti-human/mouse CD44 (IM7).

Ultra-low attachment culture dish.

2.2 Recipes of Solutions

1. Collagenase/hyaluronidase solution: Add 300 U/ml collagenase and 100 U/ml hyaluronidase in DMEM/F-12 50/50 medium supplement with 5 % FBS, 5 μg/ml insulin, 500 ng/ml hydrocortisone, 10 ng/ml EGF, 20 ng/ml cholera toxin.

2. Dispase/DNase solution: Add 5 mg/ml dispase and 0.1 mg/ml DNase in DMEM/F-12 50/50 medium supplement with 5 % FBS, 5 μg/ml insulin, 500 ng/ml hydrocortisone, 10 ng/ml EGF, 20 ng/ml cholera toxin.

3. Hemolysis Buffer: 8.26 g crystalline NH_4Cl (0.15 M), 1 g crystalline $KHCO_3$ (10 mM), 0.037 g crystalline Na_2EDTA (0.1 mM), add deionized water to 1 l, autoclave.

4. Mammosphere-forming medium: DMEM/F12 supplement with 1/50B-27, 20 ng/ml EGF, 20 ng/ml FGF, and 4 μg/ml heparin.

5. PBS, 8 g NaCl (0.137 M), 0.2 g KCl (2.7 mM), 1.44 g Na_2HPO_4 (10 mM), 0.24 g KH_2PO_4 (1.8 mM), solve in

800 ml deionized water and adjusted pH to 7.4, fill with deionized water to 1 l, autoclave.

6. Complete Cell Culture Medium: DMEM/F-12 50/50 medium supplement with 5 % FBS, 5 µg/ml insulin, 500 ng/ml hydrocortisone, 10 ng/ml EGF, 20 ng/ml cholera toxin.

Note: All solutions unless specified otherwise are sterilized by passing through a 0.45 µm filter.

3   Methods

3.1   Dissociation Human or Mouse Mammary Tumor Tissue

1. Human mammary tumor tissue is transported from the operating room in sterile specimen cups on ice. Mouse mammary tumor tissue is dissected in a biosafety hood.

2. The tissue is transferred to a 6-cm cell culture dish and chopped into 1 mm^3 pieces with a scalpel.

3. The tissue slurry is digested with 5 ml of collagenase/hyaluronidase solution for 1 h in a 37 °C CO_2 incubator.

4. The collagenase/hyaluronidase digested tissue slurry is pipetted repeatedly with a 1 ml pipette until there is no visible tissue clumps remaining.

5. All of the tissue suspensions are transferred to 15 ml conical tubes, filled with 5 ml PBS, and centrifuged at $700 \times g$ for 5 min.

6. The supernatant is discarded and the pellet is resuspended with 1 ml of 0.25 % trypsin–2.1 mM EDTA for 2 min at 37 °C. Repeat **steps 4** and **5** (**Note 1**).

7. The pellet is resuspended in dispase/DNase solution and incubated for 5 min at 37 °C. Repeat **steps 4** and **5**.

8. The pellet is resuspended in 1 ml Hemolysis buffer and incubated on ice for 5 min.

9. The cell suspension is centrifuged at $700 \times g$ for 5 min and the pellet is washed with PBS twice.

10. The pellet is resuspended in 1 ml of DMEM/F-12 complete medium. Filter through a 40 µm cell mesh and keep on ice.

3.2   Cell Culture

1. Usually, breast cancer cells are adherent. The cells should be cultured based on their standard protocol in 10 cm cell culture dishes.

2. The cells are harvested at 80 % confluency. The cell culture media is removed and the culture dish is washed once with 10 ml of PBS.

3. The cells are treated with 1 ml of 0.25 % trypsin–EDTA for 5–10 min at 37 °C.

4. Trypsin is inactivated by addition of 5 ml of FBS containing cell culture medium and the cell suspension is transferred to a 15 ml conical tube.

5. Centrifuge at $400 \times g$ for 5 min.

6. Remove the supernatant and resuspend the cells in 1 ml cell culture media. Keep the cell suspension on ice prior to proceeding further.

3.3 Mammosphere-Formation Assay

1. Cells dissociated from mammary tumors or from cell culture can be used for the mammosphere assay.

2. Cells are washed once with PBS and then resuspended in 10 ml PBS.

3. 50 µl of cells is mixed with an equal amount of 0.4 % trypan blue solution. The cells are counted with hemocytometer and the concentration is adjusted to 1×10^5 cell/ml (**Note 2**).

4. 4000–10,000 cells per ml are seeded in mammosphere-forming medium inCorning ultra-low attachment culture dish. 4 ml of the medium is used for 6 cm and 10 ml for 10 cm dishes.

5. The cells are cultured in 37 °C CO_2 incubator for 7–14 days to form the mammosphere.

6. The mammospheres are transferred to 15 ml conical tube and centrifuged at $300 \times g$ for 5 min.

7. The mammospheres are resuspended in 400 µl PBS.

8. A 96-well plate is gridded as showed in Fig. 1. 100 µl of suspended mammospheres is added into the grid-well of the 96-well plate. The number of mammospheres is counted under a 10× objective lens on an inverted microscope (**Note 3**).

9. The pictures of mammosphere are taken randomly. The diameter of mammosphere is measured by Zeiss Axiovert software (**Note 4**).

10. The typical result is shown in Fig. 2 [28].

11. For second or more generation of mammosphere formation, the mammospheres are digested from **step 6** with 500 µl of 0.25 % trypsin–EDTA for 10 min at 37 °C and washed once with full cell culture media.

12. The trypsinized mammosphere is further digested with dispase/DNase solution for 5 min at 37 °C, washed with full cell culture media and PBS sequentially.

13. The cells are filtered through a 40 µm mesh if necessary.

14. Cell number is counted and 1000–5000 cells/ml is seeded in an ultra-low attachment culture dish as **step 4**.

3.4 ALDEFLUOR Assay

The ALDEFLUOR kit from STEMCELL Technologies can be used in this assay. The protocol is based on the manufacturer's manual.

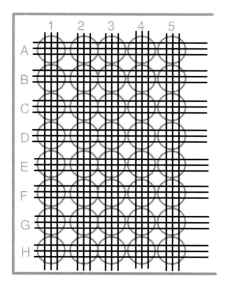

Fig. 1 The diagraph of a grid 96-well plate for mammosphere counting. A 96-well plate was reversed and gridded with a ruler and an ultrafine point permanent marker pen

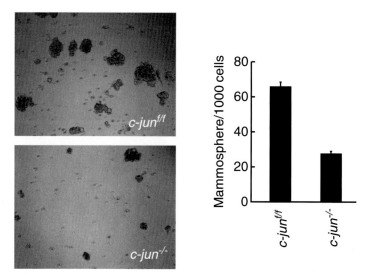

Fig. 2 Comparison of mammosphere formation between *c-jun*^{+/+} and *c-jun*^{-/-} of mouse ErbB2 mammary tumor cells. (**a**) The photograph and (**b**) the number of mammospheres formed from these two cell lines. The results are adopted from Ref. [28]

1. ALDEFLUOR reagents are activated and assembled by adding 25 μl DMSO to the tube containing dry ALDEFLUOR reagents, mixed well. 25 μl of 2 N HCl is added to the tube. The tube is vortexed and kept for 15 min at room temperature. 360 μl of ALDEFLUOR assay buffer is added to the tube and

the tube is vortexed. ALDEFLUOR substrate should be aliquoted and kept at –20 °C.

2. One aliquot of ALDEFLUOR substrate from –20 °C is thawed on ice.

3. The cells from Subheading 3.1 or 3.2 are counted and adjusted to the concentration of 1×10^6 cell/ml with ALDEFLUOR assay buffer.

4. Two FACS tubes are used for each sample. Both tubes are labeled with sample ID and one of them with DEAB as well. 5 μl of DEAB solution is added to each tube labeled with DEAB.

5. 1 ml of each sample is added to the tube labeled with sample ID only. 5 μl of ALDEFLUOR substrate is added to the sample and mixed by pipetting up and down.

6. 500 μl of sample is transferred to DEAE tube and mixed.

7. All of the samples are incubated at 37 °C for 45–60 min. For the sample from cell culture, go to **step 9** directly. The cells from this step also can be used for CD44+/CD24- cell sorting.

8. The samples are centrifuge at $400 \times g$ for 5 min and the supernatants are discarded.

9. For samples derived from tumor tissue, 50 μl of APC-labeled antibody of either human or mouse lineage marker is added based on sample species with appropriate dilution. The samples are incubated on ice for 30–60 min.

10. Repeat **step 8**.

11. Cell pellets are resuspended in 0.5 ml of ALDEFLUOR assay buffer. The samples are kept on ice before FACS sorting.

12. Set up the flow cytometer. The cell population is defined with FSC and SSC plot. The data are acquired using FITC channel with DEAB sample as control. For the sample from tumor tissue, both APC and FITC channel are used.

13. FACS sorting data are analyzed in FlowJo.

14. The typical ALDEFLUOR assay results are shown in Fig. 3 [28].

3.5 CD44+/CD24- Cell Sorting

1. The cells from Subheading 3.1.are counted and the cell concentration is adjusted to 1×10^6 cell/ml with full cell culture media.

2. 5 ml of cells is transferred to a 15 ml conical tube and the cells are pelleted at $400 \times g$ for 5 min. The supernatant is discarded.

3. The cells are blocked with 250 μl of full cell culture media containing normal IgG (1/50) and purified anti-human or mouse Fcγ III/II receptor antibody (1/50) for 45–60 min on ice.

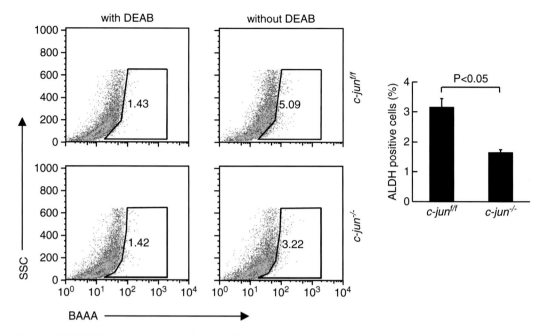

Fig. 3 ALDEFLUOR assay showed that the ALDH⁺ cells decreased upon c-Jun knockout, which means c-Jun increases stem cell population. The results are adopted from Ref. [28]

4. 5 FACS tubes are labeled for each sample with 0, PE, Cy5, APC, and test. 50 µl blocked cell suspension is added into each tube.

5. The samples in FACS tube are centrifuged at $800 \times g$ and the supernatants are carefully aspirated.

6. Four eppendorf tubes are labeled with PE, Cy5, APC, and test. 10 µl PE-mouse anti-human CD24 (1/5) and 40 µl media are added to PE-tube. 0.25 µl PE/Cy5-rat anti-human/mouse CD44 (1/200) and 50 µl media are added to Cy5-tube, and 25 µl APC-labeled human Lineage cocktail (1/2) and 25 µl full cell culture media are added to APC-tube. The antibody cocktail is made by mixing 10 µl PE-mouse anti-human CD24, 0.25 µl PE/Cy5-rat anti-human/mouse CD44, 25 µl APC-human lineage cocktail, and 15 µl media in a test tube (**Note 6**).

7. The antibody diluents are transferred to the FACS tube containing the cells with same labeling. 50 µl of media is added to the tube labeled with 0 as unstained control. All of the samples are incubated on ice for 60 min.

8. After incubation, the cells are washed with 3–4 ml PBS and resuspended in 700 µl PBS. The cell suspensions are kept on ice until acquisition of data on flow cytometer.

9. The cell population is defined with FSC (forward scatter channel) and SSC (side scatter channel) plot. The detector and

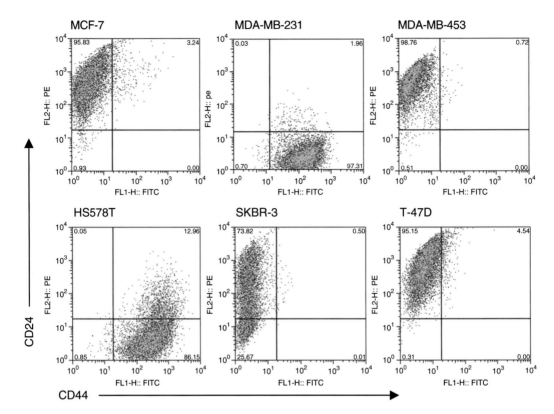

Fig. 4 FACS analysis of breast cancer cell lines of MCF-7, MDA-MB-231, MDA-MB-453, HS578T, SKBR-3, and T-47D. It was showed that the majority of basal type breast cancer cell lines (MDA-MB-231 and HS578T) is CD24$^-$CD44$^+$ and the luminal type breast cancer lines (MCF-7, MDA-MB-453, and T-47D) is CD24$^+$CD44$^-$. SKBR-3 is CD44$^-$ but with both CD24$^+$ and CD24$^-$ cells. The result are adopted from Ref. [29]

compensation of PE, PE/Cy5, and APC channel are set up with unstained and single staining control. The data are collected from test samples.

10. FACS sorting data are analyzed in FlowJo.

11. For tumor tissue from mouse, use APC-rat anti-mouse CD31 (1/100), APC-rat anti-mouse CD45 (1/100) and APC-rat anti-mouse Ter119 (1/100) as lineage markers. Use PE-rat anti-mouse CD24 (1/50) instead of PE-mouse anti-human CD24 (**Note 5**).

12. For culture cells from Subheading 3.2, the lineage marker staining can be omitted.

13. Figure 4 is the CD24 and CD44 staining of MCF-7, SKBR3, T-47D, MDA-MB-453, MDA-MB-231, and HS578A breast cancer cell lines [29].

3.6 In Vivo Transplantation Assay

1. CD44$^+$CD24$^-$ cells are FACS sorted based on the methods in Subheading 3.5. The sorted cells are washed twice with PBS.

2. The cells are counted with a hemocytometer. Trypan blue staining is used to exclude the dead cells.

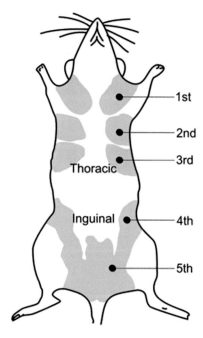

Fig. 5 The location of the fourth pair of mammary fat pad of mouse. The diagraph is adopted from Ref. [34]

3. Cell concentration is adjusted to 2×10^5 cell/ml with PBS. The following serial dilutions of 4×10^4, 1×10^4, and 2×10^3 cells/ml are made.

4. Make 40 % Matrigel solution in PBS. The cell dilutions are mixed with an equal volume of 40 % Matrigel solution. The cells in Matrigel solution are withdrawn into a 1 ml TB syringe.

5. The mouse is held with the thumb and index and pinky fingers. The fourth pair of mammary fat pads is located by the nipple.

6. The injection area is disinfected with 70 % ethanol. 100 μl of the cells in 20 % Matrigel/PBS is injected into one side of the fourth pair mammary fat pad.

7. Tumor size is measured with a caliper weekly (**Note 7**).

8. The results from a typical experiment are shown in Figs. 5 and 6 [29].

4   Notes

1. In Subheading 3.1, after 0.25 % trypsin–2.1 mM EDTA treatment, the cells become very sticky and form a very loose pellet. Instead of aspiration, use 1-ml Pipetman to carefully remove the supernatant in this step.

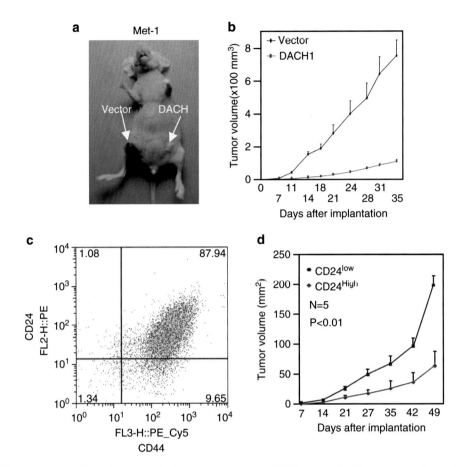

Fig. 6 The tumor formation of transplanted breast cancer cells. (**a**) Photograph of tumor formation in the nude mouse transplanted with Dach-1 overexpressing Met-1 mouse breast cancer cells and its vector control. (**b**) Tumor growth curve of Dach-1 overexpressing Met-1 cells vs vector control in transplanted mice. (**c**) Met-1 cells were FACS sorted to CD24−CD44+ and CD24+CD44+ populations. Both populations were transplanted into the mice. (**d**) The tumor formation of CD24−CD44+ vs. CD24+CD44+ Met-1 cells. CD24-CD44+ Met-1 cells was more tumorigenic than CD24+CD44+ Met-1 cells. The results are adopted from Ref. [29]

2. When counting the cells dissociated from mammary tumor, trypan blue must be used to exclude the dead cells. Automatic cell counter should not be used.

3. In Subheading 3.3, all of the mammospheres in the well should be counted. If too many mammospheres are formed, the mammosphere suspension can be diluted before adding to the grid 96-well plate.

4. For measuring the size of mammosphere, the pictures must be taken randomly. If fewer mammospheres are formed, the pictures of all mammospheres should be taken.

5. In Subheading 3.5, different breast cancer stem cell markers are used in different mouse models by different groups, except

Lin⁻CD24⁻/CD44⁺, which is used in brac1 transgenic mouse model [30] and Met-1 [29] xenograft mouse model. Other markers like CD133⁺ [30], Lin⁻CD24⁺CD29ʰⁱ [31], Lin⁻CD24⁺CD61⁺ [32], and Sca-1⁺ [33] cells were also reported to represent mouse breast cancer stem cells. One possibility is that these cells with different signature present the different origination of the breast cancer stem cells.

6. For multiple samples, the cocktail of antibodies should be made in order to keep the same staining condition. For some FACS machines, such as FACSCalibur, each sample should have its own single staining control to set up the machine.

7. In Subheading 3.6, the breast cancer cells can be genetically labeled with luciferase and then bioluminescence imaging can be used to evaluate tumor formation in vivo.

Acknowledgements

This work was supported in part by grants from NIH R01CA70896, R01CA75503, R01CA86072 (R.G.P.), the Breast Cancer Research Foundation (R.G.P.), and the Department of Defense Concept Award W81XWH-1101-0303. The Sidney Kimmel Cancer Center was supported by the NIH Cancer Center Core Grant P30CA56036 (R.G.P). This project is funded in part by the Dr. Ralph and Marian C. Falk Medical Research Trust (R.G.P.), grants from the Pennsylvania Department of Health (R.G.P.). The Department specifically disclaims responsibility for analyses, interpretations, or conclusions. There are no conflicts of interest associated with this manuscript.

Conflicts of Interest: RGP is the founder of ProstaGene, LLC and owns patents related to prostate cancer cell lines and their uses thereof.

References

1. Al-Hajj M, Wicha MS, Benito-Hernandez A, Morrison SJ, Clarke MF (2003) Prospective identification of tumorigenic breast cancer cells. Proc Natl Acad Sci U S A 100(7):3983–3988

2. Ricardo S, Vieira AF, Gerhard R, Leitao D, Pinto R, Cameselle-Teijeiro JF, Milanezi F, Schmitt F, Paredes J (2011) Breast cancer stem cell markers CD44, CD24 and ALDH1: expression distribution within intrinsic molecular subtype. J Clin Pathol 64(11):937–946. doi:10.1136/jcp.2011.090456.jcp.2011.090456 [pii]

3. Tanei T, Morimoto K, Shimazu K, Kim SJ, Tanji Y, Taguchi T, Tamaki Y, Noguchi S (2009) Association of breast cancer stem cells identified by aldehyde dehydrogenase 1 expression with resistance to sequential Paclitaxel and epirubicin-based chemotherapy for breast cancers. Clin Cancer Res 15(12):4234–4241. doi:10.1158/1078-0432.CCR-08-1479, 1078-0432.CCR-08-1479 [pii]

4. Ginestier C, Hur MH, Charafe-Jauffret E, Monville F, Dutcher J, Brown M, Jacquemier J, Viens P, Kleer CG, Liu S, Schott A, Hayes D, Birnbaum D, Wicha MS, Dontu G (2007) ALDH1 is a marker of normal and malignant human mammary stem cells and a predictor of poor clinical outcome. Cell Stem Cell 1(5):555–567. doi:10.1016/j.stem.2007.08.014, S1934-5909(07)00133-6 [pii]

5. Naor D, Sionov RV, Ish-Shalom D (1997) CD44: structure, function, and association with the malignant process. Adv Cancer Res 71:241–319

6. Zoller M (2011) CD44: can a cancer-initiating cell profit from an abundantly expressed molecule? Nat Rev Cancer 11(4):254–267. doi:10.1038/nrc3023, nrc3023 [pii]

7. Sackstein R, Merzaban JS, Cain DW, Dagia NM, Spencer JA, Lin CP, Wohlgemuth R (2008) Ex vivo glycan engineering of CD44 programs human multipotent mesenchymal stromal cell trafficking to bone. Nat Med 14(2):181–187. doi:10.1038/nm1703, nm1703 [pii]

8. Draffin JE, McFarlane S, Hill A, Johnston PG, Waugh DJ (2004) CD44 potentiates the adherence of metastatic prostate and breast cancer cells to bone marrow endothelial cells. Cancer Res 64(16):5702–5711. doi:10.1158/0008-5472.CAN-04-0389, 64/16/5702 [pii]

9. Ponta H, Sherman L, Herrlich PA (2003) CD44: from adhesion molecules to signalling regulators. Nat Rev Mol Cell Biol 4(1):33–45. doi:10.1038/nrm1004, nrm1004 [pii]

10. Lee HJ, Choe G, Jheon S, Sung SW, Lee CT, Chung JH (2010) CD24, a novel cancer biomarker, predicting disease-free survival of non-small cell lung carcinomas: a retrospective study of prognostic factor analysis from the viewpoint of forthcoming (seventh) new TNM classification. J Thorac Oncol 5(5):649–657. doi:10.1097/JTO.0b013e3181d5e554

11. Jaggupilli A, Elkord E (2012) Significance of CD44 and CD24 as cancer stem cell markers: an enduring ambiguity. Clin Dev Immunol 2012:708036. doi:10.1155/2012/708036

12. Fang X, Zheng P, Tang J, Liu Y (2010) CD24: from A to Z. Cell Mol Immunol 7(2):100–103. doi:10.1038/cmi.2009.119, cmi2009119 [pii]

13. Chute JP, Muramoto GG, Whitesides J, Colvin M, Safi R, Chao NJ, McDonnell DP (2006) Inhibition of aldehyde dehydrogenase and retinoid signaling induces the expansion of human hematopoietic stem cells. Proc Natl Acad Sci U S A 103(31):11707–11712. doi:10.1073/pnas.0603806103, 0603806103 [pii]

14. Charafe-Jauffret E, Ginestier C, Bertucci F, Cabaud O, Wicinski J, Finetti P, Josselin E, Adelaide J, Nguyen TT, Monville F, Jacquemier J, Thomassin-Piana J, Pinna G, Jalaguier A, Lambaudie E, Houvenaeghel G, Xerri L, Harel-Bellan A, Chaffanet M, Viens P, Birnbaum D (2013) ALDH1-positive cancer stem cells predict engraftment of primary breast tumors and are governed by a common stem cell program. Cancer Res 73(24):7290–7300. doi:10.1158/0008-5472.CAN-12-4704,0008-5472.CAN-12-4704 [pii]

15. Wu S, Xue W, Huang X, Yu X, Luo M, Huang Y, Liu Y, Bi Z, Qiu X, Bai S (2015) Distinct prognostic values of ALDH1 isoenzymes in breast cancer. Tumour Biol 36(4):2421–2426. doi:10.1007/s13277-014-2852-6

16. Osta WA, Chen Y, Mikhitarian K, Mitas M, Salem M, Hannun YA, Cole DJ, Gillanders WE (2004) EpCAM is overexpressed in breast cancer and is a potential target for breast cancer gene therapy. Cancer Res 64(16):5818–5824. doi:10.1158/0008-5472.CAN-04-0754, 64/16/5818 [pii]

17. Baeuerle PA, Gires O (2007) EpCAM (CD326) finding its role in cancer. Br J Cancer 96(3):417–423. doi:10.1038/sj.bjc.6603494, 6603494 [pii]

18. Sears HF, Atkinson B, Mattis J, Ernst C, Herlyn D, Steplewski Z, Hayry P, Koprowski H (1982) Phase-I clinical trial of monoclonal antibody in treatment of gastrointestinal tumours. Lancet 1(8275):762–765

19. Reynolds BA, Weiss S (1992) Generation of neurons and astrocytes from isolated cells of the adult mammalian central nervous system. Science 255(5052):1707–1710

20. Dontu G, Al-Hajj M, Abdallah WM, Clarke MF, Wicha MS (2003) Stem cells in normal breast development and breast cancer. Cell Prolif 36(Suppl 1):59–72, doi: 274 [pii]

21. Pastrana E, Silva-Vargas V, Doetsch F (2011) Eyes wide open: a critical review of sphere-formation as an assay for stem cells. Cell Stem Cell 8(5):486–498. doi:10.1016/j.stem.2011.04.007, S1934-5909(11)00172-X [pii]

22. Stingl J (2009) Detection and analysis of mammary gland stem cells. J Pathol 217(2):229–241. doi:10.1002/path.2457

23. Baccelli I, Schneeweiss A, Riethdorf S, Stenzinger A, Schillert A, Vogel V, Klein C, Saini M, Bauerle T, Wallwiener M, Holland-Letz T, Hofner T, Sprick M, Scharpff M, Marme F, Sinn HP, Pantel K, Weichert W, Trumpp A (2013) Identification of a population of blood circulating tumor cells from breast cancer patients that initiates metastasis in a xenograft assay. Nat Biotechnol 31(6):539–544. doi:10.1038/nbt.2576, nbt.2576 [pii]

24. Theodoropoulos PA, Polioudaki H, Agelaki S, Kallergi G, Saridaki Z, Mavroudis D, Georgoulias V (2010) Circulating tumor cells with a putative stem cell phenotype in peripheral blood of patients with breast cancer. Cancer Lett 288(1):99–106. doi:10.1016/j.canlet.2009.06.027, S0304-3835(09)00453-4 [pii]

25. Baumann M, Krause M, Hill R (2008) Exploring the role of cancer stem cells in

radioresistance. Nat Rev Cancer 8(7):545–554. doi:10.1038/nrc2419, nrc2419 [pii]

26. Suzuki M, Mose ES, Montel V, Tarin D (2006) Dormant cancer cells retrieved from metastasis-free organs regain tumorigenic and metastatic potency. Am J Pathol 169(2):673–681. doi:10.2353/ajpath.2006.060053,doi:S0002-9440(10)62746-0 [pii]

27. Gao H, Chakraborty G, Lee-Lim AP, Mo Q, Decker M, Vonica A, Shen R, Brogi E, Brivanlou AH, Giancotti FG (2012) The BMP inhibitor coco reactivates breast cancer cells at lung metastatic sites. Cell 150(4):764–779. doi:10.1016/j.cell.2012.06.035, S0092-8674(12)00872-0 [pii]

28. Jiao X, Katiyar S, Willmarth NE, Liu M, Ma X, Flomenberg N, Lisanti MP, Pestell RG (2010) c-Jun induces mammary epithelial cellular invasion and breast cancer stem cell expansion. J Biol Chem 285(11):8218–8226. doi:10.1074/jbc.M110.100792, M110.100792 [pii]

29. Wu K, Jiao X, Li Z, Katiyar S, Casimiro MC, Yang W, Zhang Q, Willmarth NE, Chepelev I, Crosariol M, Wei Z, Hu J, Zhao K, Pestell RG (2011) Cell fate determination factor Dachshund reprograms breast cancer stem cell function. J Biol Chem 286(3):2132–2142. doi:10.1074/jbc.M110.148395, M110.148395 [pii]

30. Wright MH, Calcagno AM, Salcido CD, Carlson MD, Ambudkar SV, Varticovski L (2008) Brca1 breast tumors contain distinct CD44+/CD24- and CD133+ cells with cancer stem cell characteristics. Breast Cancer Res 10(1):R10. doi:10.1186/bcr1855, bcr1855 [pii]

31. Zhang M, Behbod F, Atkinson RL, Landis MD, Kittrell F, Edwards D, Medina D, Tsimelzon A, Hilsenbeck S, Green JE, Michalowska AM, Rosen JM (2008) Identification of tumor-initiating cells in a p53-null mouse model of breast cancer. Cancer Res 68(12):4674–4682. doi:10.1158/0008-5472.CAN-07-6353, 68/12/4674 [pii]

32. Vaillant F, Asselin-Labat ML, Shackleton M, Forrest NC, Lindeman GJ, Visvader JE (2008) The mammary progenitor marker CD61/beta3 integrin identifies cancer stem cells in mouse models of mammary tumorigenesis. Cancer Res 68(19):7711–7717. doi:10.1158/0008-5472.CAN-08-1949, 68/19/7711 [pii]

33. Grange C, Lanzardo S, Cavallo F, Camussi G, Bussolati B (2008) Sca-1 identifies the tumor-initiating cells in mammary tumors of BALB-neuT transgenic mice. Neoplasia 10(12):1433–1443

34. Eriksson PO, Aaltonen E, Petoral R Jr, Lauritzson P, Miyazaki H, Pietras K, Mansson S, Hansson L, Leander P, Axelsson O (2014) Novel nano-sized MR contrast agent mediates strong tumor contrast enhancement in an oncogene-driven breast cancer model. PLoS One 9(10):e107762. doi:10.1371/journal.pone.0107762, PONE-D-14-18269 [pii]

Part IV

In Vitro Experimental Models for Breast Cancer

Chapter 11

Cellular Apoptosis Assay of Breast Cancer

Yu Sun and Wei-Xing Zong

Abstract

Apoptosis is an energy-dependent enzymatic cell suicide process. It almost always involves the activation of caspases. In this chapter, we systemically introduce methodologies to assay caspases dependent biochemical and morphological changes in vitro breast cancer cell lines and in vivo breast cancer tissues. In addition, mitochondrial involvement is crucial to distinguish two different apoptotic pathways. Methodology to assay dissipation of mitochondrial transmembrane potential, an early event of mitochondrial involvement, is also included. Of note, since apoptotic features may not appear to the same extent depending on the context of cell types and the death-inducing insults, a common practice is to use more than one method to assess apoptosis, qualitatively and quantitatively.

Key words Apoptosis assay, Breast cancer, Caspase activation, Annexin V/PI staining, DNA laddering assay, In situ TUNEL staining for apoptotic cells, JC-1 staining

1 Introduction

Apoptosis, or programmed cell death, plays critical roles in breast cancer development and in cancer cell response to therapeutic treatment. Its characteristic morphological changes, such as cell body shrinkage, plasma membrane blebbing, and chromatin condensation, distinguish apoptosis from other types of cell death, and are often used to qualitatively define apoptosis. Quantification of apoptosis with these morphological changes is rather difficult, insufficient, and often too subjective. More sophisticated quantification methods have been developed based on other physiological and biochemical features of apoptotic cells.

Apoptosis is an energy-dependent enzymatic cell suicide process. It almost always involves the activation of caspases. Caspases, cysteine-aspartic proteases or cysteine-dependent aspartate-directed proteases, are a family of cysteine proteases which are normally expressed as inactive zymogen and proteolytically processed to active forms following apoptotic stimuli [1]. Two apoptotic signaling pathways, namely extrinsic and intrinsic pathways, have

Jian Cao (ed.), *Breast Cancer: Methods and Protocols*, Methods in Molecular Biology, vol. 1406,
DOI 10.1007/978-1-4939-3444-7_11, © Springer Science+Business Media New York 2016

been well studied in breast cancer cells [2–4]. Extrinsic pathway is activated by the engagement of extracellular cell death-inducing ligands with death receptors on the plasma membrane that leads to the formation of the death-inducing signaling complex (DISC) and the activation of initiator caspases (caspases 8 and 10), which in turn cleave and activate effector caspases (caspases 3 and 7). The intrinsic pathway is initiated by disruption of mitochondrial outer membrane and release of apoptogenic factors, such as cytochrome c. Together with Apaf-1, cytochrome c recruits the initiator caspase (caspase 9) to form apoptosome that in turn cleaves and activates downstream effector caspases (caspases 3 and 7) [1]. In both pathways, effector caspases cleave their substrates in different subcellular compartments, leading to irreversible molecular changes which can be utilized as quantification targets. These include the exposure of phospholipids that are recognized by their specific ligand Annexin V, or DNA fragmentation that can be visualized by electrophoresis or in situ enzymatic mediated direct/indirect staining [5–7]. Disruption of mitochondria at early stage differentiates intrinsic pathway from extrinsic pathway. Therefore, method measuring mitochondrial transmembrane potential (delta psi, $\Delta\psi$) is often used for distinguishing these two pathways and for quantifying cells undergoing intrinsic apoptosis. However, since many of the apoptotic features may not appear to the same extent depending on the context of cell types and the death-inducing insults, a common practice is to use more than one method to assess apoptosis, qualitatively and quantitatively.

1.1 Immunoblotting to Detect Caspase Activation and Cleavage of Caspase Substrates

Caspases are a family of cysteine proteases that play essential role in cell death and inflammation. As many other proteases, caspases are synthesized as inactive proenzymes that require proteolytic cleavage for activation. Apoptotic caspases have been categorized as initiator caspases (caspase 2, 8, and 9), and executioner caspases (caspases 3, 6, and 7) [8].

The mechanism of caspase activation is different between initiator and executioner caspases. Initiator caspases need dimerization for activation. For example, following death receptor ligation, caspase 8 monomers are recruited to adaptor molecule, FAS-associated death domain protein (FADD), to form dimers through their prodomains. Dimerization and inter-domain cleavage are needed for the activation and stabilization of mature caspase 8 [9–11]. For caspase 9, dimerization is required for activation and inter-domain cleavage is involved in attenuation rather than promotion of its activity [10, 11]. Different from initiator caspases, the cleavage of executioner caspases between large and small subunits of procaspases leads to conformational change and brings two active sites dimerize to become the mature form of caspase [11–13].

Initiator caspases have limited substrates including itself, BID and executioner caspases. Executioner caspases have more varieties of substrates and are largely responsible for the phenotypic changes when cells undergo apoptosis. For example, caspase 3 is the major executioner caspase with more than 300 substrates including cytoskeleton proteins, such as lamin, α-fodrin, actin, and proteins involved in DNA repair and cell cycle regulation, such as polyADP-ribose polymerase (PARP) et al. [8, 14]. Therefore, biochemical methodologies, such as immunoblotting, for detection of cleavage of caspases, especially executioner caspases and their substrates are often used to determine apoptosis.

1.2 Annexin V/PI Staining

Phosphatidylserine (PS) and phosphatidylethanolamine (PE), a class of acidic phospholipids, normally reside on the inner surface of the plasma membrane, resulting from aminophospholipid translocase selectively pumping PS and PE from the outer to inner plasma membrane. When cells undergo apoptosis, activation of caspase 3 impairs aminophospholipid translocase activity, leading to the externalization of PS and PE [7], which can be recognized and bound by Annexin V. By conjugating Annexin V with fluorescent substrates, such as FITC, apoptotic cells can be labeled and quantified by using flow cytometry. The binding affinity of Annexin V to PS or PE is calcium dependent. Removing calcium from binding assay buffer will eliminate the binding of Annexin V to PS or PE. During early stage of apoptosis, cellular membrane is intact. Combined staining apoptotic cells with cell membrane impermeable dye propidium iodide (PI) can distinguish viable cells from apoptotic and necrotic cells.

1.3 DNA Laddering Assay

Apoptotic DNA fragmentation is due to caspase-dependent activation of endonucleases that cleave chromatin DNA. The responsive enzyme is Caspases Activated DNase (CAD). The activity of CAD is normally inhibited by forming complex with inhibitor of CAD (ICAD). During apoptosis, activated effector caspases cleaves ICAD to dissociate the CAD: ICAD complex, allowing CAD to migrate into the nucleus and cleave genomic DNA into internucleosomal fragments of ~180 base pairs (bp) and its multiples (360, 540 bp, etc.). The fragmented DNA can be visualized by electrophoresis or using terminal deoxynucleotidyl transferase (TdT)-mediated dUTP nick end labeling (TUNEL) assay.

1.4 In Situ TUNEL Staining for Apoptotic Cells

Another methodology to detect DNA fragmentation is Terminal deoxynucleotidyl transferase dUTP nick end labeling (TUNEL) assay that labels the terminal end of nucleic acids. The nicks in damaged DNA can be recognized by terminal deoxynucleotidyl transferase (TdT), an enzyme that catalyzes the addition of dUTPs to the DNA ends. Labeling dUTPs with a marker, such as fluorophores or haptens, including biotin, can be detected directly in the case of a fluorescent modified nucleotide (i.e., fluorescein-dUTP), or indirectly with streptavidin or antibodies, if biotin-dUTP or

BrdUTP are used, respectively [15]. TUNEL assay is advantageous in detecting apoptosis in situ, meaning to determine apoptotic cells in tissues, such as paraffin-embedded tissues. However, same as DNA ladder assay, several scenarios, such as necrosis, rather than apoptosis may generate fragmented DNA which could be recognized by TdT [16].

1.5 JC-1 Staining

Mitochondrial transmembrane potential ($\Delta\psi$) is the difference in electric potential between the interior and the exterior of mitochondrial inner membrane. $\Delta\psi$ is negative inside about 180–200 mV, and a proton gradient of about one unit, able to drive ATP synthesis. During early apoptosis, the reduction of $\Delta\psi$, depolarization of membrane, precedes nuclear chromatin condensation and DNA fragmentation. Fluorescence probes targeting $\Delta\psi$ have been used for detection of apoptotic cells. JC-1 (5,5',6,6' tetrachloro-1,1',3,3'-tetraethylbenzimidazol-carbocyanine iodide) is lipophilic cation that selectively enters mitochondria. It changes reversibly color from green to orange as membrane potentials increase. When JC-1 accumulated in healthy mitochondria, it forms so called J-aggregates and the fluorescence emission shifts from green (~530 nm, emission of JC-1 monomeric form) to orange (~590 nm, emission of J-aggregate) [17, 18], When cells undergo apoptosis, JC-1 aggregates turn to monomer due to the dissipation of membrane potential, fluorescence emission then shifts from orange (J-aggregates) to green (monomer).

2   Materials

2.1 Immunoblotting to Detect Caspase Activation and Cleavage of Caspase Substrates

Laemmli's buffer, 62.5 mM Tris–HCl, pH 6.8, 10 % glycerol, 2 % SDS, 5 % 2-mercaptoethanol, PBST, 137 mM NaCl, 2.7 mM KCl, 10 mM Na_2HPO_4, 2 mM KH_2PO_4, pH 7.2, Tween 20, 0.05 %.

2.2 Annexin V/PI Staining

Staining buffer: 10 mM HEPES, pH 7.4; 140 mM NaCl; 2.5 mM $CaCl_2$. Keep staining buffer at 4 °C prior to use. 100× Annexin V-FITC and propidium iodide are kept in dark and stored at 4 °C prior to use. PBS: 137 mM NaCl, 2.7 mM KCl, 10 mM Na_2HPO_4, 2 mM KH_2PO_4, pH 7.2.

2.3 DNA Laddering Assay

Agarose, 6× gel loading buffer, ethidium bromide (EtBr), Proteinase K (20 mg/ml), RNase cocktail (RNase A, 500 units/ml and RNase T1, 20,000 U/ml), 50× TAE buffer: Prepare a 50× stock solution in 1 l of H_2O: 242 g of Tris base, 57.1 ml of acetic acid (glacial), 100 ml of 0.5 M EDTA (pH 8.0). TES lysis buffer: 100 mM Tris, pH 8.0, 20 mM EDTA, 0.8 % (w/v) SDS.

2.4 In Situ TUNEL Staining for Apoptotic Cells

1. PBS, pH 7.4; Xylene, 100, 95, 90, 80, 70, and 50 % ethanol; 3 % H_2O_2; Hematoxylin solution; Aqua-Poly/Mount mounting medium.

2. Proteinase K 20 μg/ml, stored at –20 °C.

3. TdT buffer: 30 mM Tris base, pH 7.2, 140 mM, sodium cacodylate, 1 mM cobalt chloride, 0.25 mg/ml bovine serum albumin (BSA). Aliquot and store at –20 °C.

4. TdT reaction solution: Terminal deoxynucleotidyl transferase (TdT), biotinylated dUTP: 0.3 EU/μL of TdT enzyme and 40 pmol/μL biotinylated-dUTP in TdT buffer, aliquot and store at –20 °C.

5. TdT stop solution: 300 mM sodium chloride, 30 mM sodium citrate.

6. Avidin/biotinylated peroxidase complex (Vector Laboratories, ABC kit).

7. 3,3′-diaminobenzidine (DAB) solution: 0.03 % (g/100 ml) DAB in PBS, prepare fresh solution prior to use. Optional, add 20–40 μl H_2O_2 into 100 ml DAB solution to enhance the peroxidase activity.

2.5 JC-1 Staining

Staining buffer: medium or PBS. JC-1 stock solution, 200 μM in DMSO. Positive control: CCCP (carbonyl cyanide 3-chlorophenylhydrazone) stock solution, 50 mM in DMSO. Keep both stock solutions at –20 °C and protect JC-1 from light. Before harvesting cells, bring all reagents to room temperature.

3 Methods

3.1 Immunoblotting to Detect Caspase Activation and Cleavage of Caspase Substrates

1. Collect cells after treated with apoptotic stimuli. Centrifuge at $600 \times g$ for 5 min at 4 °C and remove supernatant by aspiration. Wash cells with cold PBS for two times.

2. Prepare cell lysates. Add Laemmli's buffer, three volumes of the size of cell pellet, to resuspend the cell pellet and incubate at room temperature for 15 min. Vortex samples every 5 min (*see* **Note 1**).

3. After determine protein concentration using BCA reagent, prepare 30 μg total protein from each sample and mix with 5× sample buffer. Samples are then boiled for 5 min.

4. Load samples in freshly made 12 % SDS-Page gel. After electrophoresis, proteins are transferred from SDS-Page gel onto nitrocellulose membrane by electroblotting.

5. Nitrocellulose membranes are blocked in PBST containing 5 % milk for 1 h and probed with the antibodies against cleaved caspase 3 or anti-PARP antibody overnight at 4 °C.

6. After wash membranes 3× 10 min with PBST, membranes are probed with horseradish peroxidase (HRP) conjugated secondary antibodies for 1 h.

7. Wash membranes 3× 10 min with PBST, membranes are incubated with enhanced chemiluminescence (ECL) for 5 min and ECL substrates can be captured on X-Ray film.

3.2 Annexin V/PI Staining

1. Treatment: $1-3\times 10^5$ breast cancer cells are seeded in 6-well plate, allow cells to grow in the plate for 16–24 h, then treat cells with apoptotic stimuli of interested, such as irradiation, DNA damaging reagents, anti-hormonal treatment or tyrosine kinase inhibitors.

2. Before harvesting cells, bring all reagents to room temperature. To make the staining solution, dilute 100× Annexin V and PI in staining buffer, protect it from light and keep it at 37 °C until use.

3. After treatment, harvest both floating and attached cells into 5 ml FACS tube (BD biosciences). Centrifuge cells at 500–600×g for 5 min and aspirate carefully to remove supernatant. Resuspend cell pellet with 2 ml PBS and repeat the centrifugation to remove supernatant.

4. Resuspend cell pellet with 500 μl prewarmed staining solution. Stain cells in cell culture incubator for 15 min.

5. Read Annexin V-FITC and PI by using flow cytometer (BD biosciences). Set FL-1 for Annexin V-FTIC and FL-3 for PI. Dot-plot 10,000 events to analyze the intensity of fluorescence. Use quadrant to separate the events (Fig. 1). Dot events in lower left quadrant (Annexin V-/PI-represent alive cells, in lower right quadrant represent (Annexin V+/PI-early apoptotic cells, upper right quadrant (Annexin V+/PI+) late apoptotic cells, and upper left quadrant (Annexin V-/PI+) secondary necrotic cells.

3.3 DNA Laddering Assay

1. Collect cells to 1.5 ml Eppendorf tubes. Centrifuge at 500×g in Eppendorf table top centrifuge for 5 min at 4 °C and remove supernatant by aspiration. Wash cells with cold PBS for two times (*see* **Note 2**).

2. Add 20 μl of TES lysis buffer and mix cell pellet with TES lysis buffer by stirring with a pipette tip or gently pipetting. Add 10 μl of RNase cocktail and mix well gently. Incubate lysates for 30–120 min at 37 °C (*see* **Note 3**).

3. Add 10 μl of proteinase K, mix gently, and incubate at 50 °C for 2 h to overnight.

4. Prepare 1.5–2 % agarose gel in TAE buffer containing 0.5 μg/ml EtBr. After gel solidified, add 6× loading buffer in DNA samples.

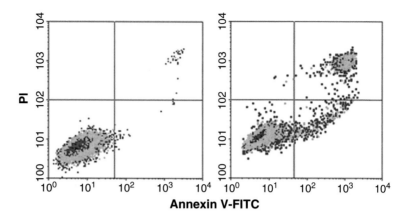

Fig. 1 MCF-7 cells were treated with phospho-ibuprofen for 24 h to induce apoptosis. Cell death was analyzed using Annexin V/PI staining. *Left*: control cells, *right*: phospho-ibuprofen treated cells

Add DNA samples into dry wells first, then add gel running buffer, TAE buffer, to cover agarose gel with samples.

5. Run the gel at a low voltage, which improves resolution of DNA fragments.

6. Visualize and photograph DNA ladders using UVP gel documentation system (UVP Inc.). DNA from apoptotic cells will form distinct ladder, whereas DNA from necrotic cells often appears to be a smear and shows no DNA ladder. DNA from viable cells will hardly migrate in the gel as a high-molecular-weight band (*see* **Note 4**).

3.4   In Situ TUNEL Staining for Apoptotic Cells

1. Tissue section preparation. Paraffin-embedded samples are sectioned into 4–6 μm tissue sections, followed by hydration by transferring the slides through the following solutions twice to xylene bath, twice in 100 % ethanol, followed by once in 95, 90, 80, 70, and 50 % ethanol and double distilled water (DDW).

2. Dry the slides at room temperature; use a hydrophobic barrier pen to draw a circle around tissue section carefully (*see* **Note 5**).

3. Rehydrate the sections in DDW for 5 min, tap away excess water on paper towel. Pipet 20 μg/ml proteinase K solution to cover sections and incubate for 15 min. Wash the slides 3× 5 min with DDW (*see* **Note 6**).

4. Inactivate endogenous peroxidases by covering the tissue sections with 3 % H_2O_2 for 10 min. Wash the slides 3× 5 min with DDW.

5. The sections are then covered by TdT buffer for 10 min. Remove TdT buffer and cover the sections with TdT reaction solution. Incubate the slides in a humidified chamber for 30 min at 37 °C (*see* **Note 7** for positive and negative controls).

6. Stop reaction by incubating the slides for 15 min in TdT stop solution. Rinse the slides with PBS.

7. Pipet 2 % BSA solution to cover the tissue sections and incubate for 30 min. Rinse the slides with PBS.

8. Incubate the tissue sections with Avidin/Biotinylated peroxidase complex for 30 min. Wash the slides 3× 5 min with PBS.

9. Incubate the sections with DAB solution. Monitor color development under microscopy until desired level of staining is achieved (typically 5–10 min). Stop the reaction by transferring the slides into DDW.

10. Lightly counterstain with hematoxylin. Monitor color development and contrast with DAB under microscopy until desired level is achieved (typically 1–2 min). Stop the reaction by transferring the slides into DDW. Wash the slides 3× 5 min with DDW.

11. Dehydrate the slides by transferring them through the following solutions: 70, 80, 90, 95, 100, 100 % ethanol, each step for 5 min. Incubate the slides with xylene 2× 5 min.

12. Cover the tissue sections with coverslips using Aqua-Poly/Mount mounting medium. Observe TUNEL (+) cells under microscopy (Fig. 2).

3.5 JC-1 Staining

1. Seed $1–3 \times 10^5$ cells in 6-well plate for experiment and one more well for positive control. Allow cells to attach and grow for 16–24 h before treatment (see **Note 8**).

2. Before harvesting, bring JC-1 stock solution to room temperature. Dilute JC-1 stock solution in PBS or cell culture media, mix by vortex the solution thoroughly. Protect the staining solution from light and keep it at 37 °C until use (see **Note 9**).

3. Harvest cells by collecting all the suspended and attached cells into a 5 ml flow cytometry tube. Centrifuge cells at 300–500×g for 5 min. Carefully remove the supernatant by aspiration. Add 2 ml PBS to resuspend cell pellet and repeat centrifugation to remove supernatant (see **Note 10**).

4. Resuspend cell pellet in 500 µl of JC-1 staining buffer, and incubate cells in cell culture incubator for 15–30 min.

5. After staining, wash cells thoroughly with PBS twice as in **step 3**. Resuspend cell pellet in 500 µl of PBS, protect from light until analyze using flow cytometry (BD biosciences). Acquire 50,000 cells for total events. Dot-plot 10,000 events for FL1-H versus FL2-H. FL1-H indicates fluorescent intensity of JC-1 monomer; FL2-H indicates fluorescent intensity of J-aggregates (Fig. 3).

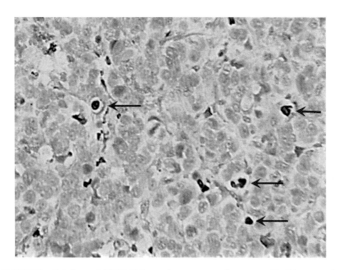

Fig. 2 TUNEL staining of MCF-7 breast cancer cell xenografted tumor. *Arrows* indicate TUNEL (+) apoptotic cells

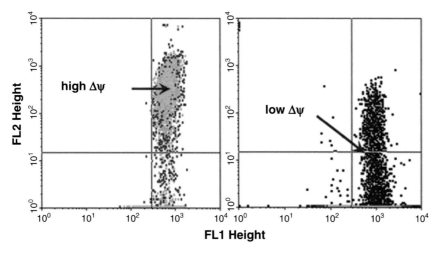

Fig. 3 MCF-7 cells were treated with phospho-ibuprofen for 3 h to induce apoptosis. $\Delta\psi$ was analyzed using JC-1. *Left*: control alive cells with high $\Delta\psi$ as high FL-2/low FL-1, right, phospho-ibuprofen treated cells undergo apoptosis with low $\Delta\psi$ as low FL-2/high FL-1

4   Notes

1. After adding Laemmli's buffer, samples can become viscous due to genomic DNA. To reduce the viscousness cell lysates can be briefly sonicated on ice to break DNA.

2. Do not use over 1–2×10^6 cells to prevent insufficient enzymatic digestion resulting in very viscous DNA solutions.

3. Do not vortex or extensive pipet to prevent damage of high molecular weight DNA.

4. Apoptosis does not always exhibit DNA fragmentation. For example, cells with resistant mutant ICAD do not show DNA fragmentation during apoptosis, although they do show some other features of apoptosis, such as chromatin condensation [19]. On the other hand, several scenarios rather than apoptosis may generate fragmented DNA. For instance, necrosis can, in some cases, produce randomly fragmented DNA [20]. Therefore, DNA fragmentation must be carefully assessed together with other features of apoptotic cells.

5. Water needs to be completely removed, as it can damage the hydrophobic barrier pen.

6. From this step, the slides need to be kept in a wet box.

7. Positive Control: incubate the sections with DNase I (3000 U/ml in 50 mM Tris–HCl, pH 7.5, 1 mg/ml BSA) for 10 min at 15–25 °C to induce DNA strand breaks, prior to the labeling procedure.
Negative Control: incubate the sections with TdT buffer only (without TdT enzyme and biotinylated dUTP) instead of TdT reaction solution.

8. Positive control: before harvesting cells, add 2 μl of CCCP stock solution into 2 ml media of positive control well and incubate for 5 min.

9. JC-1 concentration is cell line dependent. It should be optimized from 2 to 10 μM.

10. Centrifugation speed is cell-dependent. Try to avoid using high speed to centrifuge down all the cells.

References

1. Johnstone RW, Ruefli AA, Lowe SW (2002) Apoptosis: a link between cancer genetics and chemotherapy. Cell 108:153–164

2. Chinnaiyan AM et al (2000) Combined effect of tumor necrosis factor-related apoptosis-inducing ligand and ionizing radiation in breast cancer therapy. Proc Natl Acad Sci U S A 97:1754–1759

3. Gasco M, Shami S, Crook T (2002) The p53 pathway in breast cancer. Breast Cancer Res 4:70–76

4. Lowe SW, Ruley HE, Jacks T, Housman DE (1993) p53-dependent apoptosis modulates the cytotoxicity of anticancer agents. Cell 74:957–967

5. Janicke RU, Sprengart ML, Wati MR, Porter AG (1998) Caspase-3 is required for DNA fragmentation and morphological changes associated with apoptosis. J Biol Chem 273:9357–9360

6. Liu X, Zou H, Slaughter C, Wang X (1997) DFF, a heterodimeric protein that functions downstream of caspase-3 to trigger DNA fragmentation during apoptosis. Cell 89:175–184

7. Mandal D, Moitra PK, Saha S, Basu J (2002) Caspase 3 regulates phosphatidylserine externalization and phagocytosis of oxidatively stressed erythrocytes. FEBS Lett 513:184–188

8. Crawford ED, Wells JA (2011) Caspase substrates and cellular remodeling. Annu Rev Biochem 80:1055–1087

9. Martin DA, Siegel RM, Zheng L, Lenardo MJ (1998) Membrane oligomerization and cleavage activates the caspase-8 (FLICE/MACHalpha1) death signal. J Biol Chem 273:4345–4349

10. Boatright KM et al (2003) A unified model for apical caspase activation. Mol Cell 11:529–541

11. Tait SW, Green DR (2010) Mitochondria and cell death: outer membrane permeabilization and beyond. Nat Rev Mol Cell Biol 11:621–632

12. Chai J et al (2001) Crystal structure of a pro-caspase-7 zymogen: mechanisms of activation and substrate binding. Cell 107:399–407

13. Riedl SJ et al (2001) Structural basis for the activation of human procaspase-7. Proc Natl Acad Sci U S A 98:14790–14795

14. Chang HY, Yang X (2000) Proteases for cell suicide: functions and regulation of caspases. Microbiol Mol Biol Rev 64:821–846

15. Gavrieli Y, Sherman Y, Ben-Sasson SA (1992) Identification of programmed cell death in situ via specific labeling of nuclear DNA fragmentation. J Cell Biol 119:493–501

16. Grasl-Kraupp B et al (1995) In situ detection of fragmented DNA (TUNEL assay) fails to discriminate among apoptosis, necrosis, and autolytic cell death: a cautionary note. Hepatology 21:1465–1468

17. Reers M, Smith TW, Chen LB (1991) J-aggregate formation of a carbocyanine as a quantitative fluorescent indicator of membrane potential. Biochemistry 30:4480–4486

18. Smiley ST et al (1991) Intracellular heterogeneity in mitochondrial membrane potentials revealed by a J-aggregate-forming lipophilic cation JC-1. Proc Natl Acad Sci U S A 88:3671–3675

19. Sakahira H, Enari M, Ohsawa Y, Uchiyama Y, Nagata S (1999) Apoptotic nuclear morphological change without DNA fragmentation. Curr Biol 9:543–546

20. Didenko VV, Ngo H, Baskin DS (2003) Early necrotic DNA degradation: presence of blunt-ended DNA breaks, 3′ and 5′ overhangs in apoptosis, but only 5′ overhangs in early necrosis. Am J Pathol 162:1571–1578

Chapter 12

Assessment of Matrix Metalloproteinases by Gelatin Zymography

Jillian Cathcart

Abstract

Matrix metalloproteinases are endopeptidases responsible for remodeling of the extracellular matrix and have been identified as critical contributors to breast cancer progression. Gelatin zymography is a valuable tool which allows the analysis of MMP expression. In this approach, enzymes are resolved electrophoretically on a sodium dodecyl sulfate-polyacrylamide gel copolymerized with the substrate for the MMP of interest. Post electrophoresis, the enzymes are refolded in order for proteolysis of the incorporated substrate to occur. This assay yields valuable information about MMP isoforms or changes in activation and can be used to analyze the role of MMPs in normal versus pathological conditions.

Key words Matrix metalloproteinases, MMPs, Zymography, Zymogram

1 Introduction

The first matrix metalloproteinase, or MMP, was described by Gross and Lapière in 1962 following the observation that collagen in the extracellular matrix is drastically remodeled during normal organismal growth and development [1]. Over the course of the proceeding decades, roughly two dozen members of this family have been identified with numerous physiological functions described. These functions include proteolysis of a wide variety of extracellular matrix components, cleavage of bioactive-signaling molecules to influence the microenvironment, and proteolytic activation of other proteases [2, 3]. MMPs are now also understood to contribute to multiple pathologies; indeed, overexpression and overactivation of MMPs are characteristics of most cancers. Such dysregulation directly contributes to disease progression and patient prognosis by increasing angiogenesis, cancer cell proliferation, invasion, and metastasis. MMPs are calcium-dependent endopeptidases that coordinate a zinc ion necessary for catalysis. Generally synthesized as zymogens, most MMPs require proteolytic activation themselves in order to cleave matrix substrates [4–8].

Jian Cao (ed.), *Breast Cancer: Methods and Protocols*, Methods in Molecular Biology, vol. 1406,
DOI 10.1007/978-1-4939-3444-7_12, © Springer Science+Business Media New York 2016

Zymography is a powerful tool and was also first described by Gross and Lapière in their initial report on proteases cleaving collagen. Zymography yields qualitative information regarding changes in MMP activation, enzyme isoforms, and formation of quaternary structures. Furthermore, this technique requires minimal quantities of enzyme (in the picogram range for the gelatinases, MMP-2, and MMP-9) but does not require the special or costly equipment necessary for other techniques often employed for analyzing proteases. However, zymography cannot yield accurate quantitative information about the amount of MMP present in a sample nor the net proteolytic activity.

The current adaptation of this assay involves electrophoretic separation of protein samples from cell or tissue extracts or culture medium on a nonreducing sodium dodecyl sulfate-polyacrylamide (SDS-PA) gel with the preferred substrate for the MMP of interest incorporated. Because the SDS in the gel disrupts covalent interactions and thus denatures the proteins, after electrophoresis the gel must be washed with a nonionic detergent with a lower critical micelle concentration, typically Triton X-100, in order to remove the SDS and allow for refolding of the proteins within the gel. Incubation of the gel is then required in order for the enzymes to digest the substrate incorporated in the gel. Finally, the gel is stained. Following the appropriate destaining procedure, the potential enzyme activity of a sample can be observed by the presence of unstained bands where the substrate was degraded against the stained background of the gel where the substrate remains unproteolyzed (*see* Fig. 1).

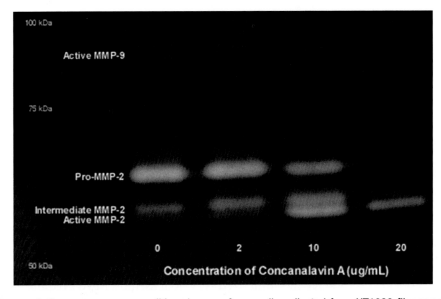

Fig. 1 Representative zymogram on conditioned serum-free media collected from HT1080 fibrosarcoma cells after 18 h treatment with increasing concentrations of Concanavalin A (Con A), a lectin known to increase transportation of MMP-14 intracellular stores to the cell surface where it can function to activate MMP-2. As the amount of endogenous MMP-2 cleaved by MMP-14 increases, a decrease in the pro-MMP-2 isoform can be observed concurrently. Note that there is no change in MMP-9 activity in response to Con A treatment until 20 μg/mL of Con A (as MMP-14 is not known to affect MMP-9 activation), at which point the cells become necrotic

2   Materials

Prepare all solutions using deionized water and analytical grade reagents. Unless indicated otherwise, all buffers and reagents should be prepared and stored at room temperature. Follow appropriate disposal regulations when discarding waste.

2.1 SDS Polyacrylamide Zymogram Gel Components

1. Resolving gel buffer: 1.5 M Tris–HCl, pH 8.8. Add 181.7 g Tris base and transfer to a beaker with 900 mL water. Stir until dissolved and adjust pH. Bring volume to 1 L total with water.

2. Stacking gel buffer: 0.5 M Tris–HCl, pH 6.8. Add 60.6 g Tris base to a beaker with 900 mL water. Stir until dissolved and adjust pH. Bring volume to 1 L total with water.

3. 30 % acrylamide/bisacrylamide (PA): Add 290 g of acrylamide and 10 g of N,N'-methylbisacrylamide in 800 mL of water. Stir until dissolved, heating to mix if necessary. pH to 7.0 or less and bring volume to 1 L total with water. Alternatively, this can be obtained commercially as acrylamide/bisacrylamide, liquid (37:5:1) containing 30 % acrylamide (w/v) and 0.8 % bisacrylamide (w/v). Store in a dark bottle, 4 °C.

4. 10 % ammonium persulfate (APS) (w/v): Dissolve 10 % APS in water.

5. 10 % sodium dodecyl sulfate (SDS) (w/v): Dissolve 10 % SDS in water.

6. N,N,N,N' -tetramethyl-ethylenediamine (TEMED).

7. Substrate (e.g., 5 % gelatin): Dissolve 50 mg of lyophilized gelatin in 800 µL of water. Bring volume to 1 mL.

8. Gel-pouring cassettes and associated gel-running apparatus.

2.2 Zymogram-Developing Components

1. 10× SDS-running buffer: Stir 30.0 g of Tris base, 144.0 g of glycine, and 10.0 g of SDS into 800 mL water. pH to 8.3 and bring volume to 1 L total with water.

2. 2.5 % Triton X-100 (v/v): Dissolve 25 mL of Triton X-100 in 975 mL water.

3. Incubation buffer: Dissolve 6.1 g of Tris base, 8.8 g of NaCl, 1.1 g of $CaCl_2$, and 20 mg of sodium azide in 800 mL of water. pH to 7.5 and bring to 1 L total with water.

4. Coomassie blue stain: Stir together 250 mg Coomassie brilliant blue R-250, 10 mL acetic acid, 45 mL methanol, and 45 mL of water. Filter through Whatman filter paper No. 1 to remove particulates.

5. Coomassie destain: Stir together 50 mL acetic acid, 50 mL methanol, and 400 mL of water.

2.3 Sample Preparation Components

1. 2× SDS sample buffer: 100 mM Tris–HCl, pH 6.8 (diluted from the 500 mM stock prepared for the SDS-PA gel), 4 % SDS (w/v), 20 % glycerol (v/v), and 0.2 % bromophenol blue (w/v).

3   Methods

3.1   10 % Sodium Dodecyl Sulfate-Polyacrylamide Gel with 0.1 % Gelatin Substrate Copolymerization Preparation

Carry out all procedures at room temperature unless noted. *See* Table 1 for a discussion on the various substrates that can be used to evaluate different MMPs.

1. Set up a gel-pouring cassette as per the manufacturer's directions to accommodate a gel with dimensions 8 cm (width) × 13 cm (height) × 1 mm (depth).

2. Prepare the separating gel: On ice, mix 3.77 mL of dH₂O, 200 µL of 5 % gelatin, 3.33 mL of acrylamide–*bis*-acrylamide stock, 2.5 mL of resolving gel buffer, 100 µL of 10 % SDS, and 100 µL of 10 % APS (see Note 1).

3. Add 8 µL of TEMED to the separating gel solution to initiate polymerization. Gently pipette to mix without causing bubble formation or aeration.

4. Immediately, pour the separating gel solution into the cassette avoiding the formation of bubbles.

5. Carefully overlay the separating gel solution with isopropanol using a transfer pipette or syringe.

6. Allow the gel to polymerize for at least 30 min at room temperature. A discernible line of separation between the separating gel and isopropanol overlay will be apparent when polymerization is complete.

7. Prepare the stacking gel: On ice, mix 3.05 mL of dH2O, 650 µL of acrylamide–bis-acrylamide stock, 1.25 mL of stacking gel buffer, 50 µL of 10 % SDS, and 50 µL of 10 % APS.

8. Decant the isopropanol overlay from the separating gel.

9. Immediately, add 8 µL of TEMED to the stacking gel solution, mix gently by pipetting, and pipette the solution on top of the separating gel until it reaches the top of the cassette.

10. Immediately insert the comb (generally ten wells) into the stacking gel, ensuring no bubbles are trapped under the comb. Allow the stacking gel to polymerize at room temperature for 30 min (*see* **Note 2**).

3.2   Sample Preparation from Serum-Free Conditioned Media

Most MMPs, including the gelatinases, are secreted enzymes and thus conditioned media collected from cell cultures are an accessible and valuable way to evaluate MMP activity. However, because serum contains proteases, cells must be cultured in serum-free media.

Table 1
Summary of the commonly used substrates for analysis of MMPs by gelatin zymography [12–15]

Substrate	Directly assessed	Notes
Gelatin	MMP-2, MMP-9	Recommended starting concentration of gelatin is 0.1 % (w/v)
		Because MMP-14 is a membrane-type MMP, its extraction and direct analysis by zymography is rife with challenges. Thus, because MMP-14 activates MMP-2, gelatin zymography is frequently used to indirectly assess changes in MMP-14 function by assessing changes in MMP-2 activation
Casein	MMP-1, MMP-7, MMP-12, MMP-13	Recommended starting concentration of casein is 0.1 % (w/v)
		Much less sensitive than gelatin zymography
		Casein migrates during electrophoresis; to reduce issues associated with this, pre-run the unloaded gel. Remaining casein is sufficient for qualitative assessment of MMP function
Collagen type I	MMP-1, MMP-13. MMP-2 and MMP-9 also detected	Recommended collagen type I starting concentration is 500 µg/mL
Heparin-enhanced	MMP-7 (most common). Bands for MMP-1 and MMP-13 can also be enhanced by heparin addition	3 µg/mL heparin is the recommended starting concentration. As the heparin runs, it will form a "heparin box" within which the MMP of interest must fall
		For MMP-7, heparin should be added directly to the enzyme sample. For MMP-1 and MMP-13, heparin should be loaded after the run has been initiated. A pre-stained marker is also added at the same time as the heparin follows its position. The 31-kDa band of the secondary marker should align with the 66-kDa band of the first protein ladder loaded. This alignment indicates the heparin box encompasses the 66 and 36 kDa region of the gel
		The signal for intermediate and active MMP-7 is increased approximately fivefold by heparin addition and pro-MMP-7 can be increased as much as 20-fold

1. Allow the cells in culture to grow to approximately 80 % confluence in the appropriate complete media.

2. Wash the monolayer with sterile PBS to remove the serum completely; three washes should be sufficient.

3. Culture the cells in a low volume of serum-free media at 37 °C for 12–16 h (*see* **Notes 3** and **4**).

4. After the appropriate incubation period, collect the media and centrifuge ($350 \times g$, 5 min at 4 °C) to remove dead cells and debris. Transfer the supernatant to a clean tube.

5. Dilute conditioned medium in SDS sample buffer to a final concentration of 1× SDS sample buffer (*see* **Note 5**).

6. Pipette to mix and allow samples to incubate at 37 °C for 20 min in order to solubilize the enzymes before loading the zymogram. Do not boil the samples (*see* **Note 6**).

3.3 Sample Preparation from Cell Lysates

Intracellular enzymes include secreted MMPs at different stages of posttranslational modification, MMPs associated with the cell surface/matrix, or MMPs which are localized intracellularly.

1. Allow the cells in culture to grow to approximately 80 % confluence in the appropriate complete media.

2. Wash the cells twice with cold sterile PBS.

3. Lyse on ice using ice-cold 2× SDS sample buffer.

4. Scrape the cells with a rubber cell scraper, transfer the lysate to a clean tube, and incubate on ice for 20 min.

5. Mix by vortexing and centrifuge $16,000 \times g$ for 20 min at 4 °C.

6. Transfer the supernatant to a clean tube and measure the protein concentration.

7. Transfer the appropriate volume of lysate to be run on the gel to a clean tube and add 1× sample buffer (diluted from 2× in water) to all samples to bring each sample to the same final volume (*see* **Note 7**).

3.4 Gel Electrophoresis and Development of Zymogram

1. Dilute the 10× running buffer to 1× in water, preparing enough volume to sufficiently fill the gel-running apparatus.

2. Carefully remove the comb from the stacking gel and place the zymogram into the gel-running apparatus. Fill the buffer chambers with 1× running buffer.

3. Load the marker and the samples and fill any empty wells with 1× sample buffer (diluted from 2× in water).

4. Run the samples through the stacking gel at constant voltage of 100 V.

5. Run the samples through the resolving gel at constant voltage of 130 V until adequate separation is achieved as indicated by

the migration of the marker (*see* **Note 8**). This will be based upon the degree of separation desired and the molecular weight of the MMPs of interest.

6. Carefully remove the gel from the running chamber and place it in plastic tray. To the tray add sufficient renaturing solution to completely submerge the zymogram (*see* **Note 9**). Incubate the gel for 30 min at room temperature, rocking gently.

7. Discard the renaturing solution and repeat **step 6**.

8. Incubate the zymogram at room temperature for an additional 30 min in developing buffer while rocking gently.

9. Discard the developing buffer and replace it with fresh developing buffer. Cover the tray with a lid or wrap tightly with plastic wrap and incubate at 37 °C for approximately 16 h (*see* **Note 10**).

10. Discard the developing buffer and stain the zymogram in staining solution until the gel is uniformly dark (*see* **Notes 11** and **12**).

11. Destain the zymogram with destaining solution until areas of proteolytic activity appear as clear bands well contrasting with the dark background of the gel (*see* **Note 13**).

4   Notes

1. The volume of water should be adjusted in the event that different substrates are used with different volumes added. The final volume of the separating gel should equal 10 mL.

2. The zymogram can be poured in advance, wrapped in a damp towel, covered with plastic wrap, and stored at 4 °C. The zymogram should be used within 2 weeks of pouring.

3. It is not advisable to culture cells beyond 24 h in serum-free media, as serum is vital to maintain cell health and viability.

4. We find the best results are obtained from cells cultured in three-centimeter dishes overlaid with 800 μL serum-free media.

5. Generally, we find 10 μL of media from cells plated in a 3-cm dish sufficient for analyzing the gelatinases on a gelatin zymogram. The 2× SDS sample buffer described above can also be prepared as a 5× stock and diluted to 1× in samples should larger volumes be necessary for adequate resolution.

6. If the intensity of the bands is too low and resolution is poor, the media can be concentrated using spin columns with the appropriate molecular weight cutoff.

7. Generally, 30 μg of total protein is sufficient to resolve the MMPs on the zymogram. It is not recommended to load more than 40 μg of protein per lane as excessive protein can negatively affect band separation and resolution. Typically, the wells of a zymogram can accommodate up to 30 μL volume.

8. Different MMP enzymes and enzyme forms are visualized on zymograms according to their molecular weights. Therefore, it is standard to run a protein marker along with your samples on the gel. While the molecular weights of each of the MMPs and their different forms (pro, intermediate, or active, as applicable) are known, the markers typically run on a zymogram are not accurately indicative of the MMP size. This is due to the fact that the markers are generally provided in reduced and denatured conditions, whereas MMPs are run on a zymogram in nonreducing buffer and then partially refolded to their native confirmation.

9. The refolding process following electrophoresis is only partial and therefore proteolytic activity can be observed for pro- and intermediate-form MMPs, with the proband appearing at a higher molecular weight. This effect is believed to be due in part to disruption of a cysteine switch that renders pro-MMPs inactive in physiological conditions by SDS. Although the pro-isoform is mostly refolded, this cysteine switch cannot be reestablished and thus the enzyme exhibits activity. Such proteolysis is merely an artifact of this technique and is not representative of the true biological activity [9, 10].

10. Incubation occurs at 37 °C in accordance with the rules of classical enzymology. Generally, an overnight incubation period of approximately 18 h is sufficient. We have observed that the longer the incubation, the greater the rate of diffusion of the MMPs away from the band and through the gel, resulting in decreased resolution. Therefore it is advisable to adjust the sample volume rather than the incubation time to improve resolution.

11. Staining solution can be collected and reused, although reused stain typically requires a longer staining time.

12. Aside from Coomassie blue, protein silver, Congo red, and Amido black stains may be used alternatively along with the associated destain.

13. MMP/inhibitor complexes (such as tissue inhibitors of metalloproteinases or TIMPs) from biological samples are dissociated by SDS during electrophoresis. Thus, the bands on the zymogram should be interpreted as potential enzymatic activity rather than net MMP activity [9, 11].

References

1. Gross J, Lapiere CM (1962) Collagenolytic activity in amphibian tissues: a tissue culture assay. Proc Natl Acad Sci 48(6):1014–1022
2. Morrison CJ et al (2009) Matrix metalloproteinase proteomics: substrates, targets, and therapy. Curr Opin Cell Biol 21(5):645–653
3. Rodríguez D, Morrison CJ, Overall CM (2010) Matrix metalloproteinases: what do they not do? New substrates and biological roles identified by murine models and proteomics. Biochem Biophys Acta 1803(1):39–54
4. Roy R, Yang J, Moses MA (2009) Matrix metalloproteinases as novel biomarkers and potential therapeutic targets in human cancer. J Clin Oncol 27(31):5287–5297
5. Bloomston M, Zervos EE, Rosemurgy AS 2nd (2002) Matrix metalloproteinases and their role in pancreatic cancer: a review of preclinical studies and clinical trials. Ann Surg Oncol 9(7):668–674
6. Egeblad M, Werb Z (2002) New functions for the matrix metalloproteinases in cancer progression. Nat Rev Cancer 2(3):161–174
7. Deryugina EI, Quigley JP (2006) Matrix metalloproteinases and tumor metastasis. Cancer Metastasis Rev 25(1):9–34
8. Overall CM, Kleifeld O (2006) Validating matrix metalloproteinases as drug targets and anti-targets for cancer therapy. Nat Rev Cancer 6(3):227–239
9. Vandooren J et al (2013) Zymography methods for visualizing hydrolytic enzymes. Nat Methods 10(3):211–220
10. Springman EB et al (1990) Multiple modes of activation of latent human fibroblast collagenase: evidence for the role of a Cys73 active-site zinc complex in latency and a "cysteine switch" mechanism for activation. Proc Natl Acad Sci 87(1):364–368
11. Hawkes SP, Li H, Taniguchi GT (2001) Zymography and reverse zymography for detecting MMPs, and TIMPs. In: Clark I (ed.) Matrix metalloproteinase protocols. Springer. p 399–410
12. Fernandez-Resa P, Mira E, Quesada AR (1995) Enhanced detection of casein zymography of matrix metalloproteinases. Anal Biochem 224(1):434–435
13. Snoek P, Von den Hoff JW (2005) Zymographic techniques for the analysis of matrix metalloproteinases and their inhibitors. BioTechniques 38(1):73–83
14. Yu WH, Woessner JF (2001) Heparin-enhanced zymographic detection of matrilysin and collagenases. Anal Biochem 293(1):38–42
15. Kleiner DE, Stetlerstevenson WG (1994) Quantitative zymography: detection of picogram quantities of gelatinases. Anal Biochem 218(2):325–329

Chapter 13

Assessment of Synthetic Matrix Metalloproteinase Inhibitors by Fluorogenic Substrate Assay

Ty J. Lively, Dale B. Bosco, Zahraa I. Khamis, and Qing-Xiang Amy Sang

Abstract

Matrix metalloproteinases (MMPs) are a family of metzincin enzymes that act as the principal regulators and remodelers of the extracellular matrix (ECM). While MMPs are involved in many normal biological processes, unregulated MMP activity has been linked to many detrimental diseases, including cancer, neurodegenerative diseases, stroke, and cardiovascular disease. Developed as tools to investigate MMP function and as potential new therapeutics, matrix metalloproteinase inhibitors (MMPIs) have been designed, synthesized, and tested to regulate MMP activity. This chapter focuses on the use of enzyme kinetics to characterize inhibitors of MMPs. MMP activity is measured via fluorescence spectroscopy using a fluorogenic substrate that contains a 7-methoxycoumarin-4-acetic acid N-succinimidyl ester (Mca) fluorophore and a 2,4-dinitrophenyl (Dpa) quencher separated by a scissile bond. MMP inhibitor (MMPI) potency can be determined from the reduction in fluorescent intensity when compared to the absence of the inhibitor. This chapter describes a technique to characterize a variety of MMPs through enzyme inhibition assays.

Key words Matrix metalloproteinases, Matrix metalloproteinase inhibitors, Fluorescence resonance energy transfer, Inhibition constant, Half maximal inhibitory concentration

1 Introduction

1.1 Matrix Metalloproteinases (MMPs)

Matrix metalloproteinases (MMPs) are a family of zinc-dependent endopeptidases that play an active role in remolding the extracellular matrix (ECM). Further classification allows MMPs to be divided into six different subcategories based on their homology and substrate specificity: collagenases (MMP-1, MMP-8, MMP-13, and MMP-18), gelatinases (MMP-2 and MMP-9), stromelysins (MMP-3, MMP-10, and MMP-11), matrilysins (MMP-7 and MMP-26), membrane-type MMPs (MMP-14, MMP-15, MMP-16, MMP-17, MMP-24, and MMP-25), and others (MMP-12, MMP-19, MMP-20, MMP-21, MMP-23, MMP-27, and MMP-28) [1]. The structural components of the MMP family typically include catalytic, pro-peptide, and hemopexin-like domains. The most important

Jian Cao (ed.), *Breast Cancer: Methods and Protocols*, Methods in Molecular Biology, vol. 1406,
DOI 10.1007/978-1-4939-3444-7_13, © Springer Science+Business Media New York 2016

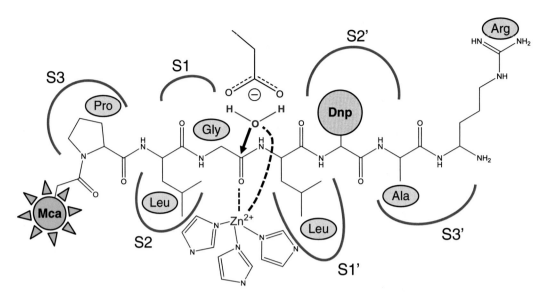

Fig. 1 Illustration of fluorogenic substrate binding and cleavage within the MMP active site. A catalytic zinc (II) held by three histidine residues utilizes a water molecule to hydrolyze the peptide bond between the glycine and leucine residues

domain, the catalytic domain, includes a five-stranded β-sheet, three long α-helices and a methionine turn that form specificity pockets for substrate binding [2]. Additionally, calcium ions and zinc ions are present in the catalytic domain for added structural and catalytic stability. Studies have shown that calcium ions within this domain stabilize MMPs, mediate signal transduction in MMP expression, and facilitate MMP activation [3]. The catalytic zinc, held in place by three histidine residues (HEXXHXXGXXH), utilizes a water molecule for the hydrolysis of a substrate's scissile peptide bond (Fig. 1) [4]. The pro-peptide domain includes a "cysteine switch" motif (PRCGXPD) that is responsible for maintaining catalytic latency, by binding to the zinc's fourth coordinate position [4]. Activation of MMPs is stimulated upon removal of the inhibitory zinc-cysteine bond through either removal of the pro-domain, chemical modification of the cysteine's thiol group, or other structural perturbations. Finally, the hemopexin domain, comprised of a four bladed β-propeller structure surrounding a central channel, facilitates binding and interactions with indigenous tissue inhibitors of MMPs (TIMPs).

Members of the MMP family play prominent roles in many physiological and pathological processes [1]. MMPs are important in development, morphogenesis, and reproduction [5]. Studies have shown that MMP-2 and MMP-9 can disrupt the blood-brain barrier inducing cerebrovascular incidents [1, 6, 13]. MMP-26 expression is upregulated in human endometrial, and ovarian, cancer cell lines, human breast ductal carcinoma in situ (DCIS), and high-grade prostatic intraepithelial neoplasia (HGPIN) [3, 7–10]. Increased expression of MMP-14 plays an important role in tumor progression and metastasis [11]. Overexpression of MMP-1, MMP-3, MMP-7, or MMP-14 in mice caused the development of hyperplastic lesions and increased the likelihood of malignant tumor formation [12]. While many MMPs have been linked to multiple diseases, others may offer protective effects, thus suggesting MMP targeting is disease dependent [1, 13–15].

1.2 Matrix Metalloproteinase Inhibitors (TIMPs and Synthetic MMPIs)

Inhibitors of matrix metalloproteinases act as post-activation regulators of MMP activity. The tissue inhibitors of metalloproteinases (TIMP-1, TIMP-2, TIMP-3, and TIMP-4) are the primary physiological inhibitors of MMPs. Previous attempts to use TIMPs as therapeutic agents to control upregulated MMP activity have been unsuccessful due to lack of MMP selectivity and involvement in other non-MMP-related processes [16]. Consequently, synthetic matrix metalloproteinase inhibitors (MMPIs) are being researched and developed to provide higher selectivity and specificity toward various MMPs [1, 6, 12, 20, 21]. Inhibitor specificity can be modulated by designing around specificity pocket interactions. For example, the S_1' site may include a hydrophobic pocket that is deep (MMP-3, MMP-12, and MMP-14), intermediate (MMP-2, MMP-8, and MMP-9), or shallow (MMP-1 and MMP-7) [1, 18–21]. Some MMPIs replace the scissile bond with a noncleavable zinc-binding group (ZBG) essentially mimicking a substrate, while others use non-peptidomimetic structures to enhance binding affinities [1]. Inhibitory potency can be tuned by utilizing different ZBGs, with the most potent inhibitors involving hydroxamic acid. However, selectivity is compromised as the hydroxamate's chelation efficacy overwhelms other binding contributors [1, 6, 21, 22]. Consequently, other ZBGs have been developed including thiols, carboxylates, phosphinates, and sulfodiimines [1, 6, 12, 21, 22].

1.3 Fluorescence Resonance Energy Transfer (FRET) Enzyme Inhibition Assay

MMP fluorescent substrates utilize fluorescence resonance energy transfer (FRET) or intramolecular fluorescence energy transfer (IFET) to assay MMP activity [16]. Resonance energy transfer depends on the existence of a fluorescent donor molecule and a quenching acceptor molecule. The interaction between the two molecules results in an energy transfer from donor to acceptor. This energy transfer can be intramolecular or intermolecular. In intramolecular energy transfer, the fluorophore and the quenching group are separated in the peptide chain at a distance which allows efficient energy transfer. Enzymatic cleavage of the substrate separates the donor from the acceptor molecules and terminates the interaction between the two. This results in an increase in fluorescence intensity which is proportional to the number of substrates cleaved [17]. This increase in fluorescent activity can be seen in the FRET substrate Mca-Pro-Leu-Gly-Leu-Dap(Dnp)-Ala-Arg-NH$_2$ (M-1895) when the Gly-Leu bond cleaves, causing the Mca group to separate from the 2,4-dinitrophenyl quencher.

The enzyme inhibition assay described in this chapter uses a FRET substrate (Mca-Pro-Leu-Gly-Leu-Dap(Dnp)-Ala-Arg-NH$_2$)

$$\frac{v_i}{v_0} = \frac{[E]_0 - [I]_0 - K_i^{app} + \sqrt{([I]_0 + K_i^{app} - [E]_0)^2 + 4[E]_0 K_i^{app}}}{2[E]_0}$$

Fig. 2 In this equation V_i and V_0 are the initial rates with and without the inhibitor, respectively. and $[E]_0$ and $[I]_0$ are the initial enzyme and inhibitor concentrations, respectively. Finally, $K_{i\,app}$ is the apparent inhibition constant between the inhibitor and the enzyme

$$K_i\,(app) = K_i \left(1 + \frac{[S]}{K_m}\right)$$

Fig. 3 For competitive inhibitors K_i apparent is equivalent to the true K_i under the substrate concentration. $[S] \ll K_m$. K_m is the Michaelis-Menten constant

$$IC_{50} = K_i \left(1 + \frac{[S]}{K_m}\right)$$

Fig. 4 For classical competitive inhibitors (inhibitors that have a K_i value significantly larger than the enzyme's concentration), the IC_{50} point is directly proportional to the K_i value multiplied by a correction factor

with tight-binding MMPIs to determine inhibition potency. One measure of inhibitor potency is the inhibition dissociation constant, K_i. The K_i value may be calculated using the Morrison equation [6] (Fig. 2). For tight-binding competitive inhibitors, a correction factor is introduced into the Morison equation to eliminate the influence of substrate concentration on the inhibition dissociation constant (Fig. 3). An alternate method for measuring the potency of an inhibitor is to determine the half maximal inhibitory concentration, or IC_{50} point, defined as the concentration of the inhibitor that yields 50 % inhibition of enzyme activity. By ascertaining the IC_{50} point, the K_i value for a competitive inhibitor can be calculated through algebraic rearrangement of the following equation (Fig. 4):

2   Materials

2.1 Luminescence Spectrometer Components

1. PerkinElmer LS 50B luminescence spectrometer or another fluorimeter with temperature control set to 25 °C.
2. Computer and monitor.
3. Fl WinLab program or similar.

2.2 Cuvette Components

1. Kinetics buffer: 50 mM 4-(2-hydroxyethyl)-1-piperazineet hanesulfonic acid (HEPES), 0.2 M NaCl, and 10 mM CaCl$_2$ * 2H$_2$O, pH 7.5, 0.01 % Brij-35 (w/v %). Add about 100 mL of

ddH$_2$O to a 500 mL beaker or glass bottle. Weigh out 5.9575 g HEPES, 5.844 g NaCl, and 0.735 g CaCl$_2$ * 2H$_2$O and transfer these components into the bottle and mix. Add ddH$_2$O until volume is at 400 mL and continue mixing until pH 7.5 is reached. Fill it to 495 mL with ddH$_2$O. Finally add 5 mL of 1 % Brij-35 solution to 500 mL bottle containing buffer components yielding a 0.01 % Brij-35 (w/v %) buffer solution. Store buffer in 4 °C fridge (*see* **Note 1**).

2. One percent Brij-35 solution: in a 50 mL tube, prepare a 1 % Brij-35 (w/v %) solution by mixing 500 mg of Brij-35 with 50 mL ddH$_2$O; mix well.

3. 20 Mm tris(2-carboxyethyl)phosphine (TCEP): 20 mM TCEP, kinetics buffer. Weigh out 8.6 mg TCEP and add to a 1.5 mL tube. Fill tube to 1.5 mL with kinetics buffer. Store in 4 °C fridge.

4. Twenty-five percent glycerol solution: glycerol, kinetics buffer. In a 15 mL tube, mix 3.75 mL of kinetics buffer with 1.25 mL of glycerol.

5. Substrate: Stock substrate Mca-Pro-Leu-Gly-Leu-Dap(Dnp)-Ala-Arg-NH$_2$ (M-1895), cold ddH$_2$O, dimethyl sulfoxide (DMSO). In a 2 mL tube, mix 20 μL stock substrate, 400 μL cold ddH$_2$O, and 380 μL DMSO; quickly place solution on ice.

2.3 SigmaPlot Components

1. Computer and monitor.

2. SigmaPlot 2000 program or similar.

3 Methods

3.1 Preparation of Assay Buffer

1. Remove kinetics buffer from 4 °C fridge and allow it to warm to room temperature.

2. Dilute the 20 mM TCEP solution to 2 mM TCEP by adding to a 0.5 mL tube 36 μL kinetics buffer and 4 μL 20 mM TCEP; mix.

3. Prepare final assay buffer by adding 5.5 μL 2 mM TCEP and 2194.5 μL kinetics buffer to a 15 mL tube; mix well (*see* **Note 2**).

3.2 Enzyme Dilution

1. Turn on fluorimeter and set temperature control to 25 °C.

2. Obtain stock enzyme from −80 °C freezer and allow it to thaw.

3. Begin testing the slope of the stock enzyme by adding 176 μL of kinetics buffer in a 0.5 mL tube. Add 10 μL of DMSO to the bottom of the tube. Next, add 10 μL of stock enzyme to the tube and mix by rapidly pipetting several times (*see* **Notes 3** and **4**).

4. Incubate the tube at room temperature for 30 min.

5. After 15 min of incubation, turn on fluorimeter and computer. Once the computer is on, open the program "Fl WinLab" and select "Time Drive."

6. Once the tube has incubated for 30 min, add 4 μL of substrate and mix by rapidly pipetting several times (*see* **Note 4**).

7. Quickly pipette the contents of the tube into a cuvette and place it in the fluorimeter (*see* **Note 5**).

8. Click the "Start" button on Fl WinLab and let the sample run for 15 min.

9. Once the sample has finished running, click the "Stop" button on Fl WinLab.

10. Click "Data Handling" on Fl WinLab. Select the sample and the slope option and set the range from 6 to 10 min. Click calculate.

11. Measure fluorescence versus time for 15 min and then find the slope over the desired period (typically between 6 and 10 min).

12. If the stock concentration is too high, dilute the enzyme with the 25 % glycerol solution (*see* **Note 6**).

13. Repeat **steps 3–12** until the slope of the enzyme is approximately 30–35.

14. Aliquot diluted enzyme into 12 μL aliquots and store at −80 °C (*see* **Note 7**).

3.3   Inhibitor Dilution

1. Remove 40 μL inhibitor stock from storage and dilute with 40 μL of appropriate solvent, such as DMSO (*see* **Note 8**).

2. Next, perform serial dilution until all inhibitor dilutions are made. Store the inhibitor dilutions in appropriate conditions, such as 4 °C.

3.4   General Instructions for Assays

1. Turn on water bath attached to fluorimeter and set temperature to 25 °C.

2. Wash cuvettes by first rinsing them with ddH$_2$O and then by rinsing them with methanol. Dry the cuvettes using light duty tissue wipes.

3. Remove enzyme from −80 °C freezer and substrate from −20 °C. Allow all to thaw and then place on ice.

4. Begin the control run by pipetting 176 μL of assay buffer into a 0.5 mL tube. Add 10 μL of inhibitor diluent to the bottom of the tube. Next, add 10 μL of enzyme to the tube and mix by rapidly pipetting several times.

5. Allow tube to incubate at room temperature for 30 min. Be sure to record the time started and the time you finish in a lab notebook.

6. After the control has been prepared, prepare the serial dilutions for the inhibitor. The exact amount of dilutions made will depend on how well the inhibitor you are using reacts with the enzyme you are using.

7. After 15 min has passed, turn on fluorimeter and computer. Once the computer is on, set up the program Fl WinLab which will be used to measure the slope of the enzyme.

8. Begin preparation of the next sample by adding 176 µL of TCEP plus kinetics buffer solution into a 0.5 mL tube, followed by adding 10 µL of the inhibitor dilution and 10 µL of the enzyme. Once again to mix rapidly, dispense the pipette several times. Allow this sample to once again incubate for 30 min.

9. After the control has incubated for 30 min, add 4 µL of substrate to the control tube by rapidly dispensing the pipette several times. At this point the total volume inside of the tube should be 200 µL.

10. Quickly pipette the contents of the tube into a cuvette and place the cuvette inside the fluorimeter.

11. Quickly hit the start button on Fl WinLab and allow the sample to run for 15 min.

12. Every 15 min, prepare the next sample to be tested.

13. Once the sample has finished running, remove the cuvette and place the next sample in the fluorimeter.

14. Record the value of the slope between the 6 and the 10 min mark in your lab notebook.

15. Clean cuvette and repeat until all samples have been tested.

16. Once completed turn off water bath, fluorimeter, and computer.

17. Make Henderson plots or other graphs on computer using SigmaPlot 2000.

3.5 SigmaPlot 2000 Instructions

1. Turn on computer and start the SigmaPlot 2000 program. Enter data for columns 1 and 2 (Fig. 5).

2. Use quick transformations to do columns 3, 4, and 5.

3. Find the IC_{50} point (x = -log final concentration; y = % activity V_i/V_0).

1 = [T] (Nm) initial inhibitor concentration	2 = slope from inhibition assays	3 = final concentration in cuvette	4 = % activity (V_f/V_0)	5 = -log final concentration in cuvette
Enter data	Enter data	Col (1)/20 = [T] x 10/200	Col (2)/control slope	-log(col(3))

Fig. 5 SigmaPlot 2000 columns for enzyme inhibition data. In columns 1 and 2, enter the inhibitor concentration and enzyme slopes from the inhibition assays, respectively. In column 3, calculate the final concentration of the inhibitor in the cuvette by dividing the initial inhibitor concentration by 20. In column 4, calculate the percent of enzyme activity by dividing the inhibitor slope from the inhibition assays by the control slopes. Column 5 is the log of the final inhibitor concentration in the assay cuvette

(a) Graph > create   graph > scatter,   next > simple   scatter, next > XY pair, next > x = -log final concentration, y = % activity > finish.

(b) *Fit curve*—right-click data point > fit curve > equation category = sigmoidal > equation   name = sigmoidal,   4 parameter > next > regression wizard > click columns (x = 5, y = 4) > finish.

(c) Minimize graph.

(d) Add minor ticks by double-click axis; these are used to find the IC_{50} value.

(e) Use toolbar to draw line for IC_{50}; look for where the activity of the enzyme is at 50 %.

(f) Calculate the IC_{50} using the equation $IC_{50} = 10^{|x \text{-axis}|}$ (*see* **Note 9**).

4. Find K_i (x = final concentration in cuvette; y = V_i / V_0).

(a) Create graph > scatter > simple > XY pair > x = final concentration in cuvette, y = % activity > finish.

(b) *Fit curve*—right-click data point > fit curve > user defined > Morrison, edit code > variables = make sure correct (I = col(3); y = col(4)); constraints = E (enzyme concentration; E > 0, E < value slightly lgr than stock concentration/20); initial parameters (set K_i same as IC_{50}; set E = stock concentration/20, unless EDM which is then stock/20 × 0.05) > run.

5. *Arrange graphs*—arrange graphs > 2 up or 2 side-by-side > finish.

6. *Label graphs*—include titles and axis, put IC_{50} on sigmoidal graph, and put K_i and R^2 on Morrison graph.

4   Notes

1. Brij-35 concentration has been shown to affect enzyme activity. Therefore, it is important to be precise when adding it to the kinetics buffer [23].

2. The final volume of assay buffer will be around 2200 μL. This amount of buffer will be good for 8–12 assays; however, trials involving inhibitor dilutions will require at least 21 assays. For these trials you want to make double the amount of buffer. To do this mix 11 μL of 2 mM TCEP with 4389 μL kinetics buffer in a 15 mL tube.

3. It is important to add the DMSO at the bottom of the tube in order to keep the DMSO from evaporating. Additionally, it is essential not to completely mix the DMSO with the assay buffer until after the enzyme has been added. Mixing the DMSO beforehand has shown to give inconsistent slope values.

4. Before mixing you should see two separate liquids in the tube. Only stop mixing when the two liquids become completely miscible with each other.

5. When placing the cuvette in the fluorimeter, be sure that the orientation of the cuvette is consistent for every run.

6. The amount of 25 % glycerol in kinetics buffer you use to dilute the enzyme will depend on the initial concentration of the stock enzyme. It is important not to over-dilute the enzyme; therefore, it is best to start with a small dilution (10 µL enzyme in 300 µL 25 % glycerol). Continue to test and dilute the enzyme as appropriate until the slope is 30–35. Additionally, it is important that enzyme concentration is known after dilution.

7. After freezing and thawing, the enzyme's slope should decrease by roughly 20 %. Therefore, the ideal slope range after freezing is between 25 and 30.

8. The exact amount of tubes needed for inhibitor dilutions will depend on how effective the inhibitor is at inhibiting the enzyme.

9. When calculating the IC_{50}, find the point on the x-axis that corresponds to y-axis value where the enzyme activity is at 50 %. For instance, if the x-axis value that corresponds to 50 % enzyme activity is –2.4, then use the following format: $IC_{50} = 10^{|2.4|}$.

References

1. Jin Y, Roycok MD, Bosco DB et al (2013) Matrix metalloproteinase inhibitors based on the 3-mercaptopyrrolidine core. J Med Chem 56:4357–4373. doi:10.1021/jm400529f

2. Hooper NM (1994) Families of zinc metalloproteases. FEBS Lett 354:1–6

3. Lee S, Park HI, Sang Q-X A (2007) Calcium regulates tertiary structure and enzymatic activity of human endometase/matrilysin-2 and its role in promoting human breast cancer cell invasion. Biochem J 403:31–42. doi:10.1042/BJ20061390

4. Woessner J, Nagase H (2000) Matrix metalloproteinases and TIMPs. Oxford University Press, New York

5. Nagase H, Woessner FJ Jr (1999) Matrix metalloproteinases. J Biol Chem 274:21491–21494

6. Hurst DR, Schwartz MA, Jin Y et al (2005) Inhibition of enzyme activity of and cell-mediated substrate cleavage by membrane type 1 matrix metalloproteinase by newly developed mercaptosulphide inhibitors. Biochem J 392:527–536. doi:10.1042/BJ20050545

7. Park HI, Turk BE, Gerkema FE et al (2002) Peptide substrate specificities and protein cleavage sites of human endometase/matrilysin-2/matrix metalloproteinase-26. J Biol Chem 38:35168–35175. doi:10.1074/jbc.M205071200

8. Savinov AY, Remacle AG, Golubkov VS et al (2006) Matrix metalloproteinase 26 proteolysis of the NH_2-terminal domain of the estrogen receptor β correlates with the survival of breast cancer patients. Cancer Res 66:2716–2724

9. Zhao YG, Xiao AZ, Park HI et al (2004) Endometase/matrilysin-2 in human breast ductal carcinoma in situ and its inhibition of breast cancer invasion. Cancer Res 64:590–598

10. Lee S, Desai KK, Ickowski KA et al (2006) Coordinated peak expression of MMP-26 and TIMP-4 in preinvasive human prostate tumor. Cancer Res 16:750–758

11. Hurst DR, Schwartz MA, Ghaffari MA et al (2004) Catalytic- and ecto-domains of membrane type 1-matrix metalloproteinase have similar inhibition profiles but distinct endopeptidase activities. Biochem J 877:775–779

12. Sang Q-X A, Jin Y, Newcomer RG et al (2006) Matrix metalloproteinase inhibitors as prospective agents for the prevention and treatment of cardiovascular and neoplastic diseases. Curr Top Med Chem 6:289–316

13. Candelario-Jalil E, Yang Y, Rosenburg GA (2009) Diverse roles of matrix metalloproteinase and tissue inhibitors of metalloproteinases in neuroinflammation and cerebral ischemia. Neuroscience 158:983–994

14. Lehrke M, Greif M, Broedl UC et al (2009) MMP-1 serum levels predict coronary atherosclerosis in humans. Cardiovasc Diabetol 8:50

15. Tuomainen AM, Nyyssonen K, Laukkanen JA et al (2005) Serum matrix metalloproteinase-8 concentrations are associated with cardiovascular outcome in men. Arterioscler Thromb Vasc Biol 27:2722–2728

16. Baker AH, Edwards DR, Murphy G (2002) Metalloproteinase inhibitors: biological actions and therapeutic opportunities. J Cell Sci 115:3719–3727

17. Fields GB (2010) Using fluorogenic peptide substrates to assay matrix metalloproteinases. In: Clark IM, Young DA, Rowan AD (eds) Matrix metalloproteinase protocols, vol 2, Methods in molecular biology. Springer, New York, pp 393–433

18. Gershkovich AA, Kholodovych VV (1996) Fluorogenic substrates for proteases based on intramolecular fluorescence energy transfer (IFETS). J Biochem Biophys Methods 33:135–162

19. Roycik MD, Myers JS, Newcomer RG et al (2013) Matrix metalloproteinase inhibition in atherosclerosis and stroke. Curr Mol Med 13:1299–1313

20. Park HI, Jin Y, Hurst DR et al (2003) The intermediate S_1' pocket of the endometase/matrilysin-2 active site revealed by enzyme inhibition kinetic studies, protein sequence analyses, and homology modeling. J Biol Chem 51:51646–51653. doi:10.1074/jbc.M310109200

21. Hu J, Van den Steen PE, Sang Q-X A et al (2007) Matrix metalloproteinase inhibitors as therapy for inflammatory and vascular diseases. Nat Rev Drug Discov 6:480–498

22. Muroski ME, Roycik MD, Newcomer RG et al (2008) Matrix metalloproteinase-9/gelatinase B is a putative therapeutic target of chronic obstructive pulmonary disease and multiple sclerosis. Curr Pharm Biotechnol 9:34–46

23. Park HI, Lee S, Ullah A et al (2010) Effects of detergents on catalytic activity of human endometase/matrilysin 2, a putative cancer biomarker. Anal Biochem 396:262–268

Chapter 14

Determination of Breast Cancer Cell Migratory Ability

David Schmitt, Joel Andrews, and Ming Tan

Abstract

Cell migration is defined as the movement of individual cells, sheets of cells, or clusters of cells from one location to another (Friedl et al., Int J Dev Biol 48:441–449, 2004). This ability of cells to migrate is critical to a wide variety of normal and pathological processes, including embryonic development, wound healing, immune responses, and cancer (Leber et al., Int J Oncol 34:881–895, 2009). Migration of tumor cells is widely thought to be an essential component of the metastatic spread of tumor cells to new sites, and inhibiting metastasis is an important therapeutic goal in cancer treatments (Horwitz and Webb, Curr Biol 13:R756–759, 2003). Therefore, the ability to observe and quantify migration in cancer cells is critical not only for basic cancer biology but especially for drug development (Friedl and Gilmour, Nat Rev Mol Cell Biol 10: 445–457, 2009). Researchers continue to develop new techniques for measuring cell migration in vitro. This chapter will discuss two techniques commonly used to study cell migration: wound healing and Boyden chamber migration assays.

Key words Migration, Wound healing, Time-lapse microscopy

1 Introduction

1.1 Wound-Healing Assay Using Time-Lapse Microscopy

Tissue wounds initiate a complex series of cellular events that promote the proliferation and migration of cells in order to close the wound [5]. Since these changes may be observed on a microscopic scale, researchers have taken advantage of these phenomena to study the migration of cells. Such studies are especially pertinent to cancer cells, since migratory ability is a major factor in determining cancer aggressiveness [6]. Researchers believe determining migratory ability in vitro may translate to the metastatic potential the breast cancer cells possess in humans [7]. Therefore, wound-healing assays, or "scratch" assays, have been performed for many years as a way to study the migratory ability of breast cancer cells (as well as other cell types) and whether experimental treatments may be able to inhibit this ability [8].

Wound-healing assays typically involve growing a confluent monolayer of breast cancer cells. A scratch is then made in the monolayer, which disrupts cell–cell interactions and initiates a complex

Jian Cao (ed.), *Breast Cancer: Methods and Protocols*, Methods in Molecular Biology, vol. 1406,
DOI 10.1007/978-1-4939-3444-7_14, © Springer Science+Business Media New York 2016

and ordered series of events to repair the tissue. As the cells migrate in an effort to close the "wound" that was made in the cell monolayer, the scratch is visualized over time with the use of a microscope. Complete "healing" of the wound usually takes anywhere from several hours to over a day, depending on the cell type and treatments being studied, as well as the width of the initial scratch.

Although scratch assays may be imaged and analyzed manually, a number of steps can be automated, including focus, acquisition, and analysis. Automated microscopes equipped with a motorized stage can be used to collect time-lapse data of the wounds at multiple locations within the culture dish over the course of an experiment. The location of each image field is saved into a file, allowing the stage to return to that precise point for each acquisition. This has the advantage that the exact same field is imaged at each time point, and the collected images can be made into a time-lapse video. This also greatly reduces the time spent by the researcher in data collection.

Analysis of scratch assays can be performed manually, as described below, or by using an automated algorithm. Such algorithms are often included in image analysis software packages and operate by first segmenting the image into cell and non-cell areas and then measuring the changes in non-cell area between subsequent images. Although this removes some of the subjectivity in measuring scratch width, care must be taken to ensure that the segmentation of the image corresponds to the actual presence or absence of cells. Additionally, this technique may be less useful for cells that do not maintain contact with their neighbors, or migrate as a sheet. Many plate-based imaging systems, such as the Celigo (Nexelcom, Lawrence, MA), include software modules for performing this task.

The major benefit of the wound-healing assay is the relative ease with which it can be performed and quantified. This assay is simple, inexpensive, and uses common laboratory supplies. It has an additional advantage in that it preserves the context of cell–cell interactions, unlike the Boyden chamber assay discussed below, which requires the disruption of cell–cell junctions before cells can be added to the chamber.

One of the most significant drawbacks to this technique is that it is two-dimensional and thus fails to faithfully replicate the three-dimensional environment in which cells normally live. It has been established that culturing cells in a 3D environment can drastically change a number of cell characteristics, including gene expression, morphology, and drug responsiveness [9, 10]. An additional drawback to the scratch assay is that it requires the physical disruption of the monolayer, which releases the contents of the scratched cells into the media. Although the scratched monolayer is washed in an effort to remove dislodged and damaged cells, the potential still exists for damaged cells and cell contents to remain in the wells and potentially affect the remaining cells in an uncontrolled manner. To counter this, some companies offer removable or biodegradable inserts which are placed into the culture plate before cells are added, such as the Oris and Oris Pro systems (Platypus Technologies, Madison, WI). These inserts are either manually removed or simply dissolve, and cell migration into the previously occupied area is measured, thus avoiding the presence of potentially complicating factors in the media.

It should also be noted that the wound-healing assay cannot be used to study chemotaxis due to the lack of a defined chemical gradient. Therefore, for chemotaxis studies, the Boyden chamber assay would be more appropriate. Additionally, large numbers of cells are required since the wound-healing assay is most commonly performed in cell culture plates. This usually is not a hindrance for breast cancer researchers, however, considering the relative ease and rapid in vitro growth of most breast cancer cell lines.

1.2 Introduction: Boyden Chamber Assay

The Boyden chamber assay, also commonly called the trans-well migration assay or chemotaxis assay, is a tool used to study cell migration [13]. This technique was originally developed by Stephen Boyden for analyzing the chemotaxis of leukocytes [14]. The driving force behind cell migration is chemotaxis, which is cell motility distinguished by movement toward higher concentrations of chemoattractants, such as growth factors. Overall, this assay consists of a cell culture insert that is placed inside a well of a multi-well cell culture plate. The cell culture insert described here is a PET (polyethylene terephthalate) membrane, with pores of a known, uniform size. For breast cancer cells, the appropriate pore size is 8 μm. Breast cancer cells are seeded into the insert in serum-free medium. The bottom well, which is the well of the multi-well plate, contains a chemoattractant. Due to its convenience and effectiveness, a commonly used chemoattractant is 20 % fetal bovine serum (FBS) in normal cell culture medium. After a given incubation time, breast cancer cells that have migrated to the bottom side of the membrane are fixed, stained, and quantified under a light microscope.

One notable benefit of the Boyden chamber assay is that it is a time-efficient method to analyze cell motility/migration independent of the effects of cell proliferation. This benefit is based on the fact that the time required for cells to migrate through the porous membrane is less than the time required for cells to progress through the cell cycle. For this reason, cell proliferation rates are typically not a factor during the Boyden chamber assay. Boyden chamber assays also have the advantage of being three-dimensional, in contrast to the scratch assay. One consideration that the user should be aware of is that the contributions of cell–cell interactions to cell migration are not amenable to study with the Boyden chamber assay. The reasons for this are twofold; first, the cells need to be dissociated into a suspension in order to be counted and equally seeded into the wells, and second, if any cell–cell interactions do form after seeding, these cell–cell junctions must be released in order for the cells to efficiently migrate through the porous membrane used in this assay. If the user desires to consider the effect of cell–cell interactions on cell migration, the wound-healing assay also described in this chapter would be more appropriate. However, it is typically warranted to analyze cell migration using both assays, given that the specific treatments or gene manipulations under study may affect cell migration in different ways. A final benefit of the Boyden chamber assay is its ease of use. Users can quickly master the techniques involved and produce repeatable results.

Several Boyden chamber devices of various sizes and compositions are commercially available. Boyden chamber assays may also be supplemented by the addition of a layer of the basement membrane proteins collagen or laminin that must be degraded before cells can pass through. Such a setup is used to assay invasive potential along with migratory ability. Regardless of the specific chamber used, cell migration is usually quantified by counting the number of migrated cells under a light microscope. The user may also use image analysis software to detect the total area of each membrane that is covered by migrated cells. This method of quantification is especially useful for assays with high numbers of migrating cells, in which counting individual cells would be time-consuming and may introduce user error.

The Boyden chamber assay will require optimization depending on the cell type, the specific type of chamber being used, and the chemoattractants. Incubation time will also need to be determined based on cell type, cell number, membrane pore size, and the composition and concentration of the chemoattractants. The method described below is specifically designed for measuring the migration of breast cancer cells using 24-well cell culture plates and 24-well Boyden chamber cell culture inserts from BD Falcon™.

2 Materials

2.1 Wound-Healing Assay

Cell lines: Human breast cancer cell lines may be purchased from the American Type Culture Collection (ATCC, Manassas, VA). Cells should be cultured in the appropriate cell culture medium, such as DMEM/F-12 supplemented with 10 % FBS. Cells should be tested to be free of mycoplasma contamination and always cultured using proper aseptic cell culture conditions.

Other reagents: Prepare all reagents using proper sterile technique under a cell culture hood. Store reagents at 4 °C. Before use, warm cell culture reagents to 37 °C in a clean water bath. Closely follow all waste disposal regulations when disposing of waste materials.

1. Cell culture media supplemented with 10 % fetal bovine serum.

2. Dulbecco's Phosphate-Buffered Saline.

3. 0.25 % trypsin–EDTA.

4. Breast cancer cell lines.

5. 200 µL pipette tips.

6. Six-well cell culture plates.

7. Actinomycin D (optional).

2.2 Boyden Chamber Assay

1. Cell culture media supplemented with 10 % fetal bovine serum.

2. Dulbecco's Phosphate-Buffered Saline.

3. 0.25 % trypsin–EDTA.

4. Breast cancer cell lines.

5. 24-well BD Falcon™ Cell Culture Inserts—8 µm pore size.

6. 24-well BD Falcon™ Cell Culture Insert Companion Plates.

7. Methanol.

8. Solution of 0.05 % crystal violet in 20 % ethanol.

9. Cotton-tipped swabs.

10. Glass beakers (250 mL or larger).

2.3 Instruments

1. 37 °C CO_2 incubator.

2. Laminar flow hood.

3. Light microscope.

3 Methods

3.1 Wound-Healing Assay

3.1.1 Preparing the Cells

1. Seed cells in triplicate wells at approximately 80 % confluence in six-well cell culture plates in normal cell culture media. This is equal to approximately 1×10^6 cells/well; however, the exact number will vary with cell type.

2. Incubate the plates at 37 °C in a 5 % CO_2 cell culture incubator overnight.

3. The next day, observe cells under a light microscope to ensure cells have reached greater than 90 % confluence. Ideally, the cells should be 95–100 % confluent.

4. Make a "scratch" in the cell monolayer in each well of the six-well plate with a 200 µL pipette tip, making sure to hold the pipette tip fully perpendicular to the plate surface to ensure uniform scratch diameter.

5. Gently wash the cells with 1 mL of DPBS to remove the dislodged cells. If care is not used, the cell monolayer may lift or be rinsed away. The borders of the scratch are especially vulnerable to disruption.

6. Carefully suction the DPBS wash from the wells. Gently wash the cells once more with 1 mL of DPBS to remove any remaining dislodged cells and again remove the DPBS by suction.

7. Add 3 mL of fresh cell culture medium to each well. If you want to ensure that any changes in cell migration observed during this assay are due solely to differences in cell migration and not different proliferation rates, you may wish to add actinomycin D (1 μg/mL). Actinomycin D inhibits cell proliferation by inhibiting RNA synthesis.

3.1.2 Observation of Cells: Manual Method

1. On the underside of the six-well cell culture plate, use a fine point permanent marker to mark three points along the length of the scratch. These points will be the reference marks used to ensure you are imaging the same portion of the scratch at each time point.

2. Using an inverted light microscope equipped with a camera, select a 10× objective and find the first reference mark you made, focusing the image just above the reference point and making sure the reference mark is just visible in the field of view. The region of interest is the scratch in the cell monolayer, so center the image over the scratch.

3. Capture the image with the attached camera and repeat for the two other reference points per well. These images will be analyzed later to measure the width of each scratch at each point for each well at time point 0 h.

4. Return the six-well plate to the cell culture incubator.

5. Capture images in the same manner as described above in **steps 2** and **3** at the 6, 12, and 24 h time points. You may also wish to capture a number of images with a 40× or greater objective at the 6 and 12 h time points to examine changes in cell morphology at the leading and trailing edges of migrating cells.

3.1.3 Observation of Cells: Automated Method

1. Place the plate into an appropriate fixed stage insert, being sure to note the orientation of the plate. Using the software interface for a microscope with an automated stage, select three or more points within each well for imaging.

2. If using an incubated microscope stage, set the desired time interval and duration for image acquisition, and set the incubator to 37 °C and 5 % CO_2. It is advisable to include an autofocus step at the beginning of each acquisition if the hardware permits it or to use a system that maintains focus over the course of the acquisition. This will prevent focal drift due to thermal expansion.

3. Run the acquisition.

4. If no incubated stage setup is available, remove the plate from the microscope stage after the first set of images is acquired, and return the plate to the cell culture incubator.

5. Capture images in the same manner as the steps above at the 6, 12, and 24 h time points. If desired, capture a number of images with a 40× or greater objective at the 6 and 12 h time points to examine changes in cell morphology at the leading and trailing edges of migrating cells.

3.1.4   Quantitation and Analysis

1. Using image analysis software, either a commercial solution such as NIS-Elements (Nikon Instruments, Inc.) or free, open-source options such as CellProfiler [11] or ImageJ [12], analyze each image at each time point, measuring the width of the scratch. If the edge profile of the cells is uneven along the length of the scratch, you may wish to use a parallel line measurement tool, where the lines are set to the average edge of the cells on each side. Migration may be expressed either as absolute migration rate or percent wound closure. The absolute rate is equal to $[(D_0 - D_x)/2T]$, where D_0 is the distance between edges at time zero, D_x is the distance between edges measured at the time point of interest, and T is the elapsed time. The distance is divided by two because the change in distance between time points is the result of the migration of both sides of the scratch. The percent closure for each point in each well is equal to $[(D_0 - D_x)/D_0]$, where D_0 is the measurement taken at time zero and D_x is the measurement taken at the time point of interest. Either measurement may then be normalized to the value obtained from the control sample and reported as percent with respect to the control.

3.2   Methods

3.2.1   Boyden Chamber Assay

Preparation of the Cells

1. Remove the normal growth medium containing FBS from cell cultures. The cultures should be in logarithmic growth phase and less than 75 % confluent.

2. Gently rinse the cell cultures three times with DPBS.

3. Replace the culture medium with serum-free medium.

Seeding the Cells into the Chamber

1. Using sterile conditions under the laminar flow hood, dispense 750 µL of serum-free medium into each well of the 24-well BD Falcon™ Cell Culture Insert Companion Plate.

2. Gently place one BD Falcon™ Cell Culture Insert into each well (*see* **Note 1**).

3. Dispense 500 µL of serum-free medium into each BD Falcon™ Cell Culture Insert, and place the lid back onto the BD Falcon™ Cell Culture Companion Plate (*see* **Note 2**).

4. Place the plate containing the inserts into a 37 °C CO_2 incubator and incubate for 2 h. This allows the insert membranes to rehydrate.

5. Using vacuum suction, carefully remove the serum-free media from the lower well and the insert.

6. Trypsinize the cell cultures to obtain cell suspensions. Neutralize the trypsin with cell culture medium containing 10 % FBS.

7. Count the cells using a hemacytometer and light microscope to determine the cell concentration. Using this concentration, calculate how many total cells you will need in order to seed triplicate wells for each cell line or treatment group. Prepare enough cell suspension for extra wells to provide room for pipetting error. The number of cells seeded per well will depend on the cell line used and typically varies from 20,000–50,000 cells/well.

8. Centrifuge the cells at $200 \times g$ for 4 min to form a cell pellet.

9. Remove and discard the medium (supernatant).

10. Carefully rinse the cell pellet with DPBS.

11. Centrifuge the cells + DPBS at $200 \times g$ for 3 min.

12. Remove and discard the DPBS (supernatant).

13. Gently resuspend the cells in serum-free medium to provide for 500 µL of serum-free medium + cells per well.

14. Pipette 750 µL of cell culture medium containing 20 % FBS into each well of the BD Falcon™ Cell Culture Companion Plate. Ensure there are no air bubbles trapped between the insert membrane and the medium in the lower well.

15. Pipette 500 µL of serum-free medium + cells into each insert.

16. Incubate the plate in a 37 °C CO_2 incubator for 6–24 h. Incubation time will be dependent on the cell type under study.

3.2.2 Fixation and Staining of Migrated Cells

1. Prepare two separate 24-well plates. In one plate, transfer 750 µL of 100 % methanol to each well. In the other plate, transfer 750 µL of 0.05 % crystal violet solution to each well.

2. Gently remove the inserts containing cells, one at a time, and remove the serum-free media out of the insert.

3. Using a cotton-tipped swab, gently, but thoroughly, scrub the top surface of the membrane to remove any nonmigrating cells from the chamber.

4. To fix the cells that have migrated to the bottom side of the membrane, place the inserts in the wells of the plate containing 100 % methanol. Incubate at room temperature for 20 min.

5. Remove the inserts from the methanol and air-dry them for 5 min.

6. To stain the cells, place the inserts into the wells of the plate containing the 0.05 % crystal violet solution. Allow the inserts to stain at room temperature for 20 min.

7. Destain the inserts by gently dipping them into consecutive beakers of distilled H_2O until no more stain comes off into the water. Allow the inserts to thoroughly air-dry (at least 1 h).

8. The inserts may be simply placed into a clean 24-well plate and imaged, or the membranes may be carefully cut from the inserts using a razor blade and mounted onto a slide for imaging.

3.2.3 Quantification of Cells

1. Survey each membrane under a light microscope at 10× magnification.

2. Capture images using a digital camera connected to the microscope using a 10× objective. For each membrane, capture three (or more) images at random positions (*see* **Note 3**).

3. To analyze the images, count the number of migrated cells manually, or using image analysis software. This may be done using an interactive software counter, where each mouse click registers as a counted cell, or by an automated object counting algorithm. If using an automated counting function, first validate the function by comparing it to a manual count performed on the same image before processing all images.

4. Calculate the mean number of migrated cells from nine images (three images per membrane × three membranes per cell line or treatment group = nine images), and calculate standard deviation.

5. Alternatively, if counting individual cells is problematic, the area occupied by stained cells may be measured instead. This is done with image analysis software by setting a threshold that will detect the presence of the stain on a per-pixel basis, thus defining a binary that represents the area occupied by migrated cells. Then calculate the mean area occupied for each treatment from nine images (three images per membrane × three membranes per cell line or treatment group = nine images) ± standard deviation.

6. Measured cell numbers or cell area may be reported directly or normalized to the control treatment and reported as percent relative to control.

4   Notes

1. When placing the cell culture inserts into the companion plate, be careful not to trap air bubbles between the insert and the media in the bottom well. To accomplish this, slowly and gently place the insert into the well at a slight angle.

2. In preparing the cells for seeding, plan for at least three replicates within each experimental group. Additionally, prepare extra volume of cell suspension in serum-free medium to allow

for slight variations in pipetting. In this protocol, since you need enough cell suspension for three replicates, prepare enough for four replicates.

3. Do not directly count the cells under a microscope. To reduce error and retain proper records, use a camera attached to the microscope to save an image file for each field of view. Cells can be counted in the desired fields from the saved images with more accuracy and objectivity.

References

1. Friedl P, Hegerfeldt Y, Tusch M (2004) Collective cell migration in morphogenesis and cancer. Int J Dev Biol 48(5-6): 441–449

2. Leber MF, Efferth T (2009) Molecular principles of cancer invasion and metastasis. Int J Oncol 34(4):881–895

3. Horwitz R, Webb D (2003) Cell migration. Curr Biol 13(19):R756–R759

4. Friedl P, Gilmour D (2009) Collective cell migration in morphogenesis, regeneration and cancer. Nat Rev Mol Cell Biol 10: 445–457

5. Rodriguez LG, Wu X, Guan JL (2005) Wound-healing assay. Methods Mol Biol 294:23–29, Edited by Jun-lin Guan

6. Ogden A, Rida PC, Aneja R (2013) Heading off with the herd: how cancer cells might maneuver supernumerary centrosomes for directional migration. Cancer Metastasis Rev 32(1-2):269–287

7. Hulkower KI, Herber RL (2011) Cell migration and invasion assays as tools for drug discovery. Pharmaceutics 3(1):107–124

8. Eccles SA, Box C, Court W (2005) Cell migration/invasion assays and their application in cancer drug discovery. Biotechnol Annu Rev 11:391–421

9. Bissell MJ, Radisky D (2001) Putting tumours in context. Nat Rev Cancer 1:46 54

10. Pampaloni F, Reynaud EG, Stelzer EHK (2007) The third dimension bridges the gap between cell culture and live tissue. Nat Rev Mol Cell Biol 8:839–845

11. Rasband WS (1997–2014) ImageJ. U. S. National Institutes of Health, Bethesda, MD. http://imagej.nih.gov/ij/

12. Carpenter AE, Jones TR, Lamprecht MR, Clarke C, Kang IH, Friman O, Guertin DA, Chang JH, Lindquist RA, Moffat J, Golland P, Sabatini DM (2006) Cell Profiler: image analysis software for identifying and quantifying cell phenotypes. Genome Biol 7:R100

13. Chen HC (2005) Boyden chamber assay. Methods Mol Biol 294:15–22

14. Boyden S (1962) The chemotactic effect of mixtures of antibody and antigen on polymorphonuclear leukocytes. J Exp Med 115:453–466

Chapter 15

A Novel Collagen Dot Assay for Monitoring Cancer Cell Migration

Vincent M. Alford, Eric Roth, Qian Zhang, and Jian Cao

Abstract

Cell migration is a critical determinant of cancer invasion and metastasis. Drugs targeting cancer cell migration have been hindered due to the lack of effective assays for monitoring cancer cell migration. Here we describe a novel method to microscopically monitor cell migration in a quantitative fashion. This assay can be used to study genes involved in cancer cell migration, as well as screening anticancer drugs that target this cellular process.

Key words Type I collagen, Migration, Two-dimensional culture, Non-tissue culture 96-well plate, Microscope

1 Introduction

Cancer cell migration is a dynamic, multistep process, which involves the rearrangement of the cytoskeleton with adhesion protein composition to form membrane protrusions at the leading edge of the cell membrane (lamellipodia) to promote translocation through adjacent tissues and structures [1]. The migratory capacity of cancer cells is often correlated with poor prognosis in patients as this cellular process is required for metastasis, which accounts for 90 % of all human cancer-related deaths [2].

Due to the relevance of cell migration in cancer dissemination, multiple systems have been established to study this cellular process. One example is the Boyden chamber assay in which a polycarbonate membrane with a defined pore size is nested between the upper and lower quadrants of the cylindrical transwell chamber. Cells are then seeded in the top chamber in serum-free media, while a chemoattractant is placed in the bottom chamber. Cells that migrate through the pores of the membrane are then stained and quantified through microscopic approaches [3]. Although a reliable, sensitive, and useful approach, this assay has disadvantages relative to the 2-D migration assay described in this chapter such as

Jian Cao (ed.), *Breast Cancer: Methods and Protocols*, Methods in Molecular Biology, vol. 1406,
DOI 10.1007/978-1-4939-3444-7_15, © Springer Science+Business Media New York 2016

prolonged drug-screening studies which are complicated by the test agent concentrations quickly equilibrating between the two transwell compartments. Another popular method adopted by many scientists is the in vitro scratch wound assay which involves generating a "scratch" in a cell monolayer and subsequently capturing images at evenly spaced time intervals. The simulated "wound healing" is then compared between the images by quantifying the migration rates of the cells [4]. Although this assay is also a useful technique, it has disadvantages relative to the 2-D migration assay described in this chapter due to the lack of a defined wound surface between cells.

Herein, we describe a new method in which cells are mixed with a type I collagen mixture and are subsequently doted onto a non-tissue culture 96-well plate. Migrating cells at the cell–collagen interface can be microscopically counted after at least an 8 h incubation period. Furthermore, by using a tooled plate to standardize the size and shape of the cell-matrix dot, this 2-D migration assay can be used in a high-throughput screening fashion. The 2-D migration assay protocol described is a simple, reproducible, and effective way to identify compounds that inhibit cancer cell migration, study pathways relevant to this cellular process, and expand our understanding of cancer dissemination.

2   Materials

Prepare all solutions for this assay in sterile conditions. All reagents used to make the dot collagen mixture, especially the type I collagen, should be kept on ice to prevent premature solidification.

2.1 Cell Culture Components

1. 0.05 % Trypsin–EDTA (1×).
2. Dulbecco's Modified Eagle Medium (DMEM): Supplemented with 10 % fetal bovine serum (FBS) and 1 % PenStrep (P/S).
3. 96-well non-tissue culture plate.
4. 37 °C CO_2 incubator.
5. Tissue culture hood.

2.2 Dot Collagen

1. 2 N NaOH: Add 8.0 g of sodium hydroxide to 100.0 mL of water.
2. Dot Collagen (3 mg/mL): Add 70 µL of sterile water, 50 µL 5× DMEM medium, and 125 µL type I collagen 6 mg/ml (acetic acid-extracted native collagen from rat tail tendon) to a 1.5 mL microcentrifuge tube and mix until contents go into solution. Adjust the pH of the dot collagen mixture using 2 N NaOH until it reaches a neutral pH of 7.4. Make sure to check pH using a pH indicator strip as a too acidic or basic solution will kill the cells (see **Note 1**).

2.3 Cell Fixation/ Nuclei Staining Solution

1. 16 % Paraformaldehyde (PFA) Solution: Dissolve 16 g of PFA powder in 84 mL phosphate-buffered saline (PBS) in a 250 mL beaker. Place beaker on hot plate (don't bring temp above 80 °C) and let contents stir for 30 min. Allow mixture to cool to room temperature before using.

2. 1× PBS Solution: Dissolve 8.00 g NaCl, 0.20 g KCl, 1.44 g Na_2HPO_4, and 0.24 g KH_2PO_4 in 800 mL of H_2O. Adjust the solution with HCl until a pH of 7.4 is achieved. Add water until contents reach 1 L. Sterilize the solution by autoclaving for 20 min at 15 psi (1.05 kg/cm²) on liquid cycle or by filter sterilization.

3. PFA Hoechst Solution: In a 15 mL Falcon tube, add 1000 μL 16 % PFA and 998 μL 1× PBS. For every 2 mL of solution made, add 2 μL of Hoechst and/or DAPI (4′,6-diamidino-2-phenylindole) (*see* **Note 2**).

3 Methods

Before conducting this assay, it is necessary to ensure there are enough cells to make the desired number of cell–collagen dots for each condition. On average, a concentration of 5×10^4 cells/μL is optimal when performing this assay. It is important to note that although this is the optimal cell density, certain cell lines may be smaller than others so more cells might be necessary to perform this assay properly. The following steps are performed in a standard tissue culture hood.

3.1 Cell Preparation

1. Once cells reach log growth rate, carefully aspirate the medium from the cell culture plate or flask they reside in (*see* **Note 3**).

2. After removing media, wash the cells with 1× phosphate-buffered saline (PBS) to remove any remaining FBS-containing medium.

3. After the wash step, aspirate PBS from the dish and incubate the cells at 37 °C for 4 min in 0.05 % Trypsin–EDTA (1×) in order to cleave adherent proteins on the cell membrane, releasing them from the bottom of the plate or flask (*see* **Note 4**).

4. Remove the cells from the incubator and mix with equal volumes of 10 % FBS-containing medium to neutralize the Trypsin–EDTA solution.

5. Mechanically pipette the cells to further separate and transfer into a new 15 mL tube.

6. Determine the cell concentration with a hemocytometer or an automated cell counter.

7. Use the counted cell density to calculate how much of the cell solution is needed to obtain 1×10^6 cells. This will give a final cell–collagen dot density of 5×10^4 cells/μL in 20 μL (*see* **Note 5**).

8. Pipette the volume of cells required into a new 1.5 mL micro-centrifuge tube and spin down ($800 \times g$ for 5 min at room temperature) to remove the media.

9. Resuspend the pellet of cells in fresh complete medium to a volume of 10 μL (*see* **Note 6**).

10. Add 10 μL of dot collagen to the suspended cells.

11. Vigorously mix the cells by pipetting to ensure a homogenous mixture of cells: dot collagen solution (*see* **Note 7**).

3.2 Drop Cell–Collagen Dot into a 96-Well Plate

1. Pipette 1.0–1.5 μL of the cell–collagen dot mixture onto a non-tissue culture 96-well plate at the center of each well (*see* **Note 8**).

2. Incubate the 96-well plate at 37 °C for 5–10 min or until collagen has solidified (*see* **Note 9**).

3. Check cell–collagen dots under a compound light microscope. They should appear to be full of cells at both the center and the peripheral edges (*see* **Note 10**).

4. Once cell–collagen dots appear to be solidified, pipette 100 μL of complete medium onto the side of each well gently to prevent washing away the cell–collagen dot.

5. Incubate the 96-well plate at 37 °C for 8–16 h depending on the cell line used (*see* **Note 11**).

6. After incubation, add 8 % PFA/Hoechst solution to each well at room temperature to fix the cells and visualize them through nuclei staining. Wrap the 96-well plate in aluminum foil to avoid exposure to light and let it stand for at least 30 min before imaging (*see* **Note 12**).

7. Use microscope visualization software to center the cell–collagen dot and record a DAPI (Fig. 1a) and bright-field (Fig. 1b) image using both a 4× magnification lens to visualize the entire dot and a 10× lens to image the four sides of the cell–collagen dot. Count the migrating cells at the cell–collagen interface and compare between the different conditions (*see* **Note 13**).

4 Notes

1. It is generally acceptable to leave the dot collagen mixture on ice in the refrigerator for up to 2 h prior to use.

2. Hoechst and DAPI are fluorescent stains that bind and label nucleic acids (double-strand DNA) for visualization of nuclei when performing microscopy.

3. Verify the morphology and health of the cells before performing the experiment as abnormalities can lead to unexplained results or poor migratory capacity.

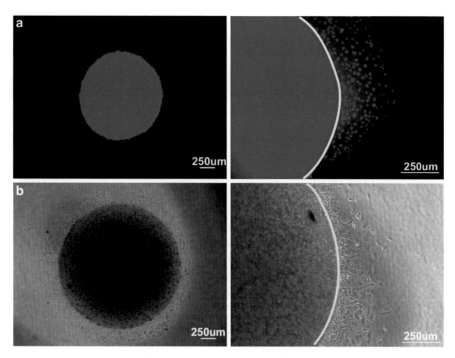

Fig. 1 2-D dot collagen migration assay using HeLA (cervical cancer) cells. (**a**) DAPI images of entire (*left* panel) and right quadrant (*right* panel) of dot collagen under 4× and 10x lens magnification respectively. (**b**) Bright-field images of entire (*left* panel) and right quadrant (*right* panel) of dot collagen under 4× and 10x lens magnification respectively.

4. The concentration and duration of treatment with Trypsin–EDTA (1×) may vary depending on the cell type.

5. It is common for small cell types to require a final cell–collagen dot density of 7×10^4 to 10×10^4 cells/μL to ensure full dots.

6. It is important to be very careful when aspirating excess media/trypsin to not disturb the pellet. It might be necessary to pipette excess liquid off mechanically to achieve a 10 μL volume when resuspending the cell pellet in fresh complete media. After resuspending the pellet, it is worth measuring the cell mixture volume by pipetting before adding the dot collagen as excessive liquid may lead to longer drying periods. Remember to account for the volume of the cells in the 10 μL resuspension mixture.

7. It is important to pipette the cell–collagen dot mixture precisely to ensure dots have a relatively even number of cells between each repeat. It is also important to not use small pipette tips (0.1–10 μL) when mixing the cell–collagen dot mixture as this may lead to mechanical disruption of the cell membrane and/or accidental cell lysis.

8. In order to optimize image quality, it is important to place dots as centered as possible. It is also important to stop pipetting the dot when the first stop is reached to avoid bubbles, as this will affect dry times and image quality. If the cell–collagen dots seem inconsistent in cell number between repeats, it may be necessary to pipette the cell–collagen dot mixture vigorously to resuspend the cells again as they may have settled. It may also help to use pipette tips that have already been pre-chilled in the refrigerator as this makes working with collagen much easier. Lastly, ensure the plate is a non-tissue culture 96-well plate as the hydrophobic surface helps maintain the collagen dot shape.

9. The drying period is highly dependent on how quickly dots are placed and will vary between experiments. The cell–collagen dots usually turn a pale opaque color when properly dried and should not be allowed to turn white in color as this means the collagen has completely dried and cells will no longer be capable of migrating.

10. Some cell lines such as HT1080 might require the addition of 2.5 % dialdehyde dextran to the final cell–collagen dot mixture (10 μL cells suspended in complete media, 8 μL dot collage mixture, 2 μL 2.5 % dextran for a 20 μL total volume) to prevent the cell–collagen mixture from contracting while drying [5–7]. This is noted as a clear empty space around the dried cell–collagen dot under microscopic visualization.

11. Note that incubation periods between cell types vary greatly, but in general shorter incubation times are best for more aggressive cell types while those less aggressive will tend to have a longer incubation period.

12. Generally, it is ok to remove the lid of the 96-well plate and invert over a designated biohazard trash can while gently tapping the side to remove old media from the wells. Instead of 8 % PFA/Hoechst solution, a 4 % PFA/Hoechst solution can be used to help remove excess background when imaging.

13. After fixation, cells should be stored in a dark place at room temperature, such as a desk drawer, and be imaged within 24–36 h.

References

1. Nagano M, Hoshino D, Koshikawa N, Akizawa T, Seiki M (2012) Turnover of focal adhesions and cancer cell migration. Int J Cell Biol 2012:310616.

2. Weigelt B, Peterse JL, Van't Veer LJ (2005) Breast cancer metastasis: markers and models. Nat Rev Cancer 5(8):591–602.

3. Chen HC (2005) Boyden chamber assay. Methods Mol Biol 294:15–22

4. Liang CC, Park AY, Guan JL (2007) In vitro scratch assay: a convenient and inexpensive method for analysis of cell migration in vitro. Nat Protoc 2(2):329–333.

5. Zhu YK, Umino T, Liu XD, Wang HJ, Romberger DJ, Spurzem JR, Rennard SI (2001) Contraction of fibroblast-containing collagen gels: initial collagen concentration regulates the

degree of contraction and cell survival. In Vitro Cell Dev Biol Anim 37(1):10–16.

6. Nien YD, Han YP, Tawil B, Chan LS, Tuan TL, Garner WL (2003) Fibrinogen inhibits fibroblast-mediated contraction of collagen. Wound Repair Regen 11(5):380–385

7. Schacht E, An Van Den Bulcke BB, De Rooze N (1997) Hydrogels prepared by crosslinking of gelatin with dextran dialdehyde. React Funct Polym 33:109–116

Chapter 16

Three-Dimensional Assay for Studying Breast Cancer Cell Invasion

Nikki A. Evensen

Abstract

Cancer cell invasion is a complex process that naturally occurs in a three-dimensional (3-D) environment comprised of tumor cells and extracellular matrix components (ECM). Therefore, examining the invasive ability of breast cancer cells in a 3-D assay is imperative to discovering novel treatment strategies aimed at preventing cancer invasion and metastasis. Here, I describe a method to quantitatively measure the number of invaded cancer cells within a 3-D microenvironment and determine the effects of potential drugs on this cellular process.

Key words Cancer cell invasion, Extracellular matrix, Three-dimensional assay

1 Introduction

Cancer cell invasion is an essential, primary step of the multistage process of cancer metastasis, which remains as the major reason for treatment failure in cancer patients. The ability of cells to break free from the primary tumor and invade through the surrounding ECM and basement membrane requires both proteolytic activity and migratory capabilities [1]. Many of the common techniques utilized to study cancer dissemination only test one of these cellular properties, such as the Transwell chamber migration assay or gelatin zymography for detecting proteolytic cleavage of matrix metalloproteinases [2]. Various modifications, such as coating the membranes used in the Transwell chamber migration assay with Matrigel matrix [3], try to circumvent this flaw. However, besides lacking the ability to simultaneously measure both of these cellular properties, these assays do not truly capture all of the interactions and obstacles that occur within the 3-D microenvironment that tumor cells are exposed to within the body. These limitations become a hindrance for screening drugs for anti-invasive potential or determining the role of genes or pathways involved in cancer cell invasion.

Jian Cao (ed.), *Breast Cancer: Methods and Protocols*, Methods in Molecular Biology, vol. 1406,
DOI 10.1007/978-1-4939-3444-7_16, © Springer Science+Business Media New York 2016

In recent years accumulating evidence demonstrates that the use of 3-D cell culture systems more closely mimics in vivo conditions than traditional 2-D cell culture. Similar gene expression profiles are observed in 3-D-cultured cancer cells and in vivo tumors, and the behavior of cancer cells cultured in 3-D more truthfully mimics the behavior of cells in vivo [4]. Phenotypic responses of cancer cells to various drugs, such as inhibition of migratory ability, are also significantly altered when cultured in a 3-D environment as compared to a 2-D environment [5]. Therefore, 3-D cell-based invasion assays are necessary for gaining a more complete understanding of this cellular process and also to obtain a higher predictive value and better optimization for future clinical efficacy of potential drugs.

Herein, we provide a protocol for a collagen gel spheroid-based 3-D assay for studying breast cancer cell invasion. Cancer cells are mixed with collagen, a component of the ECM, dotted into the center of a 96-well plate, embedded in a layer of collagen, covered with medium, and allowed to invade for a period of time. Cancer cell invasion into the surrounding matrix is then quantified. Furthermore, by utilizing a tooled plate, along with dialdehyde dextran mixed with collagen to prevent contraction, this 3-D invasion assay permits standardization and automated readout, therefore making high-throughput screening possible [6]. The 3-D invasion assay protocol provided here is reproducible, effective, easy and rapid to perform, and sensitive enough to identify compounds that inhibit cancer cell invasion or determine the role of genes involved in this pathway, ultimately furthering our understanding of cancer dissemination and increasing the probability of inhibiting this process.

2   Materials

2.1   Cell Culture Components

1. 0.25 % Trypsin/.53 mM EDTA (depending on cell line).

2. Cell culture medium appropriate for your cell line supplemented with 10 % fetal bovine serum.

3. Cancer cell lines of your choice.

4. 96-well tissue culture dish.

5. 1× phosphate-buffered saline (PBS): Mix 8 g NaCl, 0.2 g KCl, 1.44 g Na_2HPO_4, and 0.24 g KH_2PO_4 in 800 ml of water. pH the solution to 7.4 using hydrochloric acid. Bring the final volume to 1 L. Autoclave if sterile PBS is needed.

2.2   Collagen Matrix Components

1. Dot Collagen: High concentration Type I collagen (acetic acid-extracted native collagen from rat tail tendon) supplied as 100 mg in 0.02 N acetic acid ranging in concentration from 8

to 11 mg/ml. Add the necessary amount of 0.02 N acetic acid to make the final stock concentration 6 mg/ml of collagen. Store stock at 4 °C.

2. Cover Collagen: Type I collagen (acetic acid-extracted native collagen from rat tail tendon) supplied as 1 g in 0.02 N acetic acid ranging in concentration from 3 to 4 mg/ml. Store stock at 4 °C.

2.3 Analysis Components

1. Hoechst 33342 (25 µg/ml) live cell nuclear stain.

2. Fluorescent microscope with imaging system and analysis software.

3 Methods

3.1 Prepare Cells and Collagen for Assay

Store collagen mixtures on ice during all procedures.

1. Trypsinize cancer cell line cultured in standard 2-dimenstional tissue culture dishes. Neutralize cells with complete medium and count cells. Spin down the desired total number of cells to be used (*see* **Note 1**). Resuspend the cell pellet in half the total desired volume of complete medium (*see* **Note 2**).

2. For dot collagen, prepare a working concentration of 3 mg/ml, from 6 mg/ml stock, by combining 5× medium (to a final concentration of 1×), water, and 2 N NaOH (*see* **Note 3**) to desired total volume (*see* **Note 4**).

3. For cover collagen, prepare a working concentration of 1.5 mg/ml by combining 5× medium (to a final concentration of 1×), water, and 2 N NaOH (*see* **Note 3**) to desired total volume (*see* **Note 5**).

3.2 Plating Cell-Collagen Droplets

1. Add the dot collagen to the cells at a 1:1 volumetric ratio. Mix well by pipetting up and down (*see* **Note 6**).

2. Gently pipette the dot collagen-cell mixture into the middle of each well of a 96-well plate (*see* **Note 2** for volume of droplet) (*see* **Note 7**). Make the necessary amount of droplets for each experiment (*see* **Note 8**).

3. Incubate the plate at 37 °C for 5 min in order to allow collagen-cell droplets to solidify (*see* **Note 9**).

4. While the droplets solidify, adjust the pH of the cover collagen. Add enough cover collagen to cover the droplet and fill the bottom of the well (*see* **Note 10**).

5. Incubate the plate at 37 °C for 5 min in order to allow cover collagen to solidify (*see* **Note 11**).

6. Add complete medium to each well (*see* **Note 12**).

7. Incubate the plate for 18 h to allow cells to invade into the collagen matrix.

3.3 Analysis of Cancer Cell Invasion

1. Remove media from the plate by carefully turning it upside down (*see* **Note 13**).

2. Stain cells with Hoechst 33342 (25 μg/ml) nuclear stain diluted 1:250 in 1× PBS for 30 min at room temperature (*see* **Note 14**).

3. Wash cells with PBS two times for 15 min each.

4. Perform fluorescent microscopy to count the total number of invaded cancer cells for each droplet (*see* **Note 15**).

---

4 Notes

1. Every droplet needs at least 4×10^4 cells. This number could vary depending on the cells used due to the variance in cell size. The total number of cells per droplet can be optimized for each cell line. Each droplet needs to be completely filled with cells with no empty spaces. A good positive control for breast cancer cell invasion is the aggressive cell line MDA-MB-231.

2. Each droplet needs to be at least 1 μl per well. The volume for each droplet can be increased based on the experimental requirements. The total volume needed depends on the desired volume for each droplet and the number of droplets required.

3. Adjust the pH of the collagen mixtures to approximately 7.4 with 2 N NaOH. The pH needs to be adjusted just prior to using. Once the pH is adjusted, the collagen mixture must be kept on ice and used for that experiment. The collagen cannot be stored long term once the pH is adjusted.

4. Once prepared, the collagen mixture is added to the cells at a 1:1 (volume/volume) ratio. The total volume of dot collagen will be determined based on the desired volume for each droplet and the number of droplets required.

5. The cover collagen is used to overlay the droplets. The volume depends on the well size. For a 96-well plate, between 80 and 100 μl is needed to cover the droplet.

6. The proper mixing of the dot collagen and the cells is imperative to creating full droplets. The mixture can be gently and briefly vortexed if necessary. The mixture should then be put back on ice.

7. When pipetting the dot collagen-cell mixture, care should be taken not to produce any air bubbles, as they will interfere with the counting of invaded cells.

8. Due to the quick solidification of the collagen, the total amount of droplets that can be made before solidification needs to be tested.

9. Once at 37 °C, 1 μl droplets take approximately 5 min. However, the time for solidification can vary depending on the volume of each droplet and the speed at which they are plated. Empirical determination of the amount of time necessary for solidification is required. This can be tested by adding the cover collagen to droplets after varying amounts of time at 37 °C. If droplets are properly solidified, they will not disperse or move when the cover collagen is applied. If the droplets appear white, they have dried completely and the cells will not invade.

10. If using a 96-well plate, 80–100 μl should be used to cover the well. Add the cover collagen carefully to avoid creating air bubbles and moving the droplet.

11. If the cover collagen does not move when tilting the plate, it is solid. If upon the addition of the medium, the medium and cover collagen mix, the collagen did not solidify enough.

12. Add the same volume used for cover collagen. If the experimental design includes testing the effect of drugs on cancer cell invasion, the drugs should be added to the medium at twice the desired final concentration to account for the volume of the cover collagen. A positive control for inhibition of cancer cell invasion is an anti-βI integrin antibody.

13. Before removing the media, observe the collagen-cell droplet. Depending on the cell type used, collagen contraction can occur [7, 8]. This will lead to a change in the droplet size and make analysis inaccurate. If collagen contraction is an issue with the chosen cell line, dialdehyde dextran can be used to prevent it [6, 9]. The dialdehyde dextran (stock solution of 2.5 %) should be added at a 1:10 dilution (final .25 % dialdehyde dextran) to the dot collagen after mixing with the cells (Subheading 3.1, **step 1**). If no change in size, proceed with protocol as written. The collagen gel should not move from the plate when turned upside down. For a detailed protocol on how to prepare dialdehyde dextran, *see* Evensen et al. 2013 [6].

14. To simultaneously determine the total number of dead cells, Propidium iodide (PI) (2.5 μg/ml) can be added to the PBS as well. The fluorescence intensity can be observed after 30 min to determine if more time is needed with the dyes.

15. To automate the quantification of invaded cells using NIS-Elements Br 3.2 software, capture phase contrast and Hoechst images for the entire plate. A threshold from the phase contrast image is adjusted based on the difference in contrast

between the droplet and the area beyond the droplet to create two binary layers. The binary layer beyond the droplet is then copied to form a region of interest (ROI), which is then applied to the Hoechst images. The invaded cells within the ROI can then be automatically counted using object count.

References

1. Sahai E (2007) Illuminating the metastatic process. Nat Rev Cancer 7(10):737–749. doi:10.1038/nrc2229, [pii] nrc2229
2. Dufour A, Zucker S, Sampson NS, Kuscu C, Cao J (2010) Role of matrix metalloproteinase-9 dimers in cell migration: design of inhibitory peptides. J Biol Chem 285(46):35944–35956. doi:10.1074/jbc.M109.091769, [pii] M109.091769
3. Repesh LA (1989) A new in vitro assay for quantitating tumor cell invasion. Invasion Metastasis 9(3):192–208
4. Griffith LG, Swartz MA (2006) Capturing complex 3D tissue physiology in vitro. Nat Rev Mol Cell Biol 7(3):211–224
5. Millerot-Serrurot E, Guilbert M, Fourre N, Witkowski W, Said G, Van Gulick L, Terryn C, Zahm JM, Garnotel R, Jeannesson P (2010) 3D collagen type I matrix inhibits the antimigratory effect of doxorubicin. Cancer Cell Int 10:26. doi:10.1186/1475-2867-10-26,[pii]1475-2867-10-26
6. Evensen NA, Li J, Yang J, Yu X, Sampson NS, Zucker S, Cao J (2013) Development of a high-throughput three-dimensional invasion assay for anti-cancer drug discovery. PloS One 8(12): e82811. doi:10.1371/journal.pone.0082811, [pii] PONE-D-13-37596
7. Zhu YK, Umino T, Liu XD, Wang HJ, Romberger DJ, Spurzem JR, Rennard SI (2001) Contraction of fibroblast-containing collagen gels: Initial collagen concentration regulates the degree of contraction and cell survival. In Vitro Cell Dev Biol Anim 37(1):10–16
8. Nien YD, Han YP, Tawil B, Chan LS, Tuan TL, Garner WL (2003) Fibrinogen inhibits fibroblast-mediated contraction of collagen. Wound Repair Regen 11(5):380–385. doi:10.1046/J.1524-475x.2003.11511.X
9. Etienne Schacht BB, Van Den Bulcke A, De Rooze N (1997) Hydrogels prepared by cross-linking of gelatin with dextran dialdehyde. React Funct Polym 33:109–116

Chapter 17

A Combined Phagocytosis and Fluorescent Substrate Degradation Assay to Simultaneously Assess Cell Migration and Substrate Degradation

Ashleigh Pulkoski-Gross

Abstract

In order to more rapidly define the mechanism by which certain drugs and compounds can influence cancer cell invasion, we have combined a traditional phagokinetic gold migration assay with a fluorescent substrate. The purpose of this dual assay is to provide a platform by which to simultaneously monitor proteolytic activity and cancer cell migratory ability, both of which are required for the crucial step of cancer cell invasion during metastasis. This assay allows for delineation of potential mechanisms of action a compound of interest has, as one can determine whether or not a cancer cell that is being treated with the potential drug has changes in proteolytic activity and/or migratory ability at the same time.

Key words Phagokinetic, Fluorescent substrate, Migration, Substrate degradation, Dual assay

1 Introduction

Since the primary cause of death for patients diagnosed with cancer is metastasis [1], it is prudent to consider targeting the migration machinery that is utilized by cancer cells to escape their primary site. In order to do this, there must be a reasonably accessible way to monitor crucial stages of the metastatic cascade, such as cancer cell invasion. The ability of cancer cells to invade surrounding tissue is a critical component of cancer dissemination. The process of invasion requires proteolytic activity, as cleaving extracellular matrix (ECM) components reduces the physical restriction placed on a cell. The ability of cancer cells to invade surrounding tissue can be influenced by their level of protease expression, as proteases contribute to destruction of ECM and induce cell migration. Migration is also required for invasion; reducing ECM by hydrolysis of those surrounding substrates is not sufficient for invasion, as a cell must also leave the primary site to be considered invasive.

Jian Cao (ed.), *Breast Cancer: Methods and Protocols*, Methods in Molecular Biology, vol. 1406,
DOI 10.1007/978-1-4939-3444-7_17, © Springer Science+Business Media New York 2016

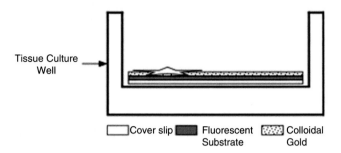

Cover slip ▮ Fluorescent ▨ Colloidal
Substrate   Gold

Fig. 1 Cartoon representation of the dual assay. A fluorescent substrate is layered on a glass coverslip, followed by a layer of colloidal gold. The coverslip rests in a tissue culture well so that cells may be plated atop the coverslip and allowed to potentially migrate across the colloidal gold layer and/or degrade the underlying fluorescent substrate

In order to develop reagents and drugs that target cell invasion, a method to monitor cell movement and proteolytic activity is required. An acceptable technique for examining only cancer cell migration is a phagokinetic gold assay first published in 1977 and further improved over the years [2–4]. Colloidal gold is layered on a gelatin-coated coverslip that cells are then placed upon. Observed areas of clearing after an incubation period are indicative of the associated cells' migratory track. While this assay is acceptable to simply monitor movement, we have modified the assay to include a fluorescent substrate in order to determine whether or not proteolytic activity is associated with a cell's migration. Combining the phagokinetic assay with a fluorescent substrate in a dual assay allows for easily testing whether or not a cell is migrating with or without involvement of proteolytic activity (Fig. 1). Equally important, this dual assay allows for easily delineating whether or not compounds shown to reduce cell migration do so by reducing protease activity. It can serve as a secondary screening method for determining whether novel compounds affect cell migration by modifying protease activity or not; in this way, the assay can directly give some insight into the mechanism of a compound that inhibits cancer cell migration and invasion. Herein we describe the method and provide an example where cells that overexpress a protease clear tracks in the colloidal gold and induce a loss of fluorescence, indicative of proteolytic activity that correlates with migration.

2 Materials

All solutions are prepared using ultrapure water (purified, deionized water with a resistance of 18.5 MΩ at 25 °C). All prepared reagents are stored at 4 °C unless otherwise noted. Follow all safety and waste disposal standards at all times.

2.1 Texas Red Gelatin-Coated Coverslip Components

1. Acid-alcohol washed coverslips: Incubate coverslips (round glass coverslip, 18 mm diameter, No.1) in a solution of 200 mL nitric acid and 100 mL hydrochloric acid for approximately 2 h, occasionally swirling the mixture and coverslips. Discard the acid carefully and wash the coverslips with ultrapure water repeatedly. Store the coverslips in 70 % ethanol, and prior to use, flame the coverslips briefly.

2. 1× phosphate-buffered saline (PBS): Mix 8 g NaCl, 0.2 g KCl, 1.44 g Na_2HPO_4, and 0.24 g KH_2PO_4 in 800 mL of water. pH the solution to 7.4 using hydrochloric acid. Bring the final volume to 1 L. The solution can be stored at room temperature after autoclaving.

3. 50 μg/mL poly-l-lysine: Dissolve poly-l-lysine (Sigma) in ultrapure water for a stock concentration of 10 mg/mL and store at −30 °C. Dilute the poly-l-lysine further in water to attain a working concentration of 50 μg/mL before each use.

4. Texas Red-labeled gelatin: Dissolve 2 mg of type B, bovine skin gelatin (Sigma-Aldrich) in 20 mL of 0.1 M $NaHCO_3$. Incubate the solution at 55 °C for 30 min to ensure the gelatin is completely dissolved (*see* **Note 1**). During the incubation time, dissolve 1 mg of Texas Red sulfonyl chloride (Molecular Probes, Life Technologies) into 100 μL of dry dimethylformamide (DMF) (Sigma, Molecular Biology Grade). After the incubation time is over, place 2 mL of the gelatin mixture into a small tube and slowly add the Texas Red-DMF mixture dropwise. Rock the Texas Red gelatin at 4 °C for at least 30 min. Add the gelatin mixture to a pre-equilibrated dialysis bag that has been clamped at one end. After clamping the other end of the dialysis bag, place the bag in a beaker containing 1 L of PBS. Place the beaker containing the dialysis bag and magnetic stir bar over a stir plate at 4 °C, gently stirring for 6 h. Change the dialysis buffer to fresh PBS and continue to stir overnight. Repeat the buffer changes until the PBS remains clear. The beaker should be covered with aluminum foil to block as much light as possible. Collect the Texas Red gelatin and add an equal volume of 100 % glycerol for a final glycerol concentration of 50 %. Store the final product at −30 °C in the dark.

5. 0.5 % glutaraldehyde: Dilute 25 % glutaraldehyde (Sigma-Aldrich, Grade I) to 0.5 % in PBS.

6. 12-well dish to place the coated coverslips and eventually seed the cancer cells.

2.2 Colloidal Gold Layer Components

1. 14.5 mM $AuHCl_4$: Measure 571 mg of tetrachloroauric (III) acid (Sigma-Aldrich) and dissolve in 100 mL of water. Store the solution protected from light at 4 °C.

2. 36.5 mM Na_2CO_3: Measure 1.93 g of $NaCO_3$ and dissolve in 400 mL of ultrapure water. After the solute dissolves, bring the

volume to 500 mL with ultrapure water in a clean graduated cylinder. Store at room temperature in a sealed container after mixing thoroughly.

3. 0.1 % formaldehyde: Dilute a 37 % formaldehyde solution to 0.1 % in ultrapure water (*see* **Note 2**).

2.3 Cell Seeding and Monitoring of Cell Migration and Substrate Degradation

1. Migratory cell line of choice.

2. Complete media appropriate for your cell line of choice.

3. Trypsin.

4. Sterile PBS.

5. Glass slides and non-fluorescent mounting media (Fluoromount G, Southern Biotech).

6. Fluorescent microscope equipped with an imaging system to capture migration/degradation patterns.

3 Methods

3.1 Preparation of Texas Red Gelatin-Coated Coverslips

1. Incubate acid-alcohol washed coverslips (after flaming and cooling briefly) in 100 µL of 50 µg/mL poly-l-lysine at room temperature for 1 h (*see* **Note 3**). Gently touch the edge of the coverslip to a clean paper towel to draw off any excess.

2. Incubate the coated coverslips in 100 µL of Texas Red gelatin diluted in PBS (1:100) for 2 h at room temperature, protected from light (*see* **Note 4**). After incubation, collect the coverslips and touch the edge of the coverslip to a clean paper towel to draw off any excess. Allow the coverslips to dry at room temperature for 90 min (*see* **Note 5**). In this, and all following steps, keep the coverslips protected from light as much as possible.

3. Incubate the gelatin-coated coverslips in ice-cold glutaraldehyde (0.5 %) on ice for 10 min. Remove from ice and continue the incubation at room temperature for 30 min.

4. Wash the coverslips in excess PBS six times in a clean six-well dish at room temperature, gently agitating. These coverslips may be stored overnight at 4 °C in PBS or may be used immediately for layering with colloidal gold. After washing in the six-well dish, place the coverslips in a 12-well dish for use in colloidal gold layering and eventual tissue culture.

3.2 Preparation of Colloidal Gold Layer (Adapted from Scott et al. [4])

1. In a 100 mL beaker with a small stir bar, heat 11 mL of ultrapure water, 1.8 mL of 14.5 mM $AuHCl_4$, and 6 mL of Na_2CO_3 to 95 °C. When the solution reaches 95 °C, add 0.1 % formaldehyde. Keep the solution heated at 95 °C (no higher) and continue the gentle agitation with the stir bar for several min-

utes (*see* **Note 6**). The mixture should turn a purple-brown color if colloidal gold is successfully generated.

2. With continuous stirring, allow the gold mixture to cool down to approximately 75 °C while still stirring and then add the hot colloidal gold to each well of the plate containing the Texas Red gelatin-coated coverslips (PBS used for storage should be removed immediately prior to the addition of the colloidal gold). Incubate the coverslips with the gold for at least 2 h at room temperature resting on a flat surface to allow even distribution of colloidal gold on the coverslip (*see* **Note 7**).

3. After colloidal gold has settled, wash the coverslips gently with PBS three times at room temperature to remove any unattached particles.

4. Incubate the washed coverslips with 75 % ethanol (prepared with ultrapure water) for 15 min in order to sterilize the surface for cell culture.

5. Remove the ethanol and rinse several times with sterile PBS in a tissue culture hood and incubate the coverslips in cell culture media with penicillin/streptomycin. This plate may be stored at 4 °C until use.

3.3 Seeding Cells and Monitoring Cell Migration and Substrate Degradation

1. Wash and trypsinize the cells of interest and neutralize in complete media.

2. Remove the culture medium used to equilibrate the prepared coverslips immediately before plating the desired number of cells on the dual-assay coverslips and replace with fresh, complete media. Keep the cells plated on the coverslips in complete media during the overnight incubation (*see* **Note 8**).

3. After the overnight incubation period, remove the media, rinse the coverslips gently with PBS, and fix cells in 4 % paraformaldehyde supplemented with Hoechst (1:2000) for 20 min at room temperature.

4. Rinse the coverslips gently with PBS and mount the coverslips on a slide using non-fluorescent mounting media. Affix the coverslips to the slide using a small amount of clear nail polish at the edge of the slip; a dot of polish will suffice. Allow to dry for at least 15 min.

5. Using a fluorescence microscope, locate a field containing Hoechst-stained nuclei. Check the red channel and identify regions of proteolyzed gelatin, indicated by a reduction in red fluorescence (*see* Fig. 2). Switch to bright field and document any cleared tracks through the colloidal gold, indicating migration.

6. Quantification of migration may be achieved by outlining the area cleared on the phagokinetic gold using software such as ImageJ. If those tracks overlay the proteolyzed area, one may choose to cal-

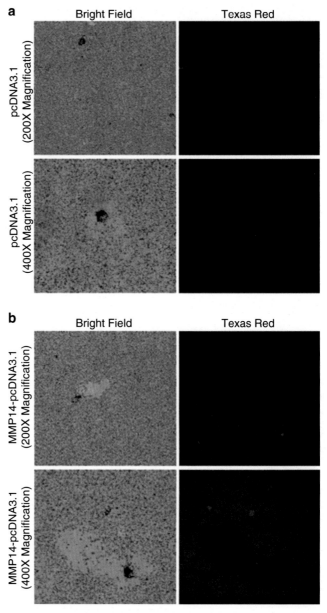

Fig. 2 Migratory cells with active protease degrade fluorescent substrate and clear tracks in the colloidal gold. (**a**) HeLa cells plated at low density on the dual-assay coverslips exhibit minimal migratory or proteolytic ability as demonstrated by lack of cleared gold in the bright field image (*left panels*) or loss of fluorescent substrate (*right panels*). (**b**) HeLa cells overexpressing the matrix metalloproteinase MMP-14 show increased migration and substrate degradation. These cells clear tracks in the colloidal gold (*left panels*) and simultaneously proteolyze the provided substrate, as demonstrated by loss of fluorescence (*right panels*)

culate the cleared area in the Texas Red gelatin and indicate it as migration as well as proteolytic activity. Alternatively, one may quantify the number of cells that have trails as opposed to those that simply cleared the area immediately surrounding themselves. Since this is a dual assay, one may determine whether cells are migrating with or without associated proteolytic activity.

4   Notes

1. The gelatin solution may be autoclaved to ensure that the substrate totally dissolves; however, incubation at 55 °C has been used successfully.

2. In order to make colloidal gold coverslips, 0.1 % formaldehyde must be made fresh before each preparation, unlike the other components. This is a crucial point, as attempts at making colloidal gold with 0.1 % formaldehyde stored at 4 °C with the other components were unsuccessful.

3. We have been successful in incubating the prepared coverslips in 100 μL of the liquid on a clean sheet of parafilm attached to a glass plate. This allows the coverslips to be coated evenly, reduces the volume of reagent required to ensure the face of the coverslip is completely covered, and allows for easy cleanup. If using this method, be careful to set the coverslips apart from one another and be sure not to force the coverslips down on the liquid and expel the liquid to the opposite side of the coverslip; gently laying them on top of the liquid bead should be sufficient. All coverslip incubations during the Texas Red gelatin-coating step are done in this way.

4. This assay is quite versatile and various fluorescent substrates may be substituted for Texas Red gelatin. We have successfully used FITC-fibronectin and FITC-collagen type IV following these procedures.

5. To facilitate coverslip drying, we use a pipette tip box equipped with several tips to use as a balance for the coverslip (Fig. 3). We place the coverslip vertically against the pipette tips in the box to allow for any potential excess to drain down and to allow the gelatin film to dry.

6. Generating colloidal gold can be very difficult and is generally the limiting factor in this assay. It is crucial to monitor and control the temperature throughout the process and to be as precise as possible with the volumes of reagents when measuring.

7. While the minimum requirement for colloidal gold settling evenly on the plate is listed above, in our experience, incubating the plate overnight with the gold particles at 4 °C can be a reasonable alternative.

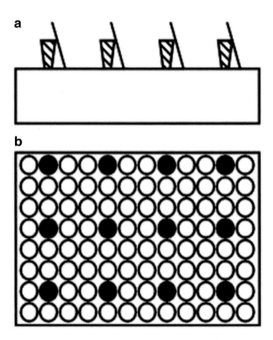

Fig. 3 Cartoon representation of the coverslip drying apparatus. (**a**) Side view of the pipette tip box and associated tips that serve as a drying rack for the coverslips after gelatin layering. Pipette tips are indicated by the *hashed trapezoid* and coverslips are indicated by the associated *black line*. (**b**) Top view of the pipette tip box with spaces indicated by *filled black circles* correlating to pipette tips. This spacing allows for the coverslips to drain and dry without contacting one another or the cover of the box

8. Cell numbers should be determined based on the cell size and migratory ability of the cells if known. Generally, plating the cells in a sparse fashion will allow for analyses of separate tracks; plating many cells will make it difficult to distinguish the tracks generated by each cell if they are sufficiently migratory. Furthermore, one may choose to include inhibitors as controls, such as protease inhibitors to reduce the degradation of substrate and potentially migration or agents that inhibit migration without directly interfering with proteolytic behavior (e.g., jasplakinolide or Taxol).

References

1. Kraljevic Pavelic S, Sedic M, Bosnjak H, Spaventi S, Pavelic K (2011) Metastasis: new perspectives on an old problem. Mol Cancer 10:22. doi:10.1186/1476-4598-10-221476-4598-10-22 [pii]
2. Albrecht-Buehler G (1977) The phagokinetic tracks of 3T3 cells. Cell 11(2):395–404, doi:0092-8674(77)90057-5 [pii]
3. Albrecht-Buehler G (1977) Phagokinetic tracks of 3T3 cells: parallels between the orientation of track segments and of cellular structures which contain actin or tubulin. Cell 12(2):333–339, doi:0092-8674(77)90109-X [pii]
4. Scott WN, McCool K, Nelson J (2000) Improved method for the production of gold colloid monolayers for use in the phagokinetic track assay for cell motility. Anal Biochem 287(2):343–344. doi:10.1006/abio.2000.4866, http://dx.doi.org/

Chapter 18

Analysis of Invadopodia Formation in Breast Cancer Cells

Ziqing Wang, Xiao Liang, Ming Cai, and Guangwei Du

Abstract

Metastasis is the major cause of breast cancer deaths. To spread from the primary tumor sites to distant tissues, solid tumor cells need to degrade the surrounding extracellular matrix (ECM). The protrusive membrane structures named invadopodia have been shown to play a critical role in the degradation of the ECM and invasion of invasive cancer cells. In this chapter, we describe a detailed protocol to examine invadopodia in human breast cancer cells.

Key words Invadopodia, Breast cancer, Invasion, Metastasis, Extracellular matrix

1 Introduction

Metastasis is a process in which cancer cells spread from their primary sites to colonize at distant organs [1, 2]. Patients with metastatic tumor usually have poor prognosis. About 90 % of breast cancer deaths are due to metastases [1, 3]. During metastasis, solid tumor cells invade the surrounding extracellular matrix (ECM), intravasate into the circulatory system through the endothelium, and then extravasate into a distant tissue, where they eventually establish secondary tumors [1, 2]. One major way of invasion requires cancer cells to degrade ECM components through matrix-digesting proteases. In recent years, increasing evidences have shown that the subcellular structures known as invadopodia are critical for the breakdown of ECM in the multiple steps of metastasis, such as local invasion, intravasation, and extravasation [4, 5].

Invadopodia are protrusive membrane structures rich in actin cytoskeleton and metalloproteases in invasive cancer cells. The primary function of invadopodia is to degrade ECM through recruiting various matrix proteases to the contacting sites of the plasma membrane and ECM, thus allowing cancer cells to leave the primary sites during metastasis. Invadopodia have been shown to be critical for the invasion of many types of cancer cells, such as melanoma, breast cancer, colon cancer, lung cancer, and prostate

Jian Cao (ed.), *Breast Cancer: Methods and Protocols*, Methods in Molecular Biology, vol. 1406,
DOI 10.1007/978-1-4939-3444-7_18, © Springer Science+Business Media New York 2016

cancer [6–10]. Disruption of the key players of invadopodia, such as Twist, cortactin, Tks5, and Tks4, blocked cancer metastasis in several cancer mouse models [11–14]. Because of the specific presence of invadopodia in invasive cancer cells and their critical roles in tumor metastasis, inhibition of key components or regulators of invadopodia has become a very attractive and unique strategy to target tumor metastasis [4, 5].

In this chapter, we describe a protocol for examining invadopodia formation in the invasive MDA-MB-231 human breast cancer cells using fluorescently labeled gelatin, one of the major components in the ECM [15]. In combination with inhibitor treatment, overexpression, and knockdown/knockout of the genes of interest, this protocol can be used to study the signaling pathways that regulate different steps of invadopodia formation.

2 Materials

The water used for all solutions in this protocol is from the purified deionized water with a sensitivity of 18 MΩ cm at 25 °C. The storage condition varies for different solutions and reagents.

2.1 Reagents for Gelatin Labeling

1. DyLight 488 labeling kits (Thermo Fisher Scientific, Rockford, IL, 53024): The DyLight Fluor is activated with an N-hydroxysuccinimide (NHS) ester, which is the most commonly used reactive group for labeling protein. NHS esters react with primary amines, forming stable, covalent amide bonds and releasing the NHS groups. The kit contains DyLight Alexa 488 labeled-NHS ester, borate buffer (0.67 M), purification resin, spin column, and microcentrifuge collection tubes (*see* **Note 1**).

2. Phosphate-buffered saline (PBS): Dissolve 8 g of NaCl, 0.2 g of KCl, 1.44 g of Na_2HPO_4, and 0.24 g of KH_2PO_4 in 800 mL distilled H_2O. Adjust the volume to 1 L with additional distilled H_2O. Adjust the pH to 7.4 with 1 M HCl. Store at room temperature.

3. Gelatin: Type A gelatin from porcine skin (Sigma, G-2500).

2.2 Reagents for Coating Coverslips with Labeled Gelatin

1. Parafilm "M" laboratory film (Bemis Flexible Packaging, Neenah, WI 54956).

2. 12 mm diameter circular microscope coverslips (Fisher, Waltham, MA, USA).

3. Poly-L-lysine solution: Dilute the stock poly-L-lysine (Sigma, P8920) to 50 µg/mL in distilled H_2O.

4. Glutaraldehyde solution: 0.5 % glutaraldehyde is diluted from 25 % glutaraldehyde stock using PBS. Store the diluted solution in refrigerator and use it in 1 month.

5. Gelatin mix solution: Mix fluorescent conjugated gelatin and 1 mg/mL unconjugated gelatin at a ratio of 1:25–1:2 (*see* **Note 2**). The optimal ratio of labeled to unlabeled gelatin is dependent on the signal intensity of labeled gelatin. We used 1:4 in this protocol (*see* **Note 3**).

6. Fluorescein-conjugated gelatin (Thermo Fisher Scientific, Rockford, IL, G13187): Fluorescent-labeled gelatin from some companies may be also used for invadopodia assay.

7. Sodium borohydride (5 mg/mL): Dissolve 10 mg of sodium borohydride in 2 mL PBS. The solution must be freshly made before use.

8. 70 % ethanol: Dilute 200 proof ethanol to 70 % with distilled H_2O.

2.3 Other Reagents for Cell Culturing and the Invadopodia Assessment

1. Cell line: MDA-MB-231 human breast cancer cells (American Type Culture Collection, ATCC).

2. DMEM: Dulbecco's Modified Eagle Medium (DMEM) (Hyclone, SH30243.01) supplemented with 10 % fetal bovine serum (FBS) from Hyclone. Store at 4 °C.

3. Enzyme-free cell dissociation solution (Millipore, S-014-B).

4. 4 % paraformaldehyde (PFA) solution: Dissolve 0.4 g paraformaldehyde powder in 10 mL PBS. Heat the solution in 60 °C water bath and shake vigorously every few minutes to help dissolve. Store at 4 °C and use within 1 week.

5. 0.1 % Triton X-100: Dilute from 10 % Triton X-100 using PBS.

6. Alexa Fluor® 594 phalloidin (Life Technologies, A22287).

7. Mounting medium: Make 4 % *n*-propyl gallate (w/v) in 90 % glycerol (v/v) and 10 % PBS. Aliquot and store at –20 °C.

8. Clear nail polish (Electron Microscopy Sciences, 72180).

3 Methods

3.1 Label Gelatin with DyLight 488

1. Prepare 2 mg/mL gelatin solution in PBS (put it in 37 °C water bath to help dissolve).

2. Add 40 μL of borate buffer from the DyLight 488 labeling kit (0.67 M) to 0.5 mL gelatin solution.

3. Take out the vial containing DyLight 488 NHS ester from freezer, and let it warm at room temperature. Add the prepared gelatin solution to the vial and vortex gently or invert ten times.

4. Briefly centrifuge the vial to collect the sample in the bottom of the tube.

5. Incubate the reaction mixture for 60 min at room temperature. Protect from light during incubation.

6. Place two spin columns in two microcentrifuge collection tubes.

7. Mix the purification resin to ensure uniform suspension and add 400 µL of the suspension into both spin columns. Centrifuge for 45 s at ~1000×g to remove the storage solution. Discard the used collection tubes and place the columns in new collection tubes.

8. Add 250–270 µL of the labeling reaction mixture to each spin column and mix the sample with the resin by pipetting up and down or briefly vortexing.

9. Centrifuge columns for 45 s at ~1000×g to collect the purified gelatin. Combine the samples from both columns (~0.5 mL total). Discard the used columns. The concentration of labeled gelatin is around 2 mg/mL.

10. Store the labeled gelatin at 4 °C for up to 1 month, protected from light. Alternatively, store labeled protein in single use aliquots at –20 °C. Avoid repeated freeze/thaw cycles.

3.2 Coat Coverslips with Fluorescently Labeled Gelatin

1. Place coverslips on a piece of parafilm with point-ended tweezers (*see* **Note 4**).

2. Pretreat coverslips with 50 µg/mL poly-L-lysine for 20 min. Wash coverslips twice with PBS, 3 min each time.

3. Cross-link coverslips with 0.5 % glutaraldehyde for 15 min. Wash three times with PBS, 3 min each time (*see* **Note 5**).

4. Cut another parafilm and put it in a humid box with nontransparent cover. Drop 40 µL of mixed gelatin solution on the parafilm (*see* **Note 6**). Invert the coverslips and place them facedown onto the gelatin solution. After 20 min incubation, wash the coverslips with PBS once for 3 min.

5. Quench the autofluorescence with 5 mg/mL sodium borohydride for 3 min. Wash three times with PBS (*see* **Note 7**).

6. Sterilize the coverslips with 70 % ethanol for 5 min (*see* **Note 8**).

7. Wash the coverslips once with PBS and once with complete medium.

8. Incubate the coverslips in complete growth medium for 1 h before use.

3.3 Invadopodia Formation Assessment

1. One or two days before the experiment, plate MDA-MB-231 cells into a 6-well plate (4×10^5 cells/well) (*see* **Note 9**). Grow cells in 10 % FBS DMEM in a 37 °C 5 % CO_2 incubator (*see* **Note 10**).

2. On the day of experiment, treat cells with appropriate inhibitors for 0.5–1 h.

3. Remove medium in the well and detach the cells with 500 µL enzyme-free cell dissociation solution. Let the plate stand for 5 min in 37 °C 5 % CO_2 incubator (*see* **Notes 11** and **12**).

4. Check the cells under an inverted microscope. After the cells are completely detached from the bottom of the well, collect the supernatant and spin at $500 \times g$ for 5 min.

5. After centrifugation, remove the supernatant and resuspend the cell pellet with 100 μL 10 % FBS DMEM.

6. Count cells and seed 1×10^5 cells per well into a 12-well plate containing the coverslips coated with fluorescent-labeled gelatin. Add appropriate inhibitors if needed.

7. Incubate cells in 37 °C 5 % CO_2 for 3–4 h (*see* **Note 13**).

8. Remove medium and fix cells with 4 % PFA for 20 min at room temperature. Wash the coverslips with PBS for 5 min (*see* **Note 14**).

9. Permeabilize the cells with 0.1 % Triton X-100 for 10 min at room temperature.

10. Wash coverslips with PBS for three times, 5 min each time.

11. Incubate coverslips with 1:200 diluted Alexa 594 phalloidin in PBS for 1 h at room temperature. Wash coverslips three times with PBS, 5 min each time.

12. After the final wash, carefully remove excess PBS by blotting the side of the coverslip with a kimwipe. Place coverslip facedown onto a drop of anti-fading mounting medium on a microscope slide (*see* **Note 15**). Avoid any air bubbles during the process (*see* **Note 16**).

13. Aspirate excess mounting medium along the coverslip. Seal them by applying nail polish along the edge of coverslips. Observe the invadopodia formation under confocal laser scanning microscope with 100× objective (Fig. 1).

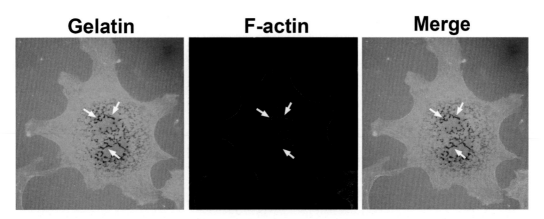

Fig. 1 A representative picture of invadopodia. MDA-MB-231 human breast cancer cells were plated on Alexa 488 gelatin-coated coverslips for 3 h. F-actin was stained with Alexa 594 phalloidin. The *arrows* show the degradation of Alexa 488 gelatin and F-actin staining, indicating the presence of active invadopodia. The images were captured by a 100× objective using a Nikon A1 laser confocal microscope

14. The formation of invadopodia will remove fluorescently labeled gelatin, leaving black spots on coverslips. Invadopodia formation can be quantified by the percentage of cells with invadopodia and/or the number of invadopodia per cell. Alternatively, the total area of degraded gelatin may also be used to indicate the invasive activity.

4   Notes

1. NHS ester-activated fluorophores are moisture sensitive. Keep them in a dark and dry environment.

2. Pre-warm the labeled gelatin and 1 mg/mL unlabeled gelatin in 37 °C water bath before mixing. Frozen gelatin must be thawed completely to prevent precipitation. Gelatin that was incompletely dissolved can cause uneven bright spots on coverslips.

3. The stiffness of ECM can directly affect the ability of breast cancer cells to form invadopodia. It is critical to use appropriate concentrations of gelatin in invadopodia experiments. The gelatin concentration (labeled and unlabeled) used for coating the slides in this protocol is about 1.2 mg/mL.

4. Coverslips with dirt can cause uneven spots with strong fluorescence. If necessary, clean the coverslips with acid alcohol (1 % HCl in 70 % ethanol).

5. Glutaraldehyde is a divalent cross-linker. It is used here to irreversibly cross-link gelatin to the glutaraldehyde-pretreated poly-L-lysine coating. The gelatin itself is not cross-linked. It binds to activated aldehyde groups associated with the poly-L-lysine after the glutaraldehyde is washed away.

6. Labeled gelatin with fluorescein or other fluorophores from commercial sources, such as Thermo Fisher Scientific, can also be used.

7. Sodium borohydride is used to reduce the autofluorescence generated by reversible Schiff's bases in the aldehyde-NH2 reaction. Prepare this solution on ice. Be cautious that sodium borohydride is highly caustic and prone to explode.

8. If the experiment lasts only for several hours, this step is not necessary.

9. Invadopodia can be also easily examined in many other metastatic cancer cell lines, e.g., breast cancer cell line 4T1, melanoma cell line RPMI7951, and lung cancer cell line A549.

10. More invadopodia can be observed when the cells are healthy.

11. Cells may be also detached by trypsin-EDTA solution. However, enzyme-free cell dissociation solution causes minimal

disruption of surface proteins and therefore greatly improves invadopodia formation.

12. The incubation time can be longer if the cells do not completely detach.

13. The time to form invadopodia varies in different cell types. An optimal plating time should be determined when a new cell line is used. For MDA-MB-231 cells, we generally analyze invadopodia plaques 3–4 h after plating. Longer incubation can increase both number and size of the invadopodia plaques. However, cell migration and merging of individual invadopodium may occur after 5 h, making the interpretation of result difficult.

14. The appropriate fixation method should be determined according to the experiment conditions, such as the antibodies used for staining. For invadopodia visualization using fluorescent microscopy, PFA is our preferred fixation reagent. Other fixation methods, such as cold methanol or glutaraldehyde, may be also used for visualizing specific antigens. However, cold methanol destroys the native quaternary structure of F-actin and hence is not suitable for actin staining with phalloidin. In addition, cold methanol removes lipids and dehydrates the cells; therefore, it may also disrupt the fine cell structures.

15. It is important to choose a correct mounting medium for certain fluorophores. For example, many commercial anti-fade mounting media do not work well for far-red fluorophores. Propyl gallate is a very general anti-fade mounting medium that is suitable for most of the commonly used fluorophores.

16. It is very common that air bubbles are formed during slide mounting. To avoid air bubbles, place one side of the coverslip on the mounting solution first, then slowly lower the tweezers, and do not release the coverslips from the tweezers until the mounting medium fully covers the coverslip.

Acknowledgment

This work was supported by a research grant RP130425 from the Cancer Prevention and Research Institute of Texas (CPRIT) and a research grant R01HL119478 from the National Heart, Lung, and Blood Institute of the National Institutes of Health to GD. The content is solely the responsibility of the authors and does not necessarily represent the official views of the CPRIT and National Institutes of Health.

References

1. Chaffer CL, Weinberg RA (2011) A perspective on cancer cell metastasis. Science 331(6024):1559–1564

2. Klein CA (2008) Cancer. The metastasis cascade. Science 321(5897):1785–1787

3. Nguyen DX, Massague J (2007) Genetic determinants of cancer metastasis. Nat Rev Genet 8(5):341–352

4. Murphy DA, Courtneidge SA (2011) The 'ins' and 'outs' of podosomes and invadopodia: characteristics, formation and function. Nat Rev Mol Cell Biol 12(7):413–426

5. Paz H, Pathak N, Yang J (2014) Invading one step at a time: the role of invadopodia in tumor metastasis. Oncogene 33(33):4193–4202

6. Nakahara H, Howard L, Thompson EW, Sato H, Seiki M, Yeh Y, Chen WT (1997) Transmembrane/cytoplasmic domain-mediated membrane type 1-matrix metalloprotease docking to invadopodia is required for cell invasion. Proc Natl Acad Sci U S A 94(15):7959–7964

7. Kelly T, Yan Y, Osborne RL, Athota AB, Rozypal TL, Colclasure JC, Chu WS (1998) Proteolysis of extracellular matrix by invadopodia facilitates human breast cancer cell invasion and is mediated by matrix metalloproteinases. Clin Exp Metastasis 16(6):501–512

8. Vishnubhotla R, Sun S, Huq J, Bulic M, Ramesh A, Guzman G, Cho M, Glover SC (2007) ROCK-II mediates colon cancer invasion via regulation of MMP-2 and MMP-13 at the site of invadopodia as revealed by multiphoton imaging. Lab Invest 87(11):1149–1158

9. Wang S, Li E, Gao Y, Wang Y, Guo Z, He J, Zhang J, Gao Z, Wang Q (2013) Study on invadopodia formation for lung carcinoma invasion with a microfluidic 3D culture device. PLoS One 8(2):e56448

10. Desai B, Ma T, Chellaiah MA (2008) Invadopodia and matrix degradation, a new property of prostate cancer cells during migration and invasion. J Biol Chem 283(20):13856–13866

11. Eckert MA, Lwin TM, Chang AT, Kim J, Danis E, Ohno-Machado L, Yang J (2011) Twist1-induced invadopodia formation promotes tumor metastasis. Cancer Cell 19(3):372–386

12. Seals DF, Azucena EF Jr, Pass I, Tesfay L, Gordon R, Woodrow M, Resau JH, Courtneidge SA (2005) The adaptor protein Tks5/Fish is required for podosome formation and function, and for the protease-driven invasion of cancer cells. Cancer Cell 7(2):155–165

13. Li CM, Chen G, Dayton TL, Kim-Kiselak C, Hoersch S, Whittaker CA, Bronson RT, Beer DG, Winslow MM, Jacks T (2013) Differential Tks5 isoform expression contributes to metastatic invasion of lung adenocarcinoma. Genes Dev 27(14):1557–1567

14. Leong HS, Robertson AE, Stoletov K, Leith SJ, Chin CA, Chien AE, Hague MN, Ablack A, Carmine-Simmen K, McPherson VA, Postenka CO, Turley EA, Courtneidge SA, Chambers AF, Lewis JD (2014) Invadopodia are required for cancer cell extravasation and are a therapeutic target for metastasis. Cell Rep 8(5):1558–1570

15. McDonald JA, Kelley DG, Broekelmann TJ (1982) Role of fibronectin in collagen deposition: Fab' to the gelatin-binding domain of fibronectin inhibits both fibronectin and collagen organization in fibroblast extracellular matrix. J Cell Biol 92(2):485–492

Chapter 19

Patient-Derived Tumor Xenograft Models of Breast Cancer

Christopher D. Suarez and Laurie E. Littlepage

Abstract

The need for model systems that more accurately predict patient outcome has led to a renewed interest and a rapid development of orthotopic transplantation models designed to grow, expand, and study patient-derived human breast tumor tissue in mice. After implanting a human breast tumor piece into a mouse mammary fat pad and allowing the tumor to grow in vivo, the tumor tissue can be either harvested and immediately implanted into mice or can be stored as tissue pieces in liquid nitrogen for surgical implantation at a later time. Here, we describe the process of surgically implanting patient-derived breast tumor tissue into the mammary gland of nonobese diabetic-severe combined immunodeficiency (NOD-SCID) mice and harvesting tumor tissue for long-term storage in liquid nitrogen.

Key words Patient-derived tumor xenograft, Breast cancer, Orthotopic transplant

1 Introduction

Patient-derived tumor xenograft (PDX) animal models are established by first transplanting human cancer tissue into a mouse host and by serially transplanting the newly formed tumor. PDX models have seen a resurgence in their use in the preclinical setting since their original development in the 1970s [1]. As increasing numbers of investigational drugs fail to receive FDA approval after costly clinical trials that do not mimic preclinical animal trials [2], researchers are using PDX models as the preferred preclinical animal model in both academic and industrial groups. Human breast cancer patients and mouse PDX models, generated from human breast tumor tissue, responded similarly to identical treatment regimens [3, 4]. In addition, PDX models have been used to conduct chemotherapy preclinical phase II animal studies that demonstrated therapeutic efficacy paralleling that observed on patients in the clinical setting [5].

Despite the significant advantages in using PDX models, these models also have challenges. Generating PDX mouse models requires the use of immune compromised host strains (e.g., NOD-SCID or

Jian Cao (ed.), *Breast Cancer: Methods and Protocols*, Methods in Molecular Biology, vol. 1406,
DOI 10.1007/978-1-4939-3444-7_19, © Springer Science+Business Media New York 2016

NOD-SCID gamma) for engraftment and propagation while preventing rejection of the human tissue [6]. Because these mice lack critical elements of their immune system, studies using PDX models to examine immunotherapeutics would be of limited value. In addition, as PDX tumors grow, they substitute the human tumor stroma for murine stroma. The integrated murine stromal components, which consist of the extracellular matrix, cancer-associated fibroblasts, blood vessels, leukocytes, and macrophages, will prevent a full evaluation of any agents that target the *human* stroma [7]. Also, the engraftment process enriches for more aggressive tumors, such as hormone receptor-negative breast tumors that have a higher take rate compared to hormone-sensitive breast tumors. Therefore, aggressive tumors are overrepresented in current PDX collections [3, 4, 8].

PDX mice currently provide one of the best model systems used to advance cancer drug discovery efforts. Here we demonstrate how to surgically implant patient-derived mammary tumor tissue into NOD-SCID mice. Engrafting ER+ patient-derived tumor tissue into mice also requires an additional surgical procedure to implant estrogen pellets. Because the estrogen pellets can be very expensive when purchased from a manufacturer, we make our own estradiol pellets regularly at a fraction of the cost of the commercial pellets. We describe our method for making these pellets. We also describe how to isolate and prepare patient-derived mammary tumor tissue that can be used for either immediate implant or long-term storage in liquid nitrogen.

2 Materials

2.1 Tissue Prep and Freezing Supplies

Patient-derived tumor tissue for implantation (fresh or frozen).

Serum-free DMEM/F12 media.

Tissue freezing media (filter sterilized): 95 % (v/v) FBS/ 5 % (v/v) DMSO.

Nalgene freezing container: (Sigma, cat. # C1562).

Eppendorf 5 ml sterile tubes.

70 % (v/v) ethanol: prepare by adding 300 ml of purified double distilled water, or Millipore water, to 700 ml of 100 % reagent grade ethanol.

Large pair of straight blunt end forceps.

Feather disposable scalpels #10 (Fisher scientific, cat. # NC9999403).

2.2 Surgery Supplies

NOD-SCID or NOD-SCID gamma mice (3–4 weeks of age). Isoflurane.

Betadine.

Hot bead sterilizer (Fine Science Tools, cat. # 18000–45).

Labsan-256CPQ disinfectant spray (Sanitation Strategies).

Surgical platform (foam board).

Bench protector (Fisher cat. # 1420641).

Compressed oxygen tank.

Anesthetic vaporizer.

Charcoal filter.

Anesthesia induction chamber.

Laboratory labeling tape.

Duct tape.

Cotton tip applicators, sterile

27-G × 1 in. needles, sterile

Noyes micro dissecting spring scissors, sterile (Roboz, cat. # RS-5676).

Knapp 4″ scissors, sterile (Roboz, cat. # RS-5960).

Blunt end Graefe micro dissecting forceps, sterile (Roboz, cat. # RS-5135).

Graefe micro dissecting forceps with teeth, sterile (Roboz, cat. # RS-5153).

Dumont #N5 cross action fine forceps (Roboz, cat. # RS-5020).

Dumont #5 fine forceps (Fine Science Tools, cat. # 11254–20).

Bovie low temperature cautery pen with micro fine tip (Fisher Scientific, cat. # NC9030107).

Sterile 9 mm E-Z clip wound closures (Stoelting, cat. # 59027).

Wound clip applicator (Fine Science Tools, cat. # 12031–09).

Wound clip remover (Fine Science Tools, cat. # 12033–00).

Heating pad.

Kendall monoject insulin syringes with 29-G × ½ in. needles, sterile.

Ketofen sterile solution (100 mg/ml) NADA 140–269: Prepare 50 mg/ml working solution in sterile 1× PBS. Store at room temperature.

2.3 Estrogen Pellets

Beeswax.

70 % ethanol.

Estradiol.

Aluminum foil.

Dry ice.

Weigh boats.

Weighing paper.

Hot plate with magnetic stirrer.

250 ml beaker.

Sterile magnetic stir bar.

Glass Pasteur pipettes.

Bunsen burner.

20 ml glass scintillation vial.

2.4 Preparation of Estrogen Pellets

1. Place aluminum foil on a flat piece of dry ice.

2. Wipe a large weigh boat with 70 % ethanol and place it on top of the aluminum foil covering the dry ice.

3. Fill 250 ml beaker with water and begin heating to 60 °C on a hot plate.

4. Weigh 1.95 g of beeswax directly into a 20 ml glass scintillation vial.

5. Weigh 50 mg estradiol powder on weighing paper.

6. Place the glass sample vial that contains the beeswax into the water-filled beaker on the hot plate. Confirm that the water level does not rise above the neck of the glass vial.

7. As soon as the wax melts, remove the glass vial from the beaker of water and place the glass vial directly onto the hot plate.

8. Add the 50 mg of estradiol powder to the glass vial.

9. Add a small, sterile magnetic stir bar to the vial to help dissolve the estradiol. It can take 15 min for the estradiol to get dissolved.

10. After the estradiol is completely dissolved, the solution will be transparent.

11. Preheat a glass Pasteur pipette using a Bunsen burner and fill the pipette with the beeswax/estradiol solution.

12. Create estradiol pellets by placing three drops of the wax mixture on the weigh boat to create a single pellet.
    Note: Each drop weighs ~10 mg, and each pellet weighs ~30 mg. Each pellet contains ~1 mg estrogen.

13. Continually flame the Pasteur pipette to prevent the wax from solidifying during this procedure.

14. After the pellets completely solidify, use sterile forceps to collect the estradiol pellets in a sterile 1.5 ml microcentrifuge tube. Pellets can be stored up to 6 months at 4 °C.

3 Methods

3.1 Thawing Cryopreserved Patient-Derived Breast Tumor Tissue

1. Thaw frozen tissue by placing the cryovials in a 37 °C water bath (*see* **Note 1**).

2. Thaw tube until a small ice pellet remains in the cryovial.

3. Spray the outside of the cryovial with 70 % ethanol and dry the tube thoroughly before transferring into a laminar flow hood.

4. After the ice in the cryovial is completely melted, allow the tissue pieces to settle to the bottom of the tube and use an aspirator to remove the media.

5. Use a sterile pipette tip or sterile forceps to transfer the tissue pieces into a new 5 ml tube (*see* **Note 2**).

6. Add 3 ml serum-free DMEM/F12.

7. Invert the 5 ml tube several times to wash the tissue pieces in the media.

8. Allow the pieces to settle to the bottom of the tube and then aspirate the media.

9. Repeat **steps 6–7** an additional two times to remove any residual DMSO from the tissue-freezing media.

10. Add 3 ml serum-free DMEM/F12 media to the tissue pieces and store the tube on ice until ready for implanting into mice.

3.2 Surgery Preparation

1. Place a recovery cage on top of a heating pad and turn heating pad on.

2. Disinfect the surgery area by cleaning the surface of the laminar flow hood with Labsan-256CPQ.

3. Secure the bench protector sheet to the foam surgery board using duct tape.

4. Place the surgery board in the hood and use duct tape to secure the nose cone to the surgery board (Fig. 1a).

5. All tools should be autoclaved prior to use in surgery and arranged on a sterile surface (Fig. 1b). Tools should also be sterilized between animals using the hot bead sterilizer.

6. Adjust the vaporizer to deliver 2 % isoflurane and oxygen at the rate of 1–2 L/min in an induction chamber to anesthetize the mice.

7. Transfer the anesthetized mouse to the surgery board.

3.3 Estrogen Pellet Implantation

1. Place anesthetized mouse in a prone position on the surgery board and insert the nose of the mouse securely into the nose cone that delivers 2 % vaporized isoflurane and oxygen at 1–2 L/min (Fig. 2a).

2. Gently apply labeling tape to secure the hind paws to the board and make sure the nose cone remains securely in place to maintain continual delivery of isoflurane and oxygen to the mouse.

3. Using cotton tipped applicators, apply betadine between the shoulder blades of the mouse.

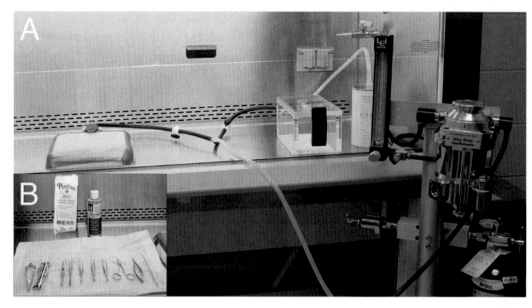

Fig. 1 Aseptic surgery setup inside a laminar flow hood. (**a**) The oxygen aids delivery of vaporized isoflurane to both the anesthesia chamber and the nose cone via Y-connection. The charcoal filter collects the waste anesthetic gas. A nitrile glove is cut and secured around the nose cone using a rubber band. Then the nose cone is secured to the surgery board using duct tape. (**b**) Surgery tools are autoclaved prior to use and maintained as sterile when placed in the laminar flow hood by using the packaging from the sterile surgical gloves. Surgery tools are also sterilized in a hot bead sterilizer between animals

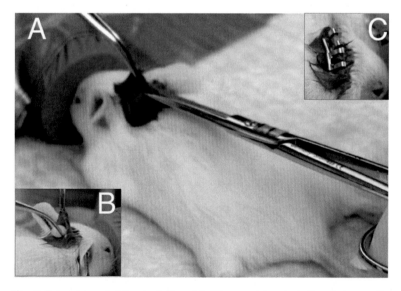

Fig. 2 Estrogen pellet implantation. (**a**) After applying betadine between the shoulder blades, the skin is lifted and small incision is made. (**b**) The pocket is opened using Graefe forceps with teeth, and the estrogen pellet is inserted using blunt-end forceps. (**c**) The incision is closed using two 9 mm staples to ensure the estrogen pellet cannot fall out of the pocket

4. Perform toe pinch to confirm mouse is completely anesthetized before making incision.

5. Using the dissection scissors, make a small incision (~3–5 mm) between the shoulder blades to create a small pocket for the estrogen pellet.

6. Using the Graefe forceps with teeth, hold open the pocket and insert an estrogen pellet into the pocket using a pair of Graefe blunt end forceps (Fig. 2b).

7. Close the incision with one or two wound clips, depending on the size of the incision that was made (Fig. 2c) (*see* **Note 3**).

*3.4   Clearing
the Epithelium
from the Mammary
Fat Pad*

1. Remove the labeling tape from the hind paws and rotate the mouse into the supine position.

2. Tape the hind paws to the surgery board and secure the upper body of the mouse to the surgery board by placing a thin piece of labeling tape loosely across the chest (Fig. 3a).

3. Using cotton tipped applicators, apply betadine to the abdomen of the mouse.

4. Toe pinch the animal to confirm it is fully anesthetized.

5. Starting just above the genitals (Fig. 3b), make a ventral midline incision (~1.5 cm), while taking care not to cut the peritoneal membrane (*see* **Note 4**).

6. Make another incision (~1 cm) down the leg of the mouse.

7. Using the Graefe forceps with teeth, grasp the skin, and using the blunt-end Graefe forceps, grasp the peritoneal membrane (Fig. 3c).

8. Gently pull the skin away from the peritoneal membrane until you expose the #4 mammary gland. You should be able to see three major blood vessels that connect to each other at the lymph node in the center of the mammary gland (Fig. 3d).

9. Use a sterile 27G needle to pin the skin to the surgery board so there is clear view of the entire mammary gland.

10. Locate the #4 mammary gland by identifying the three blood vessels that converge at the lymph node (Fig. 3d).

11. Using the Graefe blunt end forceps, grasp the lymph node, lift the tissue gently, and cauterize the #1 and #2 blood vessels (Fig. 3e).

12. Use the micro dissecting scissors to cut the mammary gland tissue located within the cauterized region (Fig. 3f) (*see* **Note 5**).

13. While gently lifting the tissue, to maintain tension on the mammary fat pad, with the blunt end forceps, cauterize the #3 blood vessel that connects the #4 and #5 mammary glands.

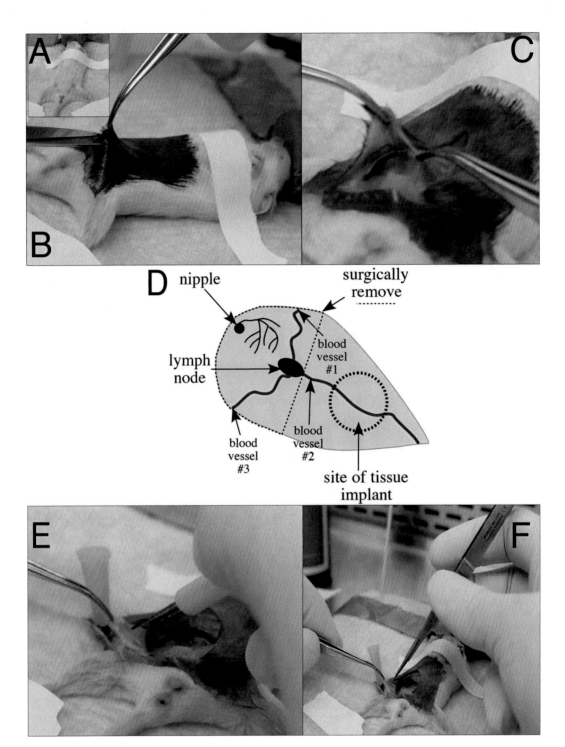

Fig. 3 Clearing the epithelium from the mammary fat pad. (**a**) A 3-week-old mouse is secured to the surgery board using labeling tape to maintain continuous delivery of isoflurane via the nosecone. (**b**) Use Graefe forceps with teeth to lift the skin of abdomen to make first small incision with Knapp 4″ scissors above the genitals. (**c**) Grasp the skin using Graefe forceps with teeth and secure the peritoneal membrane using the Graefe blunt forceps. Use the Graefe forceps with teeth to pull the skin and peel it away from the peritoneal membrane until the #4 mammary gland is visible. (**d**) The diagram shows the major features of the #4 mammary gland as viewed from the right side of the mouse. The three blood vessels converge at the lymph node and the *dashed line* shows the epithelium that is excised. (**e**) Use a low-temperature cautery pen to cauterize the blood vessels before removing the #4 mammary gland. (**f**) After cauterizing the blood vessels, excise the gland using micro dissection scissors

14. Use micro dissecting scissors to cut the tissue connecting these two glands.

15. Continue cutting toward the previously made incisions at the #1 and #2 blood vessels and remove the epithelium from the mammary fat pad.

3.5 Tissue Implant

1. Grip the remaining mammary tissue using the blunt forceps and insert the fine-tip cross action forceps to create a small pocket in the remaining mammary tissue (Fig. 4a).

2. Compress the cross action forceps to open a pocket (Fig. 4b).

3. Have an assistant pick up a tissue piece from the 5 ml vial using the straight fine-tip forceps and place it in the pocket created with the cross action forceps (Fig. 4c).

4. After placing the tissue inside the pocket, remove the cross action forceps to collapse the pocket. While the assistant removes the fine-tip forceps, use the cross action forceps to keep the tissue inside the pocket (*see* **Note 6**).

3.6 Closing the Mouse and Recovery

1. Use the Graefe blunt end forceps to align the skin where the ventral midline incision was made (Fig. 5a).

2. After confirming correct alignment, lift the skin to avoid stapling the peritoneum and close the incision by using the wound clip applicator and 9 mm E-Z clips (Fig. 5b).

3. Administer an i.p. injection of 100 μl Ketofen using the 29-G × ½ in. insulin syringe.

4. Place the mouse in the recovery cage immediately after completing the surgical procedure.

5. Allow the mouse to remain in the recovery cage until it resumes typical mobility before transferring back to its original cage.

Fig. 4 Patient-derived tissue implant. (**a**) Lift the #4 mammary gland using blunt-end forceps and insert the fine tip cross action forceps into the gland. (**b**) Compress the cross action forceps to open a small pocket. (**c**) Have an assistant use the fine tip forceps to grasp a piece of patient-derived tumor tissue and insert the tissue into the pocket you are holding open with the cross action forceps. Remove the cross action forceps first to collapse the pocket and then help the assistant remove the fine tip forceps from the pocket without removing the tissue by placing the cross action forceps between the fine tip forceps

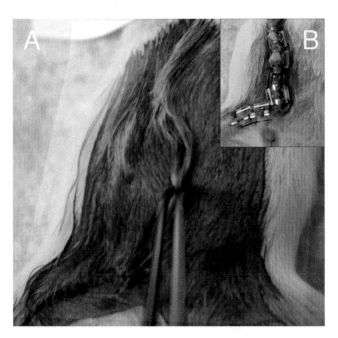

Fig. 5 Closing the incision. (**a**) Use Graefe forceps with teeth to align the skin where the incisions were made. (**b**) Use wound clip applicator and 9 mm wound closures to close the incision. Take care not to place wound closure on genitalia, which would inhibit function

3.7 Harvesting and Preparing Patient-Derived Tumor Tissue for Liquid Nitrogen Storage

1. Sacrifice the mouse that is growing the patient-derived tumor tissue.

2. Perform all of the following steps in a laminar flow hood to maintain sterility of the tissue pieces to be prepared.

3. Place the mouse in a supine position and pin each paw of the mouse to a foam board.

4. Spray the mouse with 70 % ethanol and wipe down the fur.

5. Excise the tumor using sterile dissection scissors.

6. Place the tumor in a sterile petri dish and continue to keep the tissue wet with serum-free DMEM/F12 media.

7. Using micro dissection scissors, trim away the fascia and membranous outer layer (yellow color) surrounding the tumor (Fig. 6a).

8. The tumor tissue will appear white to pink in color once the outer layer is removed (Fig. 6b).

9. Use a scalpel to section the tumor into smaller pieces, taking care to remove any necrotic tissue (*see* **Note 7**).

10. Continue to cut the tissue into smaller sections until tissue pieces are (4×2) mm in size (Fig. 6c, d) (*see* **Note 8**).

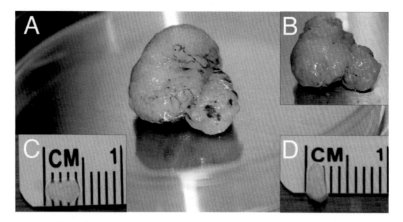

Fig. 6 Preparing patient-derived tumor tissue pieces. (**a**) After the tumor is excised, the membranous outer layer needs to be removed from the outside of the tumor. (**b**) The tumor tissue appears whiter in color once the outer layer is removed. (**c-d**) Using a disposable scalpel, the tumor tissue is cut into smaller pieces that are roughly equivalent in size at (4×2) mm. These tumor pieces are implanted into recipient mice immediately or are frozen with five pieces of tissue incubated in tissue freezing media per cryovial

11. While preparing your tissue pieces, divide them into groups of five and distribute them on a new sterile petri dish (*see* **Note 9**).

12. After all tissue is prepared as (4×2) mm pieces, place five tissue pieces inside a single cryovial containing 1 ml of tissue freezing media.

13. Freeze the tissue pieces by slowly decreasing the temperature at 1 °C/min (*see* **Note 10**).

14. Vials will be stored in the −80 °C freezer overnight before they can be stored in a liquid nitrogen cryogen tank.

4 Notes

1. When placing the tubes in the water bath, make sure that the neck of the cryovial is not submerged in the water. Submerging increases the risk of contamination for the tissue pieces because water could leak into the cryovial.

2. The 5 ml tube is recommended because it makes retrieving the tissue pieces during surgery much easier than using a longer 15 ml conical tube.

3. Use enough wound clips to prevent the estrogen pellet from escaping out of the neck pocket.

4. Take care not to make this incision too close to the genitals. Applying wound clips could inhibit function.

5. When clearing the epithelium from the mammary fat pad, be vigilant to stop any bleeding from incompletely cauterized blood vessels. If bleeding continues, utilize the cautery pen to cauterize the vessel and cotton-tipped applicators to confirm bleeding has completely stopped. Use a low temperature cautery pen and oxygen at 1–2 L/min (rather than a high temperature cautery pen and higher oxygen levels) to decrease fire risk. If you would like to verify that you remove all of the epithelium during the clearing step, then collect the surgically removed tissue for whole mount analysis.

6. The assistant should have the tissue piece grasped in the fine-tip forceps as the pocket is opened with the cross action forceps. As soon as the pocket is created, the assistant can use fine-tip forceps to place a tissue piece into the pocket.

7. Necrotic tissue from patient-derived tumors will appear white in color and is much softer than viable tumor tissue.

8. Keep the tissue moist with serum-free DMEM/F12 during this process.

9. Five tissue pieces are stored in a single cryovial to be used after thawing for transplantation into five recipient mice. After a tube is thawed, do not attempt to refreeze tissue.

10. Freeze the tissue in the appropriate freezing media using a Nalgene freezing container that will precisely decrease the temperature at a rate of 1 °C/min.

Acknowledgements

This work is supported by funding from the Indiana CTSI Young Investigator Award, the Mary Kay Foundation, the Walther Foundation, and St. Joseph's Regional Medical Center.

References

1. Shimosato Y, Kameya T, Nagai K et al (1976) Transplantation of human tumors in nude mice. J Natl Cancer Inst 56:1251–1260

2. Hay M, Thomas DW, Craighead JL et al (2014) Clinical development success rates for investigational drugs. Nat Biotechnol 32:40–51. doi:10.1038/nbt.2786

3. Marangoni E, Vincent-Salomon A, Auger N et al (2007) A new model of patient tumor-derived breast cancer xenografts for preclinical assays. Clin Cancer Res 13:3989–3998. doi:10.1158/1078-0432.CCR-07-0078

4. Zhang X, Claerhout S, Prat A et al (2013) A renewable tissue resource of phenotypically stable, biologically and ethnically diverse, patient-derived human breast cancer xenograft models. Cancer Res 73:4885–4897. doi:10.1158/0008-5472.CAN-12-4081

5. Berger DP, Fiebig HH, Winterhalter BR et al (1990) Preclinical phase II study of ifosfamide in human tumour xenografts in vivo. Cancer Chemother Pharmacol 26:S7–S11. doi:10.1007/BF00685408

6. Jin K, Teng L, Shen Y et al (2010) Patient-derived human tumour tissue xenografts in immunodeficient mice: a systematic review. Clin Transl Oncol 12:473–480. doi:10.1007/s12094-010-0540-6

7. Hidalgo M, Amant F, Biankin AV et al (2014) Patient-derived xenograft models: an emerging platform for translational cancer research. Cancer Discov 4:998–1013. doi:10.1158/2159-8290.CD-14-0001

8. DeRose YS, Wang G, Lin Y-C et al (2011) Tumor grafts derived from women with breast cancer authentically reflect tumor pathology, growth, metastasis and disease outcomes. Nat Med 17:1514–1520. doi:10.1038/nm.2454

Chapter 20

Monitoring Phosphatidic Acid Signaling in Breast Cancer Cells Using Genetically Encoded Biosensors

Maryia Lu, Li Wei Rachel Tay, Jingquan He, and Guangwei Du

Abstract

Phospholipids are important signaling molecules that regulate cell proliferation, death, migration, and metabolism. Many phospholipid signaling cascades are altered in breast cancer. To understand the functions of phospholipid signaling molecules, genetically encoded phospholipid biosensors have been developed to monitor their spatiotemporal dynamics. Compared to other phospholipids, much less is known about the subcellular production and cellular functions of phosphatidic acid (PA), partially due to the lack of a specific and sensitive PA biosensor in the past. This chapter describes the use of a newly developed PA biosensor, PASS, in two applications: regular fluorescent microscopy and fluorescence lifetime imaging microscopy-Förster/fluorescence resonance energy transfer (FLIM-FRET). These protocols can be also used with other phospholipid biosensors.

Key words Phospholipid, Phosphatidic acid, PASS, Cancer, Immunofluorescence, Fluorescence lifetime imaging microscopy, FLIM, Fluorescence resonance energy transfer, FRET

1 Introduction

Phospholipids are not only the building components of cellular membranes but can also act as signaling molecules to regulate cell proliferation, death, migration, and metabolism [1, 2]. Several phospholipids have been demonstrated to function as signaling molecules, e.g., diacylglycerol (DG), phosphatidylinositol 4,5-bisphosphate (PI4,5P$_2$), phosphatidylinostiol 3,4,5-triphosphate (PI3,4,5P$_3$), phosphatidylinositol 3-phosphate (PI3P), and phosphatidic acid (PA) [1, 2]. These phospholipids specifically bind to a variety of effector proteins, and thus modulate their enzymatic activity and subcellular localization, which are critical for normal physiological processes. The alterations of phospholipid-regulated cellular processes directly contribute to human diseases including cancer [1, 3, 4]. For example, the PI3K/PTEN pathway is one of the most mutated oncogenic pathways in breast cancer [1, 5].

Jian Cao (ed.), *Breast Cancer: Methods and Protocols*, Methods in Molecular Biology, vol. 1406,
DOI 10.1007/978-1-4939-3444-7_20, © Springer Science+Business Media New York 2016

Cellular phospholipid levels are tightly controlled by external signals. Activation of lipid metabolizing enzymes, such as phospholipases, lipid kinases, and phosphatases at different times and subcellular locations, can lead to distinct signaling outcomes [2, 6]. Therefore, faithfully monitoring where and when signaling phospholipids are generated is critical for the understanding of their functions. Biochemical methods, such as thin-layer chromatography, high-performance liquid chromatography, and mass spectrometry [7], can only measure total cellular phospholipid levels, and cannot precisely monitor the spatiotemporal production of a particular phospholipid. To circumvent these limitations, genetically encoded phospholipid biosensors have been developed by fusing green fluorescent protein (GFP) or other fluorescent proteins to protein domains or motifs that bind to a specific phospholipid, e.g., the PH domain of PLCδ for $PI4,5P_2$, the PH domain of AKT1 for $PI3,4,5P_3$, the FYVE domain of EEA1 for $PI3P_3$, and the C1 domain of PKCδ for DG [8, 9]. These biosensors have greatly advanced our knowledge of phospholipid signaling and function [2, 8].

Compared to DG and phosphoinositides, much less is known about the subcellular production and cellular functions of PA, partially due to the lack of a specific and sensitive PA biosensor in the past. Signaling PA can be produced by multiple enzymes, including two well-known families of enzymes: phospholipase D (PLD) and DG kinase (DGK) [10–12]. Several PLD and DGK family members have been demonstrated to promote cancer cell proliferation, survival, and migration [4, 12–15]. This chapter describes the use of a recently developed PA biosensor; PA biosensor with superior sensitivity (PASS) [16] in two applications: regular fluorescent microscopy and fluorescence lifetime imaging microscopy-Förster/fluorescence resonance energy transfer (FLIM-FRET). These protocols can be also used for other genetically encoded phospholipid biosensors.

2 Materials

1. PASS constructs: PASS biosensors fused with the monomeric GFP or red fluorescent protein (RFP) [16] are cloned in either a regular mammalian expression vector or a lentiviral vector (Fig. 1). The RFP-PASS-4E is a mutant version of PASS that does not bind to PA, and can be used as a control to determine whether the observed localization of the biosensor reflects lipid binding or interaction with protein molecules, as have been used for phosphoinositide biosensors [8].

2. Lentiviral packaging plasmids: pMDLg/pRRE, pRSV-Rev, and pMD2.g (Addgene).

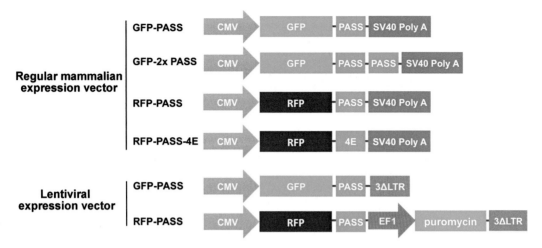

Fig. 1 Schematic maps of GFP- and RFP-PASS plasmids. GFP-PASS and RFP-PASS are cloned in a regular mammalian expression vector derived from pEGFP-C1 (Clontech), or in the lentiviral expression vectors pCDH-CMV-MCS or pCDH-CMV-MCS-EF1-Puro (System Biosciences). 4E is a PA-binding deficient mutant. GFP-2×PASS contains a tandem PASS. The plasmids, maps, and sequences will be distributed upon request

3. TLA-293T cells (Thermo Fisher).

4. MCF-7 cells (American Type Culture Collection, ATCC).

5. HCC1806 cells (American Type Culture Collection, ATCC).

6. Dulbecco's Modified Eagle Medium (DMEM) supplemented with 10 % heat-inactivated fetal bovine serum (FBS) (Hyclone).

7. RPMI-1640 medium supplemented with 10 % FBS (Hyclone).

8. 6- and 12-well cell culture plates (Greiner Bio-One).

9. Opti-MEM medium (Thermo Fisher).

10. Lipofectamine and Plus reagent (Thermo Fisher).

11. Polypropylene falcon tubes.

12. Polybrene stock (10 mg/mL).

13. Glass coverslips, 25 mm, #1 thickness (Thermo Fisher).

14. Microscope slides (Thermo Fisher).

15. Phosphate-buffered saline (PBS) without calcium and magnesium: 8 g of NaCl, 0.2 g of KCl, 1.44 g of Na_2HPO_4, and 0.24 g of KH_2PO_4 in 800 mL of distilled H_2O. Add additional distilled H_2O until the volume is 1 L. Adjust pH to 7.4 with 1 M HCl. Store at room temperature.

16. Epidermal Growth Factor (EGF, AF-100-15) from Peprotech. The stock concentration is 100 µg/mL.

17. 4 % paraformaldehyde (PFA) solution: Dissolve 0.4 g of paraformaldehyde powder in 10 mL of PBS. Heat the solution in a 60 °C water bath until the powder is completely dissolved. Store in a refrigerator and use within 1 week.

18. 0.1 % Triton X-100 in PBS: Dilute from 10 % Triton X-100 using PBS.

19. Blocking solution: 5 % normal goat serum in PBS.

20. Mounting medium: 4 % *n*-Propyl gallate (w/v) in 90 % glycerol (v/v) and 10 % PBS. Aliquot and store in a freezer.

21. Clear nail polish.

22. Straight tip forceps.

23. 50 mM ammonium chloride in PBS: Dilute from 1 M NH_4Cl in PBS stock solution. For 500 mL of 1 M stock solution, dissolve 132.5 mg of NH_4Cl in PBS. Stock solution can be stored at room temperature.

24. Mowiol mounting medium: Combine 2.4 g of Mowiol 4-88, 6 g of glycerol, 6 mL of H_2O, and 12 mL of 0.2 M Tris-Cl (pH 8.5). Store aliquots in a freezer. Mowiol can be stored in a refrigerator for less than 1 month.

25. 35 mm glass bottom dishes, #1 thickness (MatTek).

26. 1 µM fluorescein (Sigma).

3   Methods

3.1 Regular Fluorescent Microscopy

The PASS biosensor was derived from the PA-binding domain of yeast Spo20 (Spo20-PABD) by fusing Spo20-PABD to the nuclear export sequence of protein kinase A alpha (PKI-alpha) [16]. The exclusive cytoplasmic localization of PASS, in contrast to the nuclear localization of the original Spo20-PABD, significantly improves its sensitivity of PA binding.

3.1.1 Transient Transfection

1. One day before transfection, seed MCF-7 or other breast cancer cells into 6-well plates (approximately 2×10^5–5×10^5 cells per well). For optimal transfection efficiency, cells should be about 50–60 % confluent on the day of transfection. Plate cells in a 6- or 12-well plate containing coverslips if cells will be used for imaging (*see* **Note 1**).

2. On the day of transfection, add 1 µg of plasmid DNA into a 1.5 mL microfuge tube.

3. Dilute DNA with 100 µL of Opti-MEM and add 6 µL of Plus reagent. Mix well and incubate for 15 min (*see* **Note 2**).

4. Dilute 4 µL of Lipofectamine reagent in 100 µL of Opti-MEM in a new microfuge tube and mix well. Add this mixture to the tube containing DNA and mix well. Incubate for 15–30 min.

5. During incubation, aspirate media from target cells and add 800 µL of pre-warmed Opti-MEM to each well (*see* **Note 3**).

6. Add the DNA and Lipofectamine mixture onto the target cells. Incubate cells for 4–5 h in a 37 °C incubator with 5 % CO_2.

7. Replace Opti-MEM with DMEM supplemented with 10 % FBS.

8. For transient protein expression, cells can be used for image analysis 24–48 h after transfection.

3.1.2 Generation of Lentivirus and Lentiviral Infection of Target Cells

1. 24 h before transfection, seed TLA-293T cells in a 6-well plate (approximately 0.5×10^6–1×10^6 cells per well). Cells should be 40–50 % confluent on the day of transfection.

2. Transfect cells using Lipofectamine and Plus reagent or other transfection reagents as described in Subheading 3.1.1. Use 1 μg of total DNA for each well (160 ng of PASS lentiviral construct, 400 ng of pMDLg/pRRE, 200 ng of pRSV-Rev, and 240 ng of pMD2.g) (*see* **Note 4**).

3. Harvest lentivirus-containing medium and pool similar stocks if desired (*see* **Note 5**).

4. Transfer medium to a polypropylene tube, and spin at $500 \times g$ for 5 min to pellet remaining packaging cells.

5. Transfer viral supernatant to a polypropylene tube and discard remaining packaging cells.

6. Repeat viral harvesting after 24 h. If virus will not be used immediately, freeze in a –80 °C freezer.

7. Plate HCC1806 cells on coverslips in a 6- or 12-well plate 16–24 h before lentiviral infection. At the time of infection, cells should be around 40–50 % confluent.

8. Add 1–2 mL of viral supernatant to a single well. Add poly-brene to a final concentration of 1 μg/mL (*see* **Note 6**).

9. Remove and discard virus-containing medium and replace with fresh growth medium 8–24 h after infection.

10. Continue to incubate HCC1806 cells for 24–48 h to allow for expression of the sensor (*see* **Note 7**).

11. Plate cells on coverslips in a 12- or 6-well plate for imaging.

12. Cells expressing RFP-PASS can be selected with 0.1–0.2 μg/mL puromycin to generate cells stably expressing RFP-PASS (*see* **Note 8**).

3.1.3 EGF Stimulation of Cells and Slide Preparation

1. Serum starve cells expressing PASS constructs generated from Subheading 3.1.1 or 3.1.2 from 10 % FBS RPMI-1640 to 0.1 % FBS RPMI-1640. Incubate cells in 0.1 % FBS RPMI-1640 overnight (*see* **Note 9**).

2. Stimulate cells with 100 ng/mL of EGF for 0.5–30 min.

3. Wash cells with PBS once or twice to remove serum and phenol red at room temperature. This step is optional (*see* **Note 10**).

4. Fix cells with 4 % paraformaldehyde for 10 min at room temperature.

5. Wash fixed cells with PBS for 5 min, repeat two more three times.

6. Pipette 3–5 μL of anti-fade mounting medium per slide (*see* **Note 11**).

7. Using forceps, gently place the sides of coverslips on a Kimwipes to remove excess liquid.

8. Mount coverslips with cell side down onto the mounting medium.

9. Aspirate excess mounting medium and allow it to air-dry before sealing the coverslips with clear nail polish.

10. Store slides in a refrigerator protected from light.

11. Examine the slides under a regular fluorescent microscope or a laser scanning confocal microscope. A typical image of GFP-PASS before and after EGF stimulation is shown in Fig. 2.

3.1.4 Data Quantification and Analysis

Activation of many cell surface receptors changes the level of several signaling phospholipids including PA on the plasma membrane by activating their metabolizing enzymes [1, 10]. The plasma membrane translocation of PASS biosensor from cytoplasm can be used to examine the increase of PA production. One of the most common ways to evaluate the plasma membrane localization of a phospholipid biosensor is to calculate the ratio of plasma membrane to cytoplasmic intensity using a line-intensity histogram from a selected line spanning the cell with ImageJ or Fiji [8].

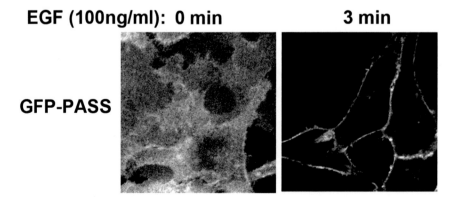

Fig. 2 Plasma membrane production of PA in EGF-stimulated HCC1806 breast cancer cells. HCC1806 cells were infected with lentivirus carrying GFP-PASS. Two days after infection, cells were serum-starved overnight, then stimulated with or without 100 ng/mL EGF for 3 min. Images were captured using a Nikon A1 laser scanning confocal microscope

1. Open the selected image in ImageJ or Fiji.

2. Using the Line Tool, draw a line through the cell, including both the plasma membrane and cytoplasm (*see* **Note 12**).

3. Save a version of this image with an arrow delineating where the line was drawn for data measurements.

4. Pixel intensity along the drawn line can be plotted using "Plot Profile" under the Analyze menu. Pixel distance along the line is plotted on the *X*-axis while pixel intensity is plotted on the *Y*-axis. These data can be saved using List or Copy and analyzed with other software such as Excel or R as desired. The Live option can be selected while drawing lines in different sections of a particular cell before choosing a final cross-section.

5. The histogram generated will resemble peaks with valleys. In the case where the PASS sensor is localized to the plasma membrane, there will be two peaks with a valley in between. The peak values will be used as plasma membrane intensity values. The cytoplasmic intensity value will be the average of the valley values. Background pixel intensity should be subtracted from average plasma membrane and cytoplasmic values.

6. To determine the change of plasma membrane localized PA, the ratio of plasma membrane pixel intensity to cytoplasmic pixel intensity can be calculated. A higher plasma membrane/cytoplasmic pixel intensity ratio can be interpreted as increased PA production on the plasma membrane. The plasma membrane/cytoplasmic pixel intensity ratio will remain unchanged if there is no increase of PA on the plasma membrane.

3.2 FLIM-FRET   Membranes comprise subdomains with different physical and signaling properties. The proper localization of signaling molecules in membrane domains such as lipid rafts determines the activity of many signaling pathways, and disruption of lipid rafts or lipid raft localization of signaling molecules has been implicated in human diseases [17, 18]. Thus, to understand how PA regulates cell functions, it is important to know where PA is generated. The precise site of lipid signaling events can be measured by the production of a specific lipid in close proximity of a membrane marker, e.g., the lipid raft marker tH (the C-terminal tail of H-Ras) or the non-lipid raft marker tK (the C-terminal tail of K-Ras) [19], using FLIM-FRET [20, 21]. FRET is a distance dependent non-radiative transfer of energy from a donor molecule in an excited state to a nearby acceptor molecule in the ground state (within 10 nm). In a FLIM setup, FRET is measured as a decrease in fluorescence lifetime of the donor fluorophore [22]. Some advantages of FLIM-FRET are that FLIM-FRET measurements are more robust and quantitative

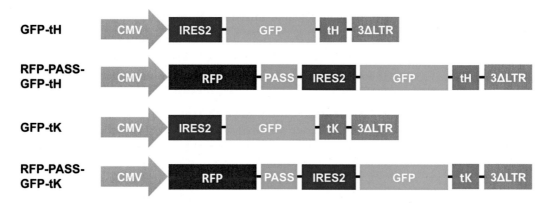

Fig. 3 Schematic maps of lentiviral expression plasmids for FLIM-FRET experiments. These plasmids contain either the donor alone (GFP-tH or GFP-tK), or both donor and acceptor in the same vector (RFP-PASS and GFP-tH, or RFP-PASS and GFP-tK). The use of a modified IRES (internal ribosome entry site), IRES2, in these constructs allows expression of both donor and acceptor in all infected cells, leading to more consistent results. The plasmids, maps, and sequences will be distributed upon request

compared to other methods to measure FRET efficiency and are independent of the expression levels of donors and acceptors [22].

To efficiently express donor–acceptor pairs in breast cancer cells, we have generated lentiviral constructs expressing donor alone (GFP-tK or GFP-tH), or bicistronic constructs expressing donor and acceptor simultaneously (GFP-tK and RFP-PASS, or GFP-tH and RFP-PASS) (Fig. 3).

3.2.1   Lentiviral Infection and Slide Preparation

1. Infect HCC1806 cells with lentivirus carrying the GFP donor, or the GFP donor plus RFP acceptor as described in Subheading 3.1.2.

2. Continue to incubate cells for 48 h to allow for expression of the sensor.

3. Switch culture medium from 10 % FBS RPMI-1640 to 0.1 % FBS RPMI-1640. Incubate cells expressing PASS constructs in 0.1 % FBS RPMI-1640 overnight (*see* **Note 9**).

4. Stimulate cells with 100 ng/mL EGF (0–30 min).

5. Fix cells with 1 mL per well of 4 % paraformaldehyde for 10 min at room temperature in the dark.

6. Wash cells twice with PBS, each time for 5 min.

7. Add 1 mL of 50 mM ammonium chloride per well to quench free aldehyde groups that can produce nonspecific fluorescent background.

8. Incubate for 10 min at room temperature in the dark.

9. Wash four times with PBS, each time for 5 min.

10. Rinse once with ddH_2O.

11. Pipette 3–5 μL of Mowiol mounting medium per slide. Avoid creating bubbles.

12. Using forceps, gently place one side of the coverslips on a Kimwipes to remove excess liquid.

13. Mount coverslips with cell side down on the Mowiol drops. Do not press down on the coverslips.

14. Dry slides in a 37 °C incubator for 1 h or at room temperature overnight. Protect from light.

15. Store slides in a refrigerator protected from light.

3.2.2 Data Acquisition and Analysis

1. Choose the 497 nm light emitting diode (LED) bulb.

2. Select the GFP filter.

3. Turn on the power strip for the computer, LIFA signal generator, and Nikon TiE widefield microscope. In this protocol, the Lambert Instruments FLIM Attachment is installed on a Nikon TiE widefield microscope.

4. Start LIFA-FLIM software.

5. Turn on LED DC and adjust the LED beam to the center (*see* **Note 13**).

6. Select the 60× oil objective. Using a transfer pipette, place a drop of 1 μM Fluorescein on the 35 mm MatTek glass bottom dish. This will be used as a lifetime reference standard, which has a lifetime of 4.0 ns (*see* **Note 14**).

7. Set exposure time at 100–150 ms, MCP at 670–700 V, frequency at 40 MHz, and phase at 330 (*see* **Notes 15** and **16**). Set up your working directory where your files will be saved.

8. Turn on LED DC, adjust focus on bright view through eyepiece or camera, select 3–4 regions of interest (ROIs), activate the FLIM button, and adjust MCP/exposure so that intensity is just below maximal (around 60,000 ADU in Statistics tab).

9. Select Idle, record Reference, and save it.

10. Set up and focus on the sample, making sure all parameters are the same as used to record reference fluorescence lifetime, activate FLIM, and adjust exposure time so that intensity is around 55,000–60,000 ADU. Do not change MCP voltage.

11. Record the FLIM analysis of the samples (*see* **Note 17**). Select ROIs for the cells in the Camera tab, select Lifetime tab and Statistics tab, and export all statistics data. Save images for both Camera and Lifetime views (*see* **Note 18**).

12. Open exported statistics data saved in Excel format and compare lifetime (phase) values (τ) between the cells expressing donor only (GFP) and donor + acceptor (GFP + RFP).

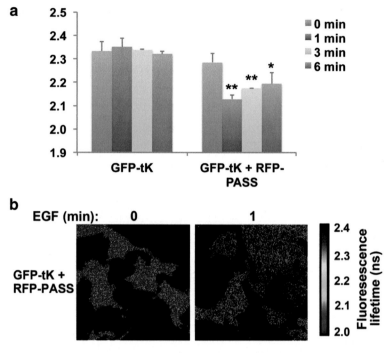

Fig. 4 Spatiotemporal production of PA in a membrane microdomain labeled by tK in EGF-stimulated HCC1806 breast cancer cells. (**a**) Changes in fluorescence lifetime in cells expressing GFP-tK alone or GFP-tK + RFP-PASS. Cells were serum-starved overnight, then stimulated with 100 ng/mL of EGF for the indicated time. Cells were imaged in the frequency domain using a wide-field FLIM-FRET microscope. A minimum of 50 cells was quantified per experiment. $N = 3$. *, $p < 0.05$; **, $p < 0.01$ versus 0 min (no EGF stimulation). (**b**) Representative fluorescence lifetime images

A minimum of 30 ROIs measured per sample group is recommended for statistical analysis of a single experiment. Fluorescence lifetimes for one sample group can be averaged and compared to other sample groups for significant differences using ANOVA. Decreases in GFP fluorescence lifetime in the GFP + RFP group versus the GFP only group reflect FRET. Alternatively, decreases in GFP fluorescence lifetime in the GFP + RFP group after EGF stimulation can be also used to reflect FRET. Using the fluorescent protein-fused biosensors and FLIM-FRET, PA is found to be produced in a membrane microdomain labeled by tK in EGF-stimulated HCC1806 breast cancer cells (Fig. 4).

4 Notes

1. Some breast cancer cell lines, e.g., HCC1806, are very difficult to transfect. If transfection efficiency is lower than 10 %, image quantification will be challenging. To more efficiently

deliver genes of interest, lentiviral infection or other transfection reagents can be used for these cell lines.

2. Both Lipofectamine and Plus reagents tend to adhere to polypropylene microfuge tubes. When adding reagents, pipette directly into the middle of the Opti-MEM medium, avoiding the sides of the tube. Alternatively, polystyrene tubes can be used to decrease the amount of DNA-lipid complexes that bind to the side of the tubes.

3. Aspirate and pipette gently when handling TLA-293T packaging cells as they detach from the plate very easily.

4. Third-generation lentiviral packaging plasmids are described in this protocol, but the second-generation packaging plasmids can also be used for PASS in the lentiviral vector. Extra care should be taken when handling second-generation lentiviral packaging plasmids.

5. BL2 safety practices should be followed when preparing and handling lentiviral particles. Personal protective clothing should be worn at all times. Use plastic pipettes in place of glass pipettes or needles. Liquid waste should be decontaminated with at least 10 % bleach. Laboratory materials that come in contact with viral particles should be treated as biohazardous waste and autoclaved. Please follow all safety guidelines from your institution and from the CDC and NIH for work in a BL2 facility.

6. Polybrene is a polycation that reduces charge repulsion between the virus and the cell membrane. The optimal final concentration of polybrene can be determined with titration experiments, but is usually between 1 and 12 μg/mL. Overexposure to polybrene (>24 h) can be toxic to cells. Certain cell types may be more sensitive to polybrene than the others. Other transduction reagents may be considered for these cell types, such as Retronectin (Clontech).

7. The fluorescent signal of GFP and RFP may be observed 24 h after infection; however, it may be very weak. The maximal fluorescent signal should be observed 48–72 h after infection.

8. Different cell lines have different sensitivities to puromycin. Additionally, the effective concentration of puromycin from different companies can vary. To induce efficient cell death of non-infected cells while reducing undesired stress responses, it is necessary to perform a titration experiment for puromycin concentrations when new cell lines are used.

9. Medium containing 0.1 % FBS can significantly reduce the basal growth factor signaling. Medium without FBS may cause unnecessary stress responses for some cells.

10. Although PBS washes can remove fluorescent background, it may also alter intracellular signaling. To faithfully monitor changes in PA levels, it is preferred to fix cells directly without PBS washes.

11. It is important to choose a correct mounting medium for certain fluorophores. Many commercial anti-fade mounting media do not work well for far-red fluorophores. *n*-propyl gallate is a very general anti-fade mounting medium that is suitable for most commonly used fluorophores.

12. At least three line-intensity histograms should be generated for a single cell to account for variation among cells.

13. The LED should be allowed to warm up for at least 30 min before imaging slides to avoid light fluctuations while acquiring data.

14. A new reference measurement should be determined for every experiment.

15. Since phase and modulation lifetimes are dependent on the excitation power level used, any changes to the power level will render any data generated unusable for comparative analysis. Do not change LED power, MCP voltage, filter sets, or the objective between measurements.

16. Phase settings will vary depending on the FLIM software used for analysis. It is highly recommended to test the FLIM setup and software with a mock experiment before conducting any important experiments.

17. These files can be opened for image analysis later.

18. Lifetime images are pseudocolored based on the timescale range chosen. The scale can be adjusted if needed to include the maximum and minimum fluorescence lifetimes measured.

Acknowledgements

This work was supported by a research grant RP130425 from the Cancer Prevention and Research Institute of Texas (CPRIT) and a research grant R01HL119478 from the National Heart, Lung, and Blood Institute of the National Institutes of Health to GD, and a UTHealth Innovation for Cancer Prevention Research Training Program Predoctoral Fellowship grant RP140103 from the CPRIT to ML. The content is solely the responsibility of the authors and does not necessarily represent the official views of the CPRIT and National Institutes of Health.

References

1. Di Paolo G, De Camilli P (2006) Phosphoinositides in cell regulation and membrane dynamics. Nature 443(7112):651–657

2. Maekawa M, Fairn GD (2014) Molecular probes to visualize the location, organization and dynamics of lipids. J Cell Sci 127(22):4801–4812

3. Wymann MP, Schneiter R (2008) Lipid signalling in disease. Nat Rev Mol Cell Biol 9(2):162–176

4. Park JB, Lee CS, Jang JH, Ghim J, Kim YJ, You S, Hwang D, Suh PG, Ryu SH (2012) Phospholipase signalling networks in cancer. Nat Rev Cancer 12(11):782–792

5. Vogelstein B, Kinzler KW (2004) Cancer genes and the pathways they control. Nat Med 10(8):789–799

6. van Meer G, Voelker DR, Feigenson GW (2008) Membrane lipids: where they are and how they behave. Nat Rev Mol Cell Biol 9(2):112–124

7. Rusten TE, Stenmark H (2006) Analyzing phosphoinositides and their interacting proteins. Nat Methods 3(4):251–258

8. Balla T, Varnai P (2002) Visualizing cellular phosphoinositide pools with GFP-fused protein-modules. Sci STKE 2002(125):pl3

9. Lemmon MA (2008) Membrane recognition by phospholipid-binding domains. Nat Rev Mol Cell Biol 9(2):99–111

10. Zhang Y, Du G (2009) Phosphatidic acid signaling regulation of Ras superfamily of small guanosine triphosphatases. Biochim Biophys Acta 1791(9):850–855

11. Peng X, Frohman MA (2012) Mammalian phospholipase D physiological and pathological roles. Acta Physiol (Oxf) 204(2):219–226

12. Shulga YV, Topham MK, Epand RM (2011) Regulation and functions of diacylglycerol kinases. Chem Rev 111(10):6186–6208

13. Dominguez CL, Floyd DH, Xiao A, Mullins GR, Kefas BA, Xin W, Yacur MN, Abounader R, Lee JK, Wilson GM, Harris TE, Purow BW (2013) Diacylglycerol kinase alpha is a critical signaling node and novel therapeutic target in glioblastoma and other cancers. Cancer Discov 3(7):782–797

14. Bruntz RC, Lindsley CW, Brown HA (2014) Phospholipase D signaling pathways and phosphatidic acid as therapeutic targets in cancer. Pharmacol Rev 66(4):1033–1079

15. Gomez-Cambronero J (2014) Phospholipase D in cell signaling: from a myriad of cell functions to cancer growth and metastasis. J Biol Chem 289(33):22557–22566

16. Zhang F, Wang Z, Lu M, Yonekubo Y, Liang X, Zhang Y, Wu P, Zhou Y, Grinstein S, Hancock JF, Du G (2014) Temporal production of the signaling lipid phosphatidic acid by phospholipase D2 determines the output of extracellular signal-regulated kinase signaling in cancer cells. Mol Cell Biol 34(1):84–95

17. Lasserre R, Guo XJ, Conchonaud F, Hamon Y, Hawchar O, Bernard AM, Soudja SM, Lenne PF, Rigneault H, Olive D, Bismuth G, Nunes JA, Payrastre B, Marguet D, He HT (2008) Raft nanodomains contribute to Akt/PKB plasma membrane recruitment and activation. Nat Chem Biol 4(9):538–547

18. Simons K, Gerl MJ (2010) Revitalizing membrane rafts: new tools and insights. Nat Rev Mol Cell Biol 11(10):688–699

19. Harding AS, Hancock JF (2008) Using plasma membrane nanoclusters to build better signaling circuits. Trends Cell Biol 18(8):364–371

20. Ariotti N, Liang H, Xu Y, Zhang Y, Yonekubo Y, Inder K, Du G, Parton RG, Hancock JF, Plowman SJ (2010) Epidermal growth factor receptor activation remodels the plasma membrane lipid environment to induce nanocluster formation. Mol Cell Biol 30(15):3795–3804

21. Zhou Y, Liang H, Rodkey T, Ariotti N, Parton RG, Hancock JF (2014) Signal integration by lipid-mediated spatial cross talk between Ras nanoclusters. Mol Cell Biol 34(5):862–876

22. Levitt JA, Matthews DR, Ameer-Beg SM, Suhling K (2009) Fluorescence lifetime and polarization-resolved imaging in cell biology. Curr Opin Biotechnol 20(1):28–36

Chapter 21

3D In Vitro Model for Breast Cancer Research Using Magnetic Levitation and Bioprinting Method

Fransisca Leonard and Biana Godin

Abstract

Tumor microenvironment composition and architecture are known as a major factor in orchestrating the tumor growth and its response to various therapies. In this context, in vivo studies are necessary to evaluate the responses. However, while tumor cells can be of human origin, tumor microenvironment in the in vivo models is host-based. On the other hand, in vitro studies in a flat monoculture of tumor cells (the most frequently used in vitro tumor model) are unable to recapitulate the complexity of tumor microenvironment. Three-dimensional (3D) in vitro cell cultures of tumor cells have been proven to be an important experimental tool in understanding mechanisms of tumor growth, response to therapeutics, and transport of nutrients/drugs. We have recently described a novel tool to create 3D co-cultures of tumor cells and cells in the tumor microenvironment. Our method utilizes magnetic manipulation/levitation of the specific ratios of tumor cells and cells in the tumor microenvironment (from human or animal origin) aiding in the formation of tumor spheres with defined cellular composition and density, as quickly as within 24 h. This chapter describes the experimental protocols developed to model the 3D structure of the cancer environment using the above method.

Key words Three-dimensional (3D) culture, 3D bioprinting, 3D cancer cell spheroids, Magnetic nanoparticles, Cytotoxicity, Co-culture

1 Introduction

For development of cancer therapeutics in the laboratory setup or pharmaceutical industry, dose evaluation and optimization are performed in vitro prior to in vivo testing in animals. These in vitro studies are currently conducted in two-dimensional (2D) cell monocultures. However, due to the limitation of planar geometry, the model can only poorly predict in vivo behaviors.

To overcome this problem, 3D in vitro cancer models are being developed emerging as a bridge between in vitro and in vivo models. With their spatial configuration, 3D structures are a more relevant in vitro model with better representation of the cell-to-cell and cell-to-matrix contact in the native microenvironment in vivo, compared to the standard 2D monolayer culture [1]. 3D in vitro

Jian Cao (ed.), *Breast Cancer: Methods and Protocols*, Methods in Molecular Biology, vol. 1406,
DOI 10.1007/978-1-4939-3444-7_21, © Springer Science+Business Media New York 2016

models also enable a more realistic simulation of the transport of nutrient, gas, and signaling molecules between the cells. Growing tumor spheroids with multiple cell types represents better the native microenvironment of the tumor. The 3D spheroid approach has been studied in cancer research and has been proven to represent the complexity of tumor microenvironment in terms of cell–cell interactions and the presence of necrotic and hypoxia regions within the center of the tumors [1, 2].

One of the important elements in the in vitro grown spheroids is the presence of a scaffold that mimics extracellular matrix (ECM) [3, 4], an important element in the tumor stroma [5]. Tumor ECM is in general denser but much less organized than normal ECM [6], and can form a physical barrier for the drug [7] as well as can cause a cell adhesion-mediated drug resistance [8] due to the change of cancer cell activity by binding to ECM [9]. In most of the developed 3D cancer models, synthetic or naturally derived polymers are used as a scaffold [4]. However, the addition of the scaffold may stunt the cell growth and affect the cell-cell interaction, and the static concentration of the scaffold can cause a misrepresentation in the growing in vivo environment with time. In our model, we have cancer spheroids that are grown with incorporation of fibroblasts as a part of the tumor stroma component. Fibroblasts produce fibronectin and collagen which can naturally form the fibrotic capsule in 3D [10]. In our in vitro system, fibronectin concentration can increase along with the growth of the spheroids containing fibroblasts, thus enabling a more realistic prognosis for the response of the tumor in vivo to the tested drugs. In addition to fibroblast, the tumor microenvironment in vivo is composed of cells from various origins, including adipocytes, endothelial cells, and inflammatory cells. Together, these supporting cells and fibers may account for up to 80–90 % of the total tumor volume in various malignancies [5, 11]. Thus, the addition of these other cells in an in vitro model significantly changes cell-cell interactions and signaling pathways within tumors.

In this chapter, we describe the use of magnetic levitation and 3D bioprinting [12, 13] to form 3D cancer cell spheroids [10], which can be designed with various cell types. Depending on tumor types, the lesion can consist of fibroblasts, adipocytes, endothelial cells, as well as immune-competent cells. This model has been utilized previously in breast cancer [10], adipose cells [14], and lung cancer [15] studies, and is currently being studied for microenvironment evaluation in co-culture with immune cells such as macrophages. This spheroid model enables the simulation of in vivo environment without using an artificial scaffold and without the need of external surface for support. Additionally, due to the ability to reach larger diameter in a short time, the 3D model is more accurate in depicting the condition in the in vivo lesions, including the presence of necrotic and hypoxia regions in the center of tumor lesion.

2   Materials

- Nanoshuttle (n3D Biosciences, Houston, TX, USA).
- Cells of interest: may consist of a combination of cancer cells, fibroblasts, myofibroblasts, immune cells, or adipocytes.
- Cell culture medium with and without fetal calf serum (FCS, 10 %).
- RPMI1640 complete medium: RPMI1640 medium, 10 % human serum, 1 % MEM vitamin solution, and 1 % penicillin–streptomycin.
- Trypsin–EDTA (0.25 %).
- Phosphate buffered saline (PBS), pH 7.4.
- Vybrant DiD/DiL/DiO cell-labeling solution (Molecular Probes, Eugene, Oregon, USA).
- T25 and T75 flasks.
- 96-well low attachment plates.
- MagPen™—Teflon pen (n3D Biosciences, Houston, TX, USA).
- Ficoll-Paque premium grade (GE Healthcare Bio-sciences AB, Uppsala, Sweden).
- Ethanol.
- Fixation solution: 4 % Paraformaldehyde in PBS.
- Permeabilizing solution: 0.1–0.2 % Triton X-100 in PBS.
- DAPI nuclear stain: 1 µg/mL solution in PBS.
- Primary and secondary antibodies (based on the desired immunohistochemistry/immunofluorescence analysis and manufacturer's protocol).
- Histogel™ (Richard-Allan Scientific, Kalamazoo, Michigan, USA).
- OCT compound—frozen tissue mounting media (Sakura Finetek, Torrance, CA, USA).
- Prolong Gold®—antifade mounting agent (Life Technologies, Eugene, Oregon, USA).
- Cytoseal™ XYL (Richard-Allan Scientific, Kalamazoo, Michigan, USA).

3   Methods

3.1   Pre-incubation of Cells with Magnetic Nanoparticles

The processes for this section and Subheading 3.2 are summarized schematically in Fig. 1.

1. Make sure that the cells used can grow in the same media.

2. Grow the cells of interest in T25 or T75 flasks (use the standard conditions).

3. Wash cells with PBS by removing medium and adding PBS from the side of the dish without disturbing the cell layer. Rinse the cells by spreading PBS all over the dish. Discard the PBS and repeat this step twice.

4. Trypsinize cells by incubation with 1 mL (T25) or 3.5 mL (T75) trypsin–EDTA for 2–5 min at 37 °C.

5. After the cells are detached from the surface, neutralize trypsin by adding 3 mL (T25) or 10.5 mL (T75) serum-containing medium (*see* **Note 1**). Harvest all the cells into a 15 mL falcon tube and centrifuge according to the procedures for the cells.

6. Remove the supernatant containing trypsin and replenish with a fresh medium. Count the cells and plate them into T25 flask.

7. Culture cells to ~70 % confluence before adding the nanoparticle assembly (*see* **Note 2**).

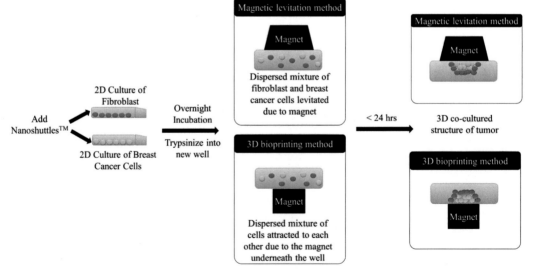

Fig. 1 Schematic presentation of the development of 3D in vitro breast tumor spheroid from a dispersed mixture of fibroblasts (in *blue*) and breast cancer cells (in *green*) after the addition of nanoshuttles. The magnet placed on top or on the bottom of the plate aids the attraction of the nanoshuttle internalized cells to form a 3D tumor mass composed of fibroblasts and breast cancer cells and can be grown for several days. Figure modified from ref. 10

8. Prepare the nanoparticles assembly by removing them from refrigerator and allow to reach room temperature (for about 15 min). Homogenize nanoparticles before using by vigorous vortexing or pipette mixing.

9. Add the nanoparticles assembly to the cells with the concentration of 2–4 μL/cm². Make sure the nanoparticles were distributed evenly throughout the flask by gently tilting the flask back and forth.

10. Incubate the cells with nanoparticle assembly overnight to allow the uptake process.

11. (optional) Cell staining for live cell tracking purposes: add 5 μL Vybrant DiD/DiI/DiO cell-labeling solution for each mL culture medium into the flask containing cells 20 min before cells are to be trypsinized.

3.2 Spheroid Formation

1. Trypsinize the cells incubated overnight with nanoshuttle (as described in Subheading 3.1, **step 3–5**).

2. After centrifugation, remove the supernatant and resuspend cells with fresh medium. (*see* **Note 3**).

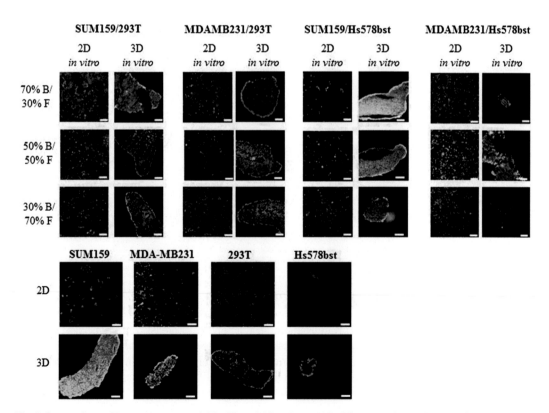

Fig. 2 Comparison of breast tumor model in 2D and 3D culture with different ratios and types of breast cancer cells (in *green*) and different types of fibroblasts (in *red*), grown for 3 days

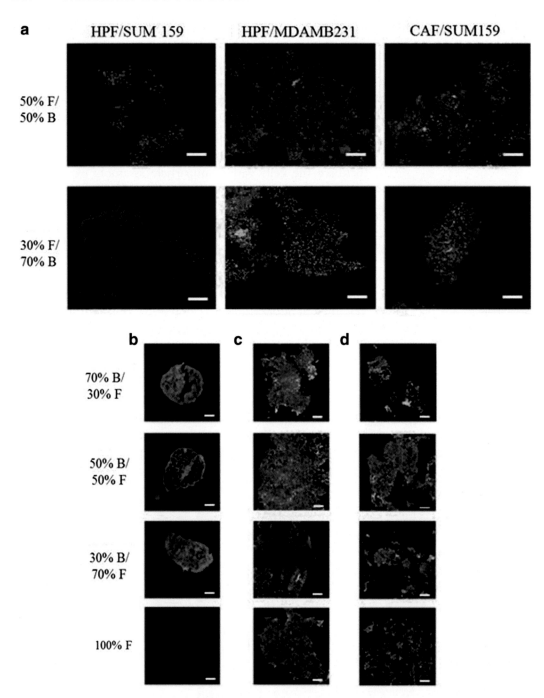

Fig. 3 3D in vitro breast tumor model with different ratios and types of breast cancer cells (in *green*) and primary fibroblasts (in *red*) co-cultured for 3 days using magnetic levitation system. All cells were counter-stained with DAPI (*blue*) for nucleus. (**a**) Confocal microscopy images of the spheres. (**b**–**d**) Fluorescent images of the spheres taken after cryo-sectioning (4 μm slices) (**b**) MDA-MB-231 and Hs578bst (**c**) MDA-MB-231 and HPF (**d**) MDA-MB-231 and CAF. Images were taken with 10× objective magnification, scale bar = 100 μm

3. (optional) For co-culture of different cell types (Figs. 2 and 3): mix the cell types needed in the model at the appropriate ratio (*see* **Note 4**).

4. Count the cell number using hemacytometer or any cell counter instrument.

5. Seed the number of cells needed to grow the spheroids in the appropriate plate (*see* **Notes 5–7**).

6. Place the lid inserts, magnet drive, and well-plate lid (when using magnetic levitation drive), or place the plate directly on top of the 96 well spheroid drive (when using bioprinting kit).

7. Place the cells at 37 °C, 5 % CO_2, cell spheres will start forming after a couple of hours of incubation. Depending on the size of spheres needed, incubate the cells for 1–5 days.

8. Replace the medium every 1–3 days depending on the cell types. This can be done by tilting the plate while keeping the drive magnet underneath the plate. The spheres will be retained by the magnet on the bottom. Aspire the medium carefully without damaging the spheres and replenish the well with fresh medium (*see* **Notes 8 and 9**).

3.3 (Optional) Harvesting Immune Cells from Human/ Mouse Blood Using Gradient Centrifugation Method

1. Obtain buffy coat from blood donation service (human) or harvest blood from mice.

2. Mix buffy coat 1:1 with PBS.

3. Prepare 20 mL of Ficoll on 50 mL falcon tubes.

4. Carefully overlay the buffy coat on top of the Ficoll layer. Make sure the layer is undisturbed.

5. Centrifuge at $1200 \times g$ and 4 °C for 30 min.

6. After centrifugation, the cells will form several layers. Mononuclear cells will form a ring between the Ficoll and the plasma layer. Remove most of the plasma to gain access to the mononuclear cells.

7. Harvest mononuclear cells and combine the cells from two tubes into one 50 mL tube. Fill the tube with fresh PBS.

8. Centrifuge the cells at $300 \times g$ for 7 min. Remove PBS and replenish with the fresh PBS. Repeat the washing step two times.

9. After the last centrifugation, resuspend cells with RPMI1640 complete medium.

10. Plate the cells at $0.5–1.5 \times 10^6$ cells/cm^2 density on the low attachment flasks. Incubate for 1 h at 37 °C and 5 % CO_2.

11. After 1 h incubation, the monocytes and macrophages are attached to the bottom of the flask, while lymphocytes are still in suspension. Remove lymphocytes for further experiments if

needed. Wash monocytes twice with PBS, and replenish with fresh medium and place the cells back into the incubator.

12. After 7 days of incubation, monocytes will be differentiated to macrophages and ready to be used for further experiments.

3.4 Experimental Design for Drug Toxicity Testing with and Without Macrophages

See **Note 10** for sphere handling.

1. Count immune cells needed for the experiment. The ratio of number of macrophages to cancer cells can range from 1:10 to 1:100, depending on the model. Carefully remove the medium on the wells containing spheroids with the magnet drive latching underneath the plate. Add the macrophages to the spheres in a lower volume than needed to allow addition of drugs onto the wells.

2. Prepare the drug in the concentration to be tested for the appropriate volume. Working volume in 96-well plates is between 100 and 200 μL.

3. Incubate spheres with the drugs for the duration of the experiment in the incubator at 37 °C and 5 % CO_2.

4. While in incubation, the growth and changes in spheroids can be followed by inverted microscopy imaging. This method can also be utilized to follow the permeation of drugs into the spheroids (Fig. 4). Hold the plate flat while transferring from incubator to microscope stage (*see* **Note 11**).

3.5 Additional Handling with Teflon Pen for Transferring Spheroids to New Containers

1. All the tools involved in transferring the spheroids need to be sterile or otherwise sterilized by treatment with 70 % (v/v) ethanol. The Teflon part can also be autoclaved. Caution: the rod magnet cannot be autoclaved.

2. Insert the rod magnet to the Teflon pen, handle the assembly either with forceps or by hand. Caution: use plastic forceps, metal forceps will be attracted to the magnet and harder to handle.

3. Remove the magnetic drive from the plate before removing the spheroids.

4. Prepare the new container with a fresh medium or fixing agent where the spheroids will be transferred to before starting the process.

5. Place the Teflon pen facing downwards in the well, reaching for the surface where the spheroids are located. The spheres will be attracted to the magnet rod and attached to the Teflon pen. Lid the Teflon pen and remove the magnet rod from the Teflon pen, the spheres will remain on the pen.

6. Place the Teflon pen downwards to the new container or well where the spheroids are to be transferred to. The spheroids will be transferred to the new container as soon as it touches the medium. Additionally, the magnetic drive plate can be placed underneath the wells to increase the attraction to the bottom

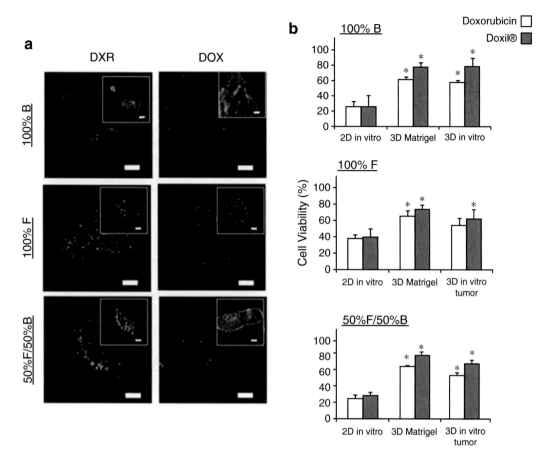

Fig. 4 Distribution and therapeutic efficacy of doxorubicin HCl and Doxil® (doxorubicin liposomes) in 3D in vitro tumors: (**a**) Fluorescent images of 3D in vitro tumors composed of monoculture or co-culture of fibroblast (in *red*) and breast cancer cells (in *green*)—inset images, comparing 72 h treatment with doxorubicin and Doxil ®, *blue*—nucleus and orange—fluorescent emission from doxorubicin, Scale bar = 100 μm, (**b**) Viability assay treated with either doxorubicin or Doxil® (100 nM) for 72 h, comparing on three different in vitro systems: (1) 2D in vitro (grown for 1 day), (2) 3D in vitro (grown for 1 day), and (3) 3D Matrigel™ (grown for 7 days) * = statistically significant difference to 2D in vitro with the same treatment, $n = 4$, $p < 0.05$

of the well and ease the transfer. In case of transferring the spheres to OCT fluid, do not let all the fluid to freeze before the transfer. Leave a fluid area where the spheroids are to be placed, as the spheres would not attach to the frozen area.

7. Close the plate.

3.6 Spheroid Fixation Process for Immunohisto-chemistry, Immunofluorescence or Fluorescence Imaging

1. Follow Subheading 3.1, **step 3** to remove medium and wash with PBS, repeat two times.

2. Add the preferred fixation agents, such as 4 % (w/v) paraformaldehyde, fix the spheres for 10 min at room temperature. (*see* **Note 12**).

3. Wash the culture three times with PBS (repeat Subheading 3.1, **step 3**).

4. The culture can be frozen in the OCT compound for further processing by transferring them to the suitable cassettes (*see* Subheading 3.4), or paraffin embedded for sectioning (Fig. 5). For paraffin embedding, it is recommended to fix the spheroids in Histogel™ before placing them in the cassettes to avoid the loss of sample due to the size. This can be done by heating the Histogel™ to 60 °C and placing the spheroids on the mold for OCT compound. Transfer the spheroids to the Histogel™ (*see* **Note 7** or Subheading 3.4), and overlay the culture with additional thin layer of Histogel™. After the Histogel™ solidified, the sample can be removed from the mold, placed in the cassettes and further processed for sectioning and staining.

5. For immunostaining of intracellular proteins, the cell membrane will need to be permeabilized by adding 0.1 % (v/v) Triton X-100 as fixation solution for 10 min. Then remove the fixing solution and wash with PBS three times (repeat Subheading 3.1, **step 3**). (*see* **Note 12**).

6. Add 1 % BSA as blocking solution and incubate for 5 min at room temperature. Increase the incubation time to 1 h if you are working with the whole spheroid instead of the cell sections.

7. Remove blocking solution, add primary antibody in appropriate concentration and incubate for 30 min at 37 °C, 4 h at room temperature, or overnight at 4 °C (*see* **Note 13**). Protect the cells from light starting from this step if fluorescent antibody is used.

8. Wash the spheroids three times with PBS, repeat Subheading 3.6, **step 7** for secondary antibody if needed. Protect the cells from light.

9. After washing three times with PBS, add nuclear stain, e.g., DAPI (1 μg/mL) for counterstaining. Incubate for 15 min at 37 °C (*see* **Note 14**).

10. Aspirate the DAPI solution and wash culture three times with PBS. The culture can be stored protected from light in PBS for several months at 4 °C.

11. If needed, the spheroids can be transferred to the slides using the Teflon pen or pipet tip (*see* Subheading 3.4).

12. Fix the spheroids with an antifade reagent (e.g., Prolong Gold®), cover the slide with coverslip and seal the sample with mounting media (e.g., CytoSeal™ XYL).

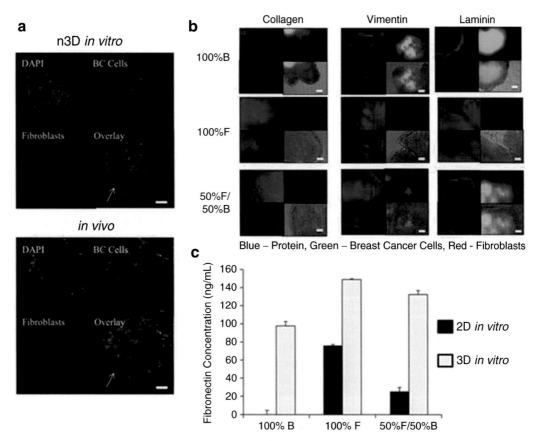

Fig. 5 Characterization of in vitro 3D co-cultures: (**a**) Fluorescent images comparing the phenotype between 3D in vitro co-culture grown in 6-well plates and in vivo tumors composed of breast cancer cells (*green* signal) and fibroblasts (*red* signal) after 7 days growth. *Blue* signal is from DAPI, staining the nucleus. *Yellow* arrows indicate the fibrotic capsule formed in both 3D in vitro and in vivo tumors. Scale bar = 100 μm. (**b**) Collagen, fibronectin, and vimentin (in *blue*) immunofluorescent whole mount staining on 7-day-old 3D in vitro monocultures and cocultures for breast cancer (in *green*) to fibroblast cells (in *red*) grown in 24 well plate, overlayed on bright-field image of tumor, Scale bar = 100 μm (**c**) Fibronectin concentration detected in 7 days grown 2D and 3D monocultures and co-cultures of breast cancer and fibroblasts cells. F = fibroblasts (293T) and B = breast cancer cells (SUM159)

4 Notes

1. For serum-sensitive cells, use trypsin neutralizing solution with medium.

2. Cells can also be plated together with nanoshuttle. Count the cells to reach at least 70 % confluency to optimize the nanoshuttle uptake and avoid overcrowding.

3. The cell pellet will appear brown due to the nanoparticles taken up by the cells. If the brown color is not homogeneous

within the cells, the incubation time should be increased to ensure the optimum nanoshuttle uptake.

4. If the 3D culture consists of different cell types, maintain the spheroids in the cell medium of the most demanding cell type, check the growth curves of other cell types in this growth medium.

5. It is recommended to plate 500–5000 cells in 50–100 μL for each well of 96-well plates or 100,000–300,000 cells in 350 μL for each well of 24-well plates.

6. Use low-attachment plates to avoid cell attachment to the bottom of the well.

7. For plates with magnetic levitation system with magnets on the lid, use 10–20 % less medium than usual to ensure that the medium does not reach the lid, which can cause the attachment of the cells to the lid.

8. Optionally, sterile gel-loading fine pipet tips can be used to avoid having the spheroids suctioned when removing the medium.

9. Remove medium delicately without disrupting the spheroids. If the cells are easily disrupted, then the spheroids may need longer time to form. Incubate the spheroid further before removing the medium.

10. The spheres can be transferred to other plates by using pipette with larger diameter (1 mL tips). In case the size of the spheres is bigger than the pipette diameter or the spheroids fall apart easily, the transfer can be done by using the Teflon pen (*see* Subheading 3.4).

11. For the magnetic levitation magnets, the lid insert is translucent and the spheroids should be easily imaged with the lid insert on. However, if the lid insert disturbs the imaging, the lid and magnetic drive magnet can be removed in sterile environment, leaving the plate only to be covered by the transparent plate lid.

12. Optionally, cells can be fixed with 100 % methanol, ethanol, or isopropyl alcohol, or 100 % acetone for 30 min at 4 °C. No permeabilization is needed for fixation with methanol and acetone, but methanol may compromise the epitope. Permeabilizing agent such as Triton X-100 is harsh to the cells and optionally, milder agents such as Tween 20, saponin, digitonin, and Leucoperm can be used in concentration of 0.2–0.5 % for 30 min. Fixation and permeabilization protocol will need to be optimized based on the cells and further processing required.

13. For immunofluorescence staining: make sure that no dyes in similar excitation and emission wavelength to the dye used in immunofluorescence were added for live cell imaging (**step 9** of Subheading 3.1).

14. DAPI is only recommended for histological slices, since the dye will not penetrate through the whole spheroids for most cell types.

Acknowledgements

The authors would like to acknowledge Nano3D Biosciences Inc for technical support and Susan G. Komen PDF12229449 Award, NIH U54CA143837, and NIH 1U54CA151668-01 for the financial support.

References

1. Vinci M, Gowan S, Boxall F, Patterson L, Zimmermann M, Court W, Lomas C, Mendiola M, Hardisson D, Eccles SA (2012) Advances in establishment and analysis of three-dimensional tumor spheroid-based functional assays for target validation and drug evaluation. BMC Biol 10:29. doi:10.1186/1741-7007-10-29
2. Yamada KM, Cukierman E (2007) Modeling tissue morphogenesis and cancer in 3D. Cell 130(4):601–610. doi:10.1016/j.cell.2007.08.006
3. Lukashev ME, Werb Z (1998) ECM signalling: orchestrating cell behaviour and misbehaviour. Trends Cell Biol 8(11):437–441. doi:10.1016/s0962-8924(98)01362-2
4. Talukdar S, Mandal M, Hutmacher DW, Russell PJ, Soekmadji C, Kundu SC (2011) Engineered silk fibroin protein 3D matrices for in vitro tumor model. Biomaterials 32(8): 2149–2159. doi:10.1016/j.biomaterials.2010.11.052
5. De Kruijf EM, Van Nes JGH, Van De Velde CJH, Putter H, Smit VTHBM, Liefers GJ, Kuppen PJK, Tollenaar RAEM, Mesker WE (2011) Tumor-stroma ratio in the primary tumor is a prognostic factor in early breast cancer patients, especially in triple-negative carcinoma patients. Breast Cancer Res Treat 125(3):687–696. doi:10.1007/s10549-010-0855-6
6. Lu P, Weaver VM, Werb Z (2012) The extracellular matrix: a dynamic niche in cancer progression. J Cell Biol 196(4):395–406. doi:10.1083/jcb.201102147
7. Dittmer J, Leyh B (2015) The impact of tumor stroma on drug response in breast cancer. Semin Cancer Biol 31:3–15. doi:10.1016/j.semcancer.2014.05.006
8. McMillin DW, Negri JM, Mitsiades CS (2013) The role of tumour-stromal interactions in modifying drug response: challenges and opportunities. Nat Rev Drug Discov 12(3): 217–228. doi:10.1038/nrd3870
9. Watt FM, Huck WTS (2013) Role of the extracellular matrix in regulating stem cell fate. Nat Rev Mol Cell Biol 14(8):467–473. doi:10.1038/nrm3620
10. Jaganathan H, Gage J, Leonard F, Srinivasan S, Souza GR, Dave B, Godin B (2014) Three-dimensional in vitro co-culture model of breast tumor using magnetic levitation. Sci Rep 4:6468. doi:10.1038/srep06468
11. Downey CL, Simpkins SA, White J, Holliday DL, Jones JL, Jordan LB, Kulka J, Pollock S, Rajan SS, Thygesen HH, Hanby AM, Speirs V (2014) The prognostic significance of tumour-stroma ratio in oestrogen receptor-positive breast cancer. Br J Cancer 110(7):1744–1747. doi:10.1038/bjc.2014.69
12. Haisler WL, Timm DM, Gage JA, Tseng H, Killian TC, Souza GR (2013) Three-dimensional cell culturing by magnetic levitation. Nat Protoc 8(10):1940–1949. doi:10.1038/nprot.2013.125
13. Souza GR, Molina JR, Raphael RM, Ozawa MG, Stark DJ, Levin CS, Bronk LF, Ananta JS, Mandelin J, Georgescu MM, Bankson JA, Gelovani JG, Killian TC, Arap W, Pasqualini R (2010) Three-dimensional tissue culture based on magnetic cell levitation. Nat Nanotechnol 5(4):291–296. doi:10.1038/nnano.2010.23
14. Daquinag AC, Souza GR, Kolonin MG (2013) Adipose tissue engineering in three-dimensional levitation tissue culture system based on magnetic nanoparticles. Tissue Eng Part C Methods 19(5):336–344
15. Tseng H, Gage JA, Raphael RM, Moore RH, Killian TC, Grande-Allen KJ, Souza GR (2013) Assembly of a three-dimensional multitype bronchiole coculture model using magnetic levitation. Tissue Eng Part C Methods 19(9):665–675. doi:10.1089/ten.tec.2012.0157

Part V

In Vivo Experimental Models for Breast Cancer

Chapter 22

Methods for Analyzing Tumor Angiogenesis in the Chick Chorioallantoic Membrane Model

Jacquelyn J. Ames, Terry Henderson, Lucy Liaw, and Peter C. Brooks

Abstract

Models of tumor angiogenesis have played a critical role in understanding the mechanisms involved in the recruitment of vasculature to the tumor mass, and have also provided a platform for testing antiangiogenic potential of new therapeutics that combat the development of malignant growth. In this regard, the chorioallantoic membrane (CAM) of the developing chick embryo has proven to be an elegant model for investigation of angiogenic processes. Here, we describe methods for effectively utilizing the preestablished vascular network of the chick CAM to investigate and quantify tumor-associated angiogenesis in a breast tumor model.

Key words Angiogenesis, Chick chorioallantoic membrane (CAM), 4T1 tumor cells, Breast tumors, Tumor angiogenesis, Immunofluorescence (IF), von Willebrand factor (vWF), Microcomputed tomography (microCT), Microfil®

1 Introduction

Angiogenesis, the formation of new blood vessels from preexisting vasculature, is well understood to be a critical contributor to tumor growth. A continuously expanding body of work is not only deepening our understanding of angiogenic pathways activated during tumor expansion but has also led to the identification of many molecules that play critical roles in the recruitment of new vessels [1–5]. These findings are rapidly driving the identification of new targets for antiangiogenic therapeutics [6, 7]. Thus, reliable and reproducible preclinical models that mimic multiple steps in new blood vessel formation are necessary for investigation of pathological angiogenic processes.

While in vitro angiogenic assays such as endothelial tube formation and the aortic ring model can provide important cellular and molecular insight [8–10], these models are limited in regard to the complex environments in which angiogenesis arises. To address

Jian Cao (ed.), *Breast Cancer: Methods and Protocols*, Methods in Molecular Biology, vol. 1406,
DOI 10.1007/978-1-4939-3444-7_22, © Springer Science+Business Media New York 2016

these limitations, numerous in vivo models of angiogenesis have been developed, including several angiogenic eye models as well as Matrigel™ plug assays [11, 12]. Tumor angiogenesis occurs in many distinct tissue microenvironments [13, 14] and is tightly regulated by a diverse set of factors, which can be specific to a particular tumor type. Thus, the use of reproducible models in which the angiogenic process can be studied within a readily accessible tumor environment is of significant benefit. The chorioallantoic membrane (CAM) of the developing chicken embryo is a highly advantageous tool to explore the cellular and molecular processes of tumor angiogenesis and test potential inhibitors and inducers of blood vessel formation [15–20]. The CAM serves as the main respiratory organ [21], responsible for gas exchange between the outside of the shell and the developing embryo, and is highly vascularized, rendering it an attractive model for angiogenic studies [22–26]. Because the embryo does not acquire a fully functional immune response until the late stages of embryonic development [27, 28], it serves as a versatile platform for the growth of whole tumor explants [29–31] and various tumor cell lines from distinct species [32, 33]. Many recent studies have utilized the chick CAM for analyzing tumor growth, metastasis, and the angiogenic capacity [34–38] of ovarian, melanoma, and osteosarcoma cancers [39–42].

In this chapter, we discuss quantitative methods for analyzing tumor angiogenesis by not only direct vessel counts, but also by Microfil® polymer perfusion followed by microCT scans. Microfil® perfusion provides information that goes beyond what we can infer from histology-based microvascular density counts alone. Given that the Microfil® perfusion generates casts of tumor-associated vessels capable of blood flow; this procedure provides additional support to confirm that the vessels quantified are indeed functional. Moreover, microCT scans obtained from Microfil® perfusion convey further information pertaining to total tumor vascular density as well as average vessel diameter, which is crucial when investigating the impact of antiangiogenic compounds.

2   Materials

All reagents should be made and stored at room temperature unless indicated otherwise. Prepare all solutions using ultrapure water (purified-deionized water to attain a sensitivity of 18 MΩ cm at 25 °C). Follow appropriate institutional disposal policies and regulations when disposing of all waste materials.

2.1 Embryo Housing and Tissue Culture Components

1. Chicken embryos (Charles River Avian Vaccine Services, North Franklin, CT, USA).

2. Embryo incubator (Brinsea, Titusville, FL, USA).

3. Model drill.

4. Model drill bit and cutting wheel.

5. Biological safety cabinet.

6. Cellulose tape.

7. Cardboard egg holder.

8. Egg candle.

9. Mineral oil.

10. 4T1 breast tumor cells.

11. Tissue culture treated plates.

12. Fetal bovine serum (FBS).

13. Complete growth media for 4T1 cells: RPMI-1640, with 10 % FBS, and final concentration of 1 % sodium pyruvate and 1 % penicillin–streptomycin.

14. 0.05 % trypsin.

2.2 immuno-fluorescence Components

1. Phosphate buffered saline (PBS).

2. Curved-tip forceps.

3. Straight-tip forceps.

4. Small dissecting scissors.

5. Tissue-Tek O.C.T. compound.

6. Tissue-Tek disposable vinyl specimen molds.

7. Cryostat.

8. Fixing solution: 50 % methanol, 50 % acetone.

9. Glass slides.

10. Bovine serum albumin (BSA).

11. Blocking buffer: 2.5 % BSA in PBS. Weigh 0.625 g of BSA and dissolve in 25 μl of PBS.

12. Primary antibody solution: 1:500 dilution of vWF antibody (polyclonal rabbit anti-human von Willebrand Factor, Dako, Carpinteria, CA, USA). 10 μl of vWF antibody in 5 ml of blocking buffer.

13. Secondary antibody solution: 1 μg/ml dilution of Alexa Fluor 594 goat anti-rabbit IgG in 2 ml of blocking buffer.

14. DAPI solution: 1:10,000 dilution of stock. 1 μl of DAPI (5 mg/ml) in 10 ml of blocking buffer.

15. Cover slips.

16. Fluoromount.

17. Clear nail polish.

18. Microscope and imaging software.

2.3 Microfil Injection and MicroCT	1. 30-gauge injection needles.
	2. 1 ml syringes.
	3. 4 % paraformaldehyde.
	4. Absolute ethanol.
	5. Microfil® MV-122.
	6. Scanco VivaCT-40 microCT and software.

3 Methods

All procedures should be conducted at room temperature unless specified otherwise.

3.1 Chicken Embryo and 4T1 Tumor Xenografts

1. Chicken embryos are obtained from Charles River Spafas at 0 days post-fertilization (dpf). Upon delivery, eggs are kept in an incubator at 37 °C with 45–60 % relative humidity, on an interval timed tilt platform that turns every 45 min. The eggs are maintained in this environment until 10 dpf (*see* **Note 1**).

2. All tissue culture work should be conducted in a sterile laminar flow hood. Grow and maintain 4T1 breast tumor cells on tissue culture treated plates with complete medium in a tissue culture incubator at 5 % CO_2, 37.5 °C, and 37 % humidity.

3. At 10 dpf the chorioallantoic membranes (CAMs) of the chicken embryo are well-developed and optimal for tumor studies (Fig. 1). The CAM should be dropped in following steps: First, remove the eggs from incubator and candle, holding the broad end of the egg to the light source to identify the location of the embryonic air sac, determine the optimal drilling position for vascular injection window and identify an area best suited to drop the CAM for tumor placement. (*See* **Note 2**). Mark these locations on the outside of the shell with a pencil.

4. All of the following steps involving the eggs should be performed in a sterilized biological safety cabinet. To separate the CAM from the shell, first sterilize the external surface area of the egg with 70 % ethanol. Begin by drilling a hole in the broad end of the egg where the air sac is located; this will relieve the internal pressure and allow for the generation of a false air sac. Next, drill a small hole through the shell in the area identified for tumor placement during candling, being diligent not to break through the white shell membrane (*see* **Note 3**). Due to the close association of the shell membrane with the CAM, if the shell membrane is compromised during this process it can lead to bleeding and rupturing of the CAM. After both holes have been drilled, holding the broad end of the egg (area of air sac) to the egg candle with the position of the drilled hole on the top side of the egg; lightly press on the exposed shell membrane with curved tip forceps to separate the CAM from the shell (*see* **Note 4**). The gentle

The Chick CAM for Tumor Angiogenesis

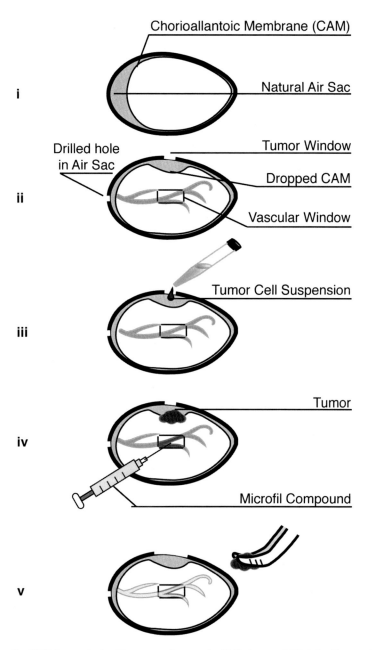

Fig. 1 Utilizing the CAM for analysis of tumor angiogenesis. (*i*) Embryos at 10 dpf with preexisting blood vessels are used for the CAM tumor-angiogenesis assay. (*ii*) Embryos are candled to identify the position of the tumor window and the vascular window (for injection of antagonist and Microfil perfusion). Holes are drilled into the natural embryonic air sac and at the location of the tumor window. The CAM is separated from the shell and a window is cut and removed from above the false air sac. The vascular window is cut, but not removed, and covered with sterile tape. (*iii*) Tumor cells are pipetted on to the dropped CAM, the tumor-window is sealed with sterile tape and the embryos are returned to the incubator. (*iv*) Seven days following tumor cell seeding, the injection window is removed and the Microfil compound is injected at a final volume of 100–500 μl. (*v*) Embryos are incubated for 2 h at 4 °C to cure the polymer. The tumor and associated CAM is removed and fixed with paraformaldehyde for 24 h prior to microCT analysis

pressure from the forceps will dissociate the CAM from the shell and the embryonic sac will fill the area of the natural air sac, creating a false air sac over the area of the CAM for tumor placement (*see* **Note 5**).

5. Once the CAM has separated from the shell, use the cutting wheel attachment to file a square around the drilled hole above the new air sac. A second window should be generated around the blood vessel identified for injections (*see* **Note 6**). Carefully drill through the shell, leaving the shell membrane intact. Do not remove the window until you are ready to proceed with the seeding of tumor cells. Place the eggs back in the humidified incubator.

6. Collect viable 4T1 cells by removing them from tissue culture plates with trypsin. Begin by removing the complete media from plates and washing the cells three times with PBS. Incubate plates with 3 ml trypsin for approximately 10 min (or until cells become liberated from the surface of the plates) at 37 °C. Add 5 ml of basal RPMI-1640 to each plate of trypsin-treated cells and use the pipette to wash any remaining semi-adherent cells off of the plate surface. Once cells are in suspension, transfer all cell suspensions to a sterile 50 ml conical tube and centrifuge for 3 min at $1500 \times g$. Proceed by removing the trypsin-containing supernatant and wash cells three times in PBS. Finally, resuspend cells in basal RPMI-1640 for a final concentration of $2.5–7.5 \times 10^7$ cells/ml (*see* **Note 7**).

7. Remove the eggs from the incubator and carefully remove the shell window; try to avoid shell pieces or dust falling inward onto the CAM (*see* **Note 8**). Next, carefully pipette 40 μl (total of $1–3 \times 10^6$ cells) of the 4T1 cell suspension onto the surface of the CAM. Take care not to make any jarring movements that could disrupt the distribution of the tumor cells on the membrane. Cover the shell window with a piece of sterile cellulose tape and return embryos to the incubator.

8. Tumors are grown for 7 days on the CAM of the developing embryo; during this time the eggs are maintained in a humidified incubator on stationary shelving. Experiments using angiogenic antagonists will need to be optimized for each particular treatment (*see* **Note 9**).

9. Tumors should be collected via careful dissection on 17 dpf and embryos immediately sacrificed in accordance with all applicable institutional IACUC polices and procedures. See Subheading 3.2 for methods necessary for preparation of tissues for immunofluorescence, and skip to Subheading 3.3 for instructions on tissue preparation for angiogenic quantification using microCT scanning.

3.2 Quantification of Tumor Angiogenesis by Microvascular Density Counts

The angiogenic capacity of 4T1 tumors can be quantified by determining the number of vessels positively stained for vWF (von Willebrand Factor) per unit area in the tumor.

1. Carefully remove tumor from the embryo by grasping an area of the CAM near the tumor and cutting the tissue surrounding the tumor with dissecting scissors. Place tumors in a petri dish and rinse once with PBS (Fig. 2). Tumor mass can be determined with a traditional weigh scale.

2. For tumors that will be used for sectioning, transfer to O.C.T.-filled cryo-molds (*see* **Note 10**). Cover any exposing tissue with O.C.T. compound and transfer immediately to dry ice to snap freeze tissue. Once tissue and O.C.T. compound is frozen, store molds at –80 °C.

3. Section frozen tumor molds using a microtome cryostat set to cut at 5 μm, and transfer sections to Superfrost Plus slides.

4. Incubate the slides in fixing solution for 1 min at –20 °C and air-dry.

5. Once the slides are dry they can be used immediately for immunofluorescence staining or stored at –80 °C for future use.

6. Wash the slides for 15 min in PBS.

7. Incubate the slides with blocking buffer for 1 h at 37 °C (*see* **Note 11**). The slides should be washed once in PBS for 5 min on shaker at room temp.

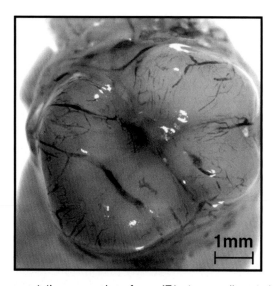

Fig. 2 Representative example of a 4T1 tumor dissected from the CAM. Representative image of a 4T1 tumor dissected from the CAM of a chicken embryo 7 days after seeding (17 dpf) exhibiting recruitment of blood vessels to the tumor mass

8. Incubate the slides with primary (vWF) antibody solution for 1 h at 37 °C. Wash slides three times with PBS for 10 min (each wash) on shaker at room temp.

9. Incubate the slides with secondary antibody solution for 1 h at 37 °C in the dark. Wash the slides three times with PBS for 15 min at room temperature while keeping the slides out of direct light. Follow by incubating the slides with the prepared DAPI solution for 3 min at room temp and perform one last wash in PBS on an orbital shaker for 5 min.

10. Remove the slides from PBS wash and allow to air-dry. Once dry, place one drop of Fluoromount mounting medium on each section of stained tissue. Cover stained sections with glass coverslip, seal with clear fingernail polish and dry completely.

11. Angiogenic vessels within the tumor can be visualized (Fig. 3a) and quantified by positive staining for vWF. An initial (×200) microscopic field is chosen within the section of the tumor. Count the number of vWF-positive vessels in the initial field and repeat for fields directly above, below and to the right and left of the initial field (Fig. 3b). Repeat by moving the field to another area of the tissue section. A total of 15–20 fields should be counted for each tumor.

3.3 Additional Method for Quantification of Tumor Angiogenesis via MicroCT Imaging

Quantitative methods of tumor angiogenesis utilizing corrosion casts have been established for murine models [43–45] and it is possible to apply a similar approach when examining tumor vascularity in the chick CAM model. The information obtained regarding tumor-associated angiogenesis through Microfil® and microCT methods is far more comprehensive than quantitation through immunofluorescence alone. With Microfil® perfusion we are able to gather not only three-dimensional image data of vascular networks but also important aspects of vessel morphology such as average vessel size and total vascular density as related to total tumor volume (Fig. 4).

1. At 17 dpf embryos are prepared for Microfil® injection and tumor harvest.

2. Prepare Microfil® medium mixture according to manufacturers instructions. Between 100 and 500 μl of the compound is required for perfusion of each embryo (*see* **Note 12**).

3. The vascular window drilled at 10 dpf can now be removed and vasculature accessed for polymer injection. Remove embryos from incubator and place in cardboard egg holder. Gently peel the shell window away from the shell membrane using fine tip curved forceps (*see* **Note 13**). Once the window is removed and the opaque shell membrane is visible, gently swab the shell membrane with mineral oil; this will render the membrane translucent and vessels will become visible.

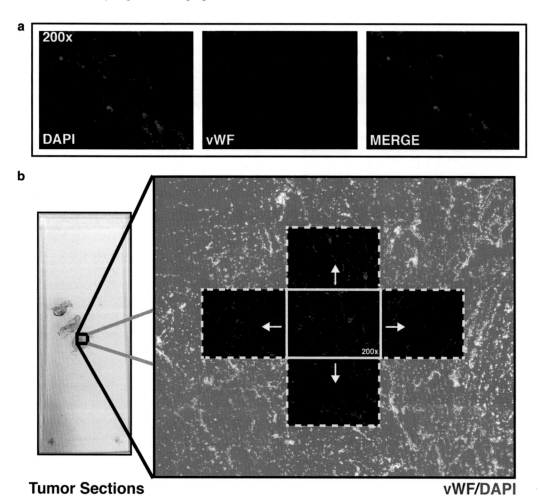

Tumor Sections   **vWF/DAPI**

Fig. 3 Angiogenic vessels within the tumor can be visualized and quantified by positive staining for von Willebrand Factor (vWF). (**a**) Frozen sections (5 μm) of 4T1 tumors are stained with antibodies directed to blood vessels (vWF/*red*) and nuclear material counterstained is with DAPI (*blue*). (**b**) An initial (×200) microscopic field (*center box*) is chosen within the section of the tumor. The number of vWF positive vessels can be determined in the initial field followed by moving the field (*dashed grid boxes*) in four separate directions (*right, left, up,* and *down*), and vessel counts made within each field. A total of 15–20 fields should be counted for each tumor

4. Once the vascular window is removed, inject between 100 and 500 μl of Microfil® into the identified blood vessel (*see* **Note 14**) using a 30-gauge needle (*see* **Note 15**). Adequate perfusion is visually assessed when all blood vessels on the CAM surrounding the tumor are fully saturated with the yellow polymer.

5. Following injections, embryos are placed at 4 °C for 2 h to set the polymer.

6. To remove tumors from the embryo, place egg on an egg holder and carefully widen the window directly over the tumor by

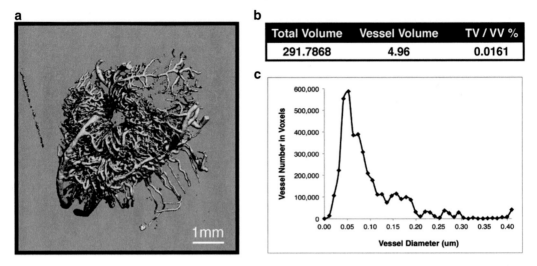

	Total Volume	Vessel Volume	TV / VV %
	291.7868	4.96	0.0161

Fig. 4 Quantification of vascular density in 4T1 xenografts by microCT. (**a**) Example of a microcomputed tomography scan from a Microfil® perfused 4T1 tumor grown on the chick CAM. (**b**) Quantitative data including total scanned volume of the tumor (TV), vessel volume (VV) of the scanned tumor, and the ratio of vessel volume divided by the total scan volume of the tumor. (**c**) Histogram plot showing the number of vessels of a given size within the scanned tumor

peeling back pieces of the shell with forceps. Next, carefully grasp the CAM in an area near the tumor with a pair of forceps, and use dissecting scissors to gently cut through the CAM tissue in a circular shape around the tumor. Transfer tumors to a 4 % paraformaldehyde solution for fixation and store at 4 °C for 24 h. After 24 h, formalin-fixed tumor tissue can be transferred to 70 % ethanol and stored at 4 °C until microCT scanning.

7. Perfused and fixed tumors should be scanned using a microCT imaging unit. Scan tumors at 10.5 μm resolution with a voltage of 55 kVp and a current of 145 μA. Resolution should be set to high (2048 × 2048 pixel image matrix) for image acquisition. Tomograms should be globally thresholded based on X-ray attenuation and used to render binarized 3D images of the tumor.

4   Notes

1. Diligence in maintaining proper conditions (temperature and humidity) is crucial to the viability and vascular development of the embryos. During the 10-day incubation period prior to xenograft implantation, if embryos are maintained in the same position the CAM will develop preferentially on one side of the embryo. Rotating the developing embryos 180° every 48 h facilitates development of the CAM on all sides of the egg

increasing the workable area of the membrane. At time points later than 10 dpf the CAM becomes more tightly associated with the shell membrane making the procedure of dropping the CAM and creating a false air sac increasingly difficult; younger embryos do not have CAMs developed enough for the assay. Plan to use 8–10 embryos per treatment group. It is possible that not all embryos received will fully develop and some may be damaged during the process of dropping the CAM. Be sure to take this into consideration when ordering embryos.

2. When candling the embryos, first mark the air sac and then identify a CAM-associated blood vessel to be utilized for injections. Small- to medium-sized vessels that transverse laterally across the egg and branch away from the natural air sac work best for agonist/antagonist/Microfil® injections. Holding the egg at the candle draw a box around the chosen blood vessel; with the vascular injection window facing you and mark the top of the egg. This will be where the CAM is dropped and the false air sac is generated. It is also important not to keep the egg on the candle for too long during this and future injection steps as the high temperature of the light can have deleterious effects on the developing embryo.

3. While drilling holes for dropping the CAM it works best to process no more than 5–6 embryos at a time. Exposure of the shell membrane/CAM for long periods will cause the tissue to undergo excessive drying rendering it difficult to separate from the shell membrane. Do not wipe the exposed shell membrane with ethanol, as this will also excessively dry the tissue.

4. When dropping the CAM, it is sometimes beneficial to gently rock the egg or softly tap the shell around the drilled hole assisting the CAM to pull away from the shell membrane.

5. Successful CAM dropping will result in a false air sac in which you lose visibility of the CAM-associated blood vessels. If an air sac or bubble is observed while blood vessels are still associated with the shell membrane, the CAM has been ruptured and the embryo cannot be used for the following tumor studies.

6. When creating the two shell windows it is helpful to drill four sides of the box and leave the corners of the shell intact cover both windows with sterile cellulose tape to prevent drying until windows are removed.

7. Be sure to collect enough cells for 8–10 embryos per treatment group. It is critical to have a uniform suspension of cells to facilitate equivalent numbers of cells grafted to each embryo. Pipetting up and down can help to create a uniform suspension,

but passing cells through an injection needle is not recommended as it is probable that cells will be lysed and not viable.

8. When removing the window from the dropped CAM, dust or larger shell pieces that fall onto the CAM can potentiate an inflammatory response and skew long-term results of the study. Shell fragments in the CAM also contribute to difficulty during the sectioning of frozen blocks by damaging microtome blades.

9. If angiogenic antagonists are to be tested, concentrations and dosing schedules should be optimized for each treatment. The amount of antagonist to be delivered and the duration of the treatment are dependent upon the specific biochemistry of the antagonist to be tested and should be determined experimentally. Systemic intravenous injections can be performed by diluting the antagonist in PBS (total volume of 100 μl) and injecting into the CAM-associated vasculature identified within the vascular window. Antagonist can also be administered topically via dilution in 50 μl of PBS by pipetting through the window above the dropped CAM and directly onto the tumor.

10. Placing tumors on edge in the O.C.T. cryo-mold will result in sections that are easily oriented through identification of the CAM.

11. Volumes of blocking buffer and antibody solutions will vary based on the size of the tissue section and the occupied area on the slide; it is imperative that all tissue area within the marked boundaries is sufficiently covered in the solutions to ensure even and adequate staining.

12. To optimize injection efficiency, mix only enough Microfil compound for injection of 3–4 embryos at a time. Once the Microfil® components have been mixed the compound will slowly begin to polymerize and extended periods of time will result in curing of the polymer leading to difficulty during injections.

13. Precautions when removing the vascular window are critical as the shell can be tightly associated with the shell membrane and CAM. Carefully remove the tape without pulling on the window and swab the window and surrounding area with mineral oil to soften the CAM before removing the window with forceps.

14. If injection through the vascular window proves challenging it is also possible to inject the Microfil® compound via blood vessels on the top of the dropped CAM. If injection from the top of the CAM is performed, widen the window above the tumor by peeling back the shell with forceps. Be sure to identify vessels

associated with the CAM as blood vessels of the embryo below the CAM are difficult to inject.

15. Adequate perfusion can be assessed by observation of the yellow Microfil® polymer replacing the embryonic blood in CAM vasculature surrounding the tumor.

Acknowledgments

This work was supported in part by a grant from the Maine Cancer Foundation (MCF) to PCB, NIH Center for Biomedical Research Excellence P30GM103392 (P.I. Robert Friesel), and institutional support from the Maine Medical Center.

References

1. Bader AG, Kang S, Vogt PK (2006) Cancer-specific mutations in PIK3CA are oncogenic in vivo. Proc Natl Acad Sci U S A 103(5): 1475–1479
2. Eliceiri BP, Klemke R, Stromblad S, Cheresh DA (1998) Integrin avb3 requirement for sustained mitogen-activated protein kinase activity during angiogenesis. J Cell Biol 140(5): 1255–1263
3. Hanahan D, Folkman J (1996) Patterns and emerging mechanisms of the angiogenic switch during tumorigenesis. Cell 86:353–364
4. Liu M, Scanlon CS, Banerjee R, Russo N, Inglehart RC, Willis AL, Weiss SJ, D'Silva NJ (2013) The histone methyltransferase EZH2 mediates tumor progression on the chick chorioallantoic membrane assay, a novel model of head and neck squamous cell carcinoma. Transl Oncol 6(3):273–281
5. Zijlstra A, Seandel M, Kupriyanova TA, Partridge JJ, Madsen MA, Hahn-Dantona EA, Quigley JP, Deryugina EI (2006) Proangiogenic role of neutrophil-like inflammatory heterophils during neovascularization induced by growth factors and human tumor cells. Blood 107(1):317–327
6. Debefve E, Pegaz B, Ballini JP, van den Bergh H (2009) Combination therapy using verteporfin and ranibizumab; optimizing the timing in the CAM model. Photochem Photobiol 85(6):1400–1408
7. Palacios-Arreola MI, Nava-Castro KE, Castro JI, Garcia-Zepeda E, Carrero JC, Morales-Montor J (2014) The role of chemokines in breast cancer pathology and its possible use as therapeutic targets. J Immunol Res 2014:849720
8. Ucuzian AA, Greisler HP (2007) In vitro models of angiogenesis. World J Surg 31(4):654–663
9. Bai Y, Zhao M, Zhang C, Li S, Qi Y, Wang B, Huang L, Li X (2014) Anti-angiogenic effects of a mutant endostatin: a new prospect for treating retinal and choroidal neovascularization. PLoS One 9(11):e112448
10. Nicosia RF (2009) The aortic ring model of angiogenesis: a quarter century of search and discovery. J Cell Mol Med 13(10):4113–4136
11. Couffinhal T, Dufourcq P, Barandon L, Leroux L, Duplaa C (2009) Mouse models to study angiogenesis in the context of cardiovascular diseases. Front Biosci 14:3310–3325
12. Norrby K (2006) In vivo models of angiogenesis. J Cell Mol Med 10(3):588–612
13. Tredan O, Lacroix-Triki M, Guiu S, Mouret-Reynier MA, Barriere J, Bidard FC, Braccini AL, Mir O, Villanueva C, Barthelemy P (2015) Angiogenesis and tumor microenvironment: bevacizumab in the breast cancer model. Target Oncol 10:189
14. Mittal K, Ebos J, Rini B (2014) Angiogenesis and the tumor microenvironment: vascular endothelial growth factor and beyond. Semin Oncol 41(2):235–251
15. Armstrong PB, Quigley JP, Sidebottom E (1982) Transepithelial invasion and intramesenchymal infiltration of the chick embryo chorioallantois by tumor cell lines. Cancer Res 42(5):1826–1837
16. Kain KH, Miller JW, Jones-Paris CR, Thomason RT, Lewis JD, Bader DM, Barnett JV, Zijlstra A (2014) The chick embryo as an expanding experimental model for cancer and cardiovascular research. Dev Dyn 243(2): 216–228
17. Nguyen M, Shing Y, Folkman J (1994) Quantitation of angiogenesis and antiangiogenesis in the chick embryo chorioallantoic membrane. Microvasc Res 47(1):31–40

18. Nowak-Sliwinska P, Segura T, Iruela-Arispe ML (2014) The chicken chorioallantoic membrane model in biology, medicine and bioengineering. Angiogenesis 17(4):779–804

19. Ribatti D (2012) Chicken chorioallantoic membrane angiogenesis model. Methods Mol Biol 843:47–57

20. Ribatti D (2014) The chick embryo chorioallantoic membrane as a model for tumor biology. Exp Cell Res 328:314

21. Ausprunk DH, Knighton DR, Folkman J (1974) Differentiation of vascular endothelium in the chick chorioallantois: a structural and autoradiographic study. Dev Biol 38:237–247

22. Borges J, Tegtmeier FT, Pardon NT, Mueller MC, Lang EM, Stark B (2003) Chorioallantoic membrane angiogenesis model for tissue engineering: a new twist on a classic model. Tissue Eng 19(3):441–450

23. Brooks PC, Montgomery AM, Cheresh DA (1999) Use of the 10-day-old chick embryo model for studying angiogenesis. Methods Mol Biol 129:257–269

24. Ribatti D, Nico B, Vacca A, Roncali L, Burri PH, Djonov V (2001) Chorioallantoic membrane capillary bed: a useful target for studying angiogenesis and anti-angiogenesis in vivo. Anat Rec 264(4):317–324

25. Tay SL, Heng PW, Chan LW (2012) The CAM-LDPI method: a novel platform for the assessment of drug absorption. J Pharm Pharmacol 64(4):517–529

26. Tay SL, Heng PW, Chan LW (2012) The chick chorioallantoic membrane imaging method as a platform to evaluate vasoactivity and assess irritancy of compounds. J Pharm Pharmacol 64(8):1128–1137

27. Janković BD, Isaković K, Lukić ML, Vujanović NL, Petrović S, Marković BM (1975) Immunological capacity of the chicken embryo. Immunology 29(3):497–508

28. Murphy JB (1914) Factors of resistance to heteroplastic tissue-grafting: studies in tissue specificity III. J Exp Med 19:513–522

29. Murphy JB (1914) The ultimate fate of mammalian tissue implanted in the chick embryo II. J Exp Med 19:181–186

30. Ausprunk D, Knighton DR, Folkman J (1975) Vascularization of normal and neoplastic tissues grafted to the chick chorioallantois. Am J Pathol 79:597–618

31. Dagg CP, Karnofsky DA, Roddy J (1956) Growth of transplantable human tumors in the chick embryo and hatched chick. Cancer Res 16(7):589–594

32. Deryugina EI, Quigley JP (2008) Chick embryo chorioallantoic membrane model systems to study and visualize human tumor cell metastasis. Histochem Cell Biol 130(6):1119–1130

33. Easty GC, Easty DM, Tchao R (1969) The growth of heterologous tumour cells in chick embryos. Eur J Cancer 5:287–295

34. Fein MR, Egeblad M (2013) Caught in the act: revealing the metastatic process by live imaging. Dis Model Mech 6(3):580–593

35. Fergelot P, Bernhard JC, Soulet F, Kilarski WW, Leon C, Courtois N, Deminiere C, Herber JM, Antczak P, Falciani F, Rioux-Leclercq N, Patard JJ, Ferriere JM, Ravaud A, Hagedorn M, Bikfalvi A (2013) The experimental renal cell carcinoma model in the chick embryo. Angiogenesis 16(1):181–194

36. Palmer TD, Lewis J, Zijlstra A (2011) Quantitative analysis of cancer metastasis using an avian embryo model. J Vis Exp (51): pii: 2815

37. van Beijnum JR, Nowak-Sliwinska P, van den Boezem E, Hautvast P, Buurman WA, Griffioen AW (2013) Tumor angiogenesis is enforced by autocrine regulation of high-mobility group box 1. Oncogene 32(3):363–374

38. Zijlstra A, Lewis J, Degryse B, Stuhlmann H, Quigley JP (2008) The inhibition of tumor cell intravasation and subsequent metastasis via regulation of in vivo tumor cell motility by the tetraspanin CD151. Cancer Cell 13(3):221–234

39. Isachenko V, Isachenko E, Mallmann P, Rahimi G (2013) Increasing follicular and stromal cell proliferation in cryopreserved human ovarian tissue after long-term precooling prior to freezing: in vitro versus chorioallantoic membrane (CAM) xenotransplantation. Cell Transplant 22(11):2053–2061

40. Lokman NA, Elder AS, Ricciardelli C, Oehler MK (2012) Chick chorioallantoic membrane (CAM) assay as an in vivo model to study the effect of newly identified molecules on ovarian cancer invasion and metastasis. Int J Mol Sci 13(8):9959–9970

41. Lopez-Rivera E, Jayaraman P, Parikh F, Davies MA, Ekmekcioglu S, Izadmehr S, Milton DR, Chipuk JE, Grimm EA, Estrada Y, Aquirre-Ghiso J, Sikora AG (2014) Inducible nitric oxide synthase drives mTOR pathway activation and proliferation of human melanoma by

reversible nitrosylation of TSC2. Cancer Res 74(4):1067–1078

42. Mu X, Sultankulov B, Agarwal R, Mahjoub A, Schott T, Greco N, Huard J, Weiss K (2014) Chick embryo extract demethylates tumor suppressor genes in osteosarcoma cells. Clin Orthop Relat Res 472(3):865–873

43. Contois LW, Akalu A, Caron JM, Tweedie E, Cretu A, Henderson T, Liaw L, Friese IR, Vary C, Brooks PC (2015) Inhibition of tumor-associated αvβ3 integrin regulates the angiogenic switch by enhancing expression of IGFBP-4 leading to reduced melanoma growth and angiogenesis *in vivo*. Angiogenesis 18:31

44. Downey CM, Singla AK, Villemaire ML, Buie HR, Boyd SK, Jirik FR (2012) Quantitative ex-vivo micro-computed tomographic imaging of blood vessels and necrotic regions within tumors. PLoS One 7(7):e41685

45. Mondy WL, Cameron D, Timmermans JP, De Clerck N, Sasov AC, Casteleyn C, Piegl LA (2009) Micro-CT or corrosion casts for use in the computer-aided design of microvasculature. Tissue Eng Part C Methods 15:729–738

Chapter 23

Pharmacokinetics and Pharmacodynamics in Breast Cancer Animal Models

Wei Wang, Subhasree Nag, and Ruiwen Zhang

Abstract

The study of pharmacokinetics (PK) and pharmacodynamics (PD) in cancer drug discovery and development is often paired and described in reciprocal terms, where PK is the analysis of the change in drug concentration with time and PD is the analysis of the biological effects of the drug at various concentrations over different time courses. While PK is defined by how a compound is absorbed, distributed, metabolized, and eliminated, PD refers to the measure of a compound's ability to interact with its intended target, leading to a biologic effect. Recent advances in anti-breast cancer drug discovery have resulted in several new drugs, but there is still a high attrition rate during clinical development. One reason for this failure is attributed to inappropriate correlation between the PK and PD parameters and subsequent extrapolation to human subjects. In this chapter, we describe the protocols of PK and PD studies in breast cancer models to assess the efficacy of an anti-breast cancer compound, noting the types and endpoints employed, and explain why it is important to link PK and PD in order to establish and evaluate dose/concentration-response relationships and subsequently describe and predict the effect-time courses for a given drug dose.

Key words Pharmacokinetics, Pharmacodynamics, Breast cancer, Animal models, PK/PD modeling

1 Introduction

Biological response of a drug is often related to its concentration at the site of action over the course of time [1]. Thus, pharmacokinetics (PK) which studies the time course of drug's fate in the body and pharmacodynamics (PD) which studies the relationship between drug concentration and biological effect become the essential components of the modern drug discovery process and crucial towards successful clinical transition of a new drug. Breast cancer, one of the most common causes of cancer related mortalities in women worldwide, has been the focus of anticancer drug discovery over the years [2, 3]. Despite considerable diversity in molecular profile, histopathology, and treatment outcome, advances in genetics and molecular/cell biology

Jian Cao (ed.), *Breast Cancer: Methods and Protocols*, Methods in Molecular Biology, vol. 1406,
DOI 10.1007/978-1-4939-3444-7_23, © Springer Science+Business Media New York 2016

have greatly extended our knowledge about the molecular pathogenesis of breast cancer, and subsequently unlocked new avenues for developing novel targeted therapies through the identification of potential therapeutic targets involved in intracellular signaling processes responsible for malignant transformation and progression [4–7]. However, anti-breast cancer drug discovery and development is associated with high attrition rates, largely, due to a lack of efficacy and unexpected toxicity of new drugs in the clinical development. Generation of high-quality, predictive PK and PD information in the early phase of the drug development process could help reduce the high attrition rates in the discovery and development pipeline [8–10].

1.1 Pharmacokinetic Considerations

New drug development is highly dependent on animal studies to provide a framework for human clinical trials [11, 12]. Current cancer drug development paradigm selects starting doses for phase I clinical trials based on prior data obtained from pharmacokinetic and toxicology studies in animals using allometric scaling based on body surface area conversions [10, 11]. Though plasma PK properties have been successfully related to pharmacological effects, the heterogeneous nature of tumor tissue often impedes drug delivery from plasma to tumor, leading to complex relationships between concentrations in plasma, tumor interstitium, and tumor cells [13]. Therefore, meaningful PK data for an anticancer drug must include tumor permeability and tumor uptake data, either in vitro or in vivo [13, 14].

Prior to advancing to in vivo testing, a number of parameters are assessed that help predict the potential of the drug candidate to reach its intended target in vivo [15, 16]. These include, but are not limited to, physicochemical characterization of the drug (i.e., determinations of molecular weight, solubility, lipophilicity), extent of plasma protein binding, interactions with cytochrome P_{450} (CYP) isoenzymes (both induction and inhibition), ability to migrate across a cell barrier (Caco-2, MDCK cell lines, and/or artificial lipid membranes are often used to address potential for absorption across the gut and blood–brain barrier), and stability (measured either as half-life of a compound mixed with human or rodent plasma and in solutions of varying pH simulating stomach or GI environments, or when incubated with liver microsomes or hepatocytes as an early metabolic stability readout) [15, 16]. An exhaustive in vivo PK profile that describes tissue distribution and accumulation over a time course is usually established before compounds undergo extensive efficacy testing in animal models [16–18].

1.2 Pharmacodynamic Considerations

A successful and meaningful PD analysis involves the measurement of biomarkers or clinical endpoints that can be quantitatively related to drug concentration in the body. In PD studies for anticancer drugs, often, four different types of outcomes are measured either, alone or in combination [19, 20]: (a) decrease in tumor size or tumor regression (this is the most frequent measure of drug efficacy); (b) modification in expression level of a biomarker which correlates with disease progression and/or survival, e.g., changes in estrogen receptor (ER) and progesterone receptor (PgR) for endocrine therapy, changes in Ki-67 levels, a cellular proliferation marker, for assessing therapeutic efficiency [21, 22]; (c) an outcome predicted based on the molecular target of the drug (e.g., an increase in apoptosis assessed by Terminal deoxynucleotidyl transferase dUTP nick end labeling (TUNEL) staining with a proapoptotic drug, or inhibition of target oncogene levels); and (d) in some cases, toxicity as a result of therapy can also be used as a clinical end-point with higher drug exposure being associated with antitumor activity and greater toxicity [10]. The concentration–response relationships that lead to the estimation of these PD parameters are both useful in preclinical drug development and in the clinic trials. Various in vitro cellular and cell free assays are often used to identify and validate molecular targets [23–31]. For further testing and confidence as a potential anti-breast cancer agent, the PD studies often include testing in animal models for target validation and therapeutic response, in relation to drug concentration in vivo.

2 Materials

Analytic grade and/or HPLC grade reagents should be used for preparation of all solutions and buffers, using ultrapure double distilled water. Prepared solutions and buffers should be filtered and stored as indicated. All cell culture equipment and biological hood and culture media should be sterile or sterilized prior to use. Gloves must be used for all procedures. Animals should be housed and treated according to regulatory guidelines and all animal study protocols must be reviewed and approved by Institutional Animal Care and Use Committee (IACUC).

2.1 Uptake and Cellular Transport Experiments

1. 60 or 100 mm dishes.

2. Cancer cell lines.

3. MDCK cell line.

4. Caco-2 cell line.

5. Cell culture medium as specified by ATCC for particular cell line.

6. FBS.

7. Penicillin–streptomycin.

8. Sterile PBS.

9. Trypsin–EDTA solution.

10. Hemocytometer for cell counting.

11. Pasteur pipettes.

12. Stop solution (200 mM KCl and 2 mM HEPES).

13. Transwell diffusion chamber.

2.2 S9 Metabolism Studies

1. 37 °C shaking water bath.

2. 25 mL round bottom glass test tube.

3. 100 mM Tris-HCl buffer (pH 7.4).

4. Methanol, UPLC grade.

5. S9 (Celsis In Vitro Technologies): CD-1 mouse S9; human S9; and Fisher 344 rat S9. All S9 fractions are in 20 mg/mL protein basis.

6. NRS-I: 2 % $NaHCO_3$; 1.7 mg/mL NADP, 7.8 mg/mL glucose-6-phosphate, 6 units/mL glucose-6-phosphate dehydrogenase.

7. NRS-II: 2 % $NaHCO_3$; 1.9 mg/mL UDPGA; 100 μg/mL PAPS.

2.3 Protein Binding Assay

1. Ultrafiltration devices (Amicon Centrifree-Amicon YM-300 filter system, Millipore Corp., Bedford, MA).

2. Water bath.

3. Centrifuge.

4. Purified mouse plasma.

5. α1-acid glycoprotein (AGP).

6. Human serum albumin (HSA).

2.4 Cytochrome P450 Enzyme Induction/Inhibition Assay

1. Fresh human hepatocytes.

2. Test compound at different concentrations.

3. CYP isoforms: CYP1A2; CYP2B6; and CYP3A4.

4. Negative control: vehicle (typically 0.1 % DMSO).

5. Positive control reagents: 3-methylchloranthrene (CYP1A2); omeprazole (CYP1A2); phenobarbital (CYP2B6); dexamethasone and rifampicin (CYP3A4).

6. Probe substrates: ethoxyresorufin (CYP1A2); bupropion (CYP2B6); midazolam (CYP3A4).

2.5 In Vitro Stability Testing

1. Mouse plasma.

2. Human plasma.

3. Water bath.

2.6 Establishment of Animal Tumor Models

1. Cell culture related materials.

2. Culture medium without FBS or antibiotics.

3. 70 % ethanol.

4. 0.9 % NaCl (physiological saline, sterile).

5. Drug preparations in appropriate vehicle.

6. 1 mL syringe with 27–30 gauge needles.

7. Sterilized animal feeding needles, 18/50 mm, curved tube style.

8. Matrigel® basement membrane matrix (Becton Dickinson Labware, Bedford, MA).

9. Xenolight RediJect d-Luciferin (850 μL of 30 mg/mL in a vial; PerkinElmer, Waltham, CA).

10. Caliper IVIS bioluminescent imaging system (Caliper, Mountain View, CA).

11. A small balance for body weight measurement.

12. Calipers for measuring tumor diameters.

13. Ear tags and ear puncher for identification of animals.

14. Disinfectant.

15. Sterile gauze.

16. Solution for sample preparation for pathology analysis: add 735 mL of absolute alcohol (100 %) into 315 mL of deionized H_2O and then add 117 mL of formalin (37 % formaldehyde).

17. Athymic nude mice, female, 4–6 weeks old, allowing 3–5 days acclimatization period.

18. Clean cages with sterile bedding, food, and water, and cage card with animal identification information.

2.7 In Vivo PK and PD Studies

1. Glass metabolic cages (for mice and rats).

2. Animal surgical instruments for blood drawing and tissue collection.

3. Diseased animal model, such as female nude mice bearing breast xenograft tumors or orthotopic tumors.

4. Normal healthy animal, such as CD-1 female mice.

5. Heparinized 1.5 mL Eppendorf tubes.

6. Tissue homogenizer.

7. Balance for tissue weight measurement.

8. Tissue storage vials and solvent.

9. Pharmacokinetic modeling program.

3 Methods

3.1 Drug Cellular Uptake Assay

1. Plate cells at a density of ~5×10^5 cells/well in 6-well plates and grow under optimum conditions.

2. Prior to the experiment, wash cells twice with PBS and incubate at 37 °C for 10 min to ensure complete removal of the medium.

3. Add 2 mL of desired concentration of drug in PBS in each well and incubate at 37 °C for 30 min to 2 h (*see* **Note 1**).

4. Following incubation, wash the cells three times with PBS and rinse the cells three times with ice-cold stop solution to stop further uptake of drug.

5. Lyse cells by incubating with PBS at –80 °C overnight and subsequent thawing at room temperature. Aliquots from each well are transferred to 1.5 mL centrifuge tubes and extracted as per established procedure.

6. Analyze with HPLC/LC-MS (*see* **Notes 2–4**).

7. Measured the protein concentration of each sample by Bradford (Bio-Rad, Hercules, CA) assay.

8. All uptake data is normalized to the protein count from each well.

3.2 Cellular Transport Studies

1. Grow cell mono-layers on the Transwell™ under appropriate conditions.

2. Prior to the experiment, wash cells twice with PBS and incubate at 37 °C for 10 min to ensure complete removal of the medium.

3. Add 200 μL of desired concentration of drug in PBS in donor chamber (*see* **Note 1**).

4. Fill receiver chamber with PBS. Sampling is carried out from the receiver chamber at predetermined time points and fresh PBS replaced in the receiver chamber to maintain sink conditions. All the experiments are performed at 37 °C.

5. The samples are analyzed by HPLC/LC-MS. The effective permeability (cm/s) is determined according to Eq. 1:

$$P = \frac{\mathrm{d}C/\mathrm{d}t}{C \times AV \times 60} \qquad (1)$$

P is the permeability in cm/s, $\mathrm{d}C/\mathrm{d}t$ denotes the slope of plot of concentration (nM) versus time (min), C represents the concentration of drug (nM/cm^3), A is the surface area of the chamber (cm^2), and V is the volume of each half-chamber (cm^3).

3.3 S9 Metabolism Studies

1. Dilute the S9 to 10 mg/mL in 100 mM Tris-HCl buffer.

2. Place the test tube into the ice bath and add 200 μL of 10 mg/mL S9.

3. Add 1280 μL of 100 mM Tris buffer.

4. Add 20 μL of 1000 μM test compound.

5. Place the test tube and the NRS separately into a 37 °C shaking water bath for 5 min.

6. Add 500 μL of 100 mM Tris-HCl or NRS I or NRS II to test tube. Start the reaction time at the addition of Tris-HCl or NRS to the samples.

7. 150 μL (duplicate) of the mixture is moved to the Eppendorf tube at 0, 15, 30, 45, and 60 min, and twofold volume (300 μL) of cold methanol is added. Vortex for 10 s. The mixture is centrifuged at 12,000 $\times g$ for 10 min and then the whole supernatant is moved to a new 100×75 mm glass tube. Air-dry using compressed clean-air. The deposit is dissolved in 150 μL of UPLC mobile phase. 100 μL of volume is injected onto the UPLC/LC-MS column.

8. The stability of the drug is determined by analysis of intact drug post incubation with S9 fractions, in comparison with negative control (without S9 fractions) (*see* **Notes 5** and **6**).

3.4 Protein Binding Assay

1. The plasma protein binding studies are carried out on whole human plasma as also specific plasma proteins such as Human serum albumin (HSA), and α1-acid glycoprotein (AGP).

2. α1-acid glycoprotein (AGP) and human serum albumin (HSA) solutions are prepared in PBS (pH 7.4). The concentrations of AGP and HSA used are 0.7 g/L and 40 g/L respectively, which is equal to the levels found in normal subjects.

3. Spike mouse plasma samples, AGP, and HSA with drug at a low and high concentration and maintain at 37 °C for 1 h. Prepare control samples using PBS instead of plasma.

4. From each of these preparations, a portion is aliquoted and placed in a sample reservoir of an Amicon Centrifree® ultrafiltration system (Millipore Co., Bedford, MA). The filter systems are centrifuged at 2000 $\times g$ until the reservoirs run dry. From each sample, triplicate portions are taken for analysis by HPLC or LC-MS. The Amicon Centrifree YM-300 filter system has a membrane molecular mass cutoff of 30,000 Da.

5. The amounts present in the filtrates are designated as "free drug" (F). The concentrations of the unfiltered solutions are also determined by triplicate analyses. This amount represents the 'total drug' concentration (T). The amount bound nonspecifically to the filter (X) also is determined. The percentage of test compound bound to plasma proteins is calculated by the following formula: Percentage bound $= [(T - F - X)/T] \times 100$ (*see* **Note 7**).

3.5 Cytochrome P450 Induction/ Inhibition Assay

1. Use fresh or platable cryopreserved hepatocytes, as either monolayer or sandwich culture, with a 1–2 day recovery period after plating.

2. Treat with test compound and positive controls for 2–3 days (changing medium with test compounds every 24 h) in media

Table 1
Assay conditions for the determination of Cytochrome P450 enzyme activities in human hepatocytes

Enzyme	Prototypical inducer	Substrate	Conc. (µM)	Incubation (min)	Marker metabolite
CYP1A2	3-methylcholanthrene	Phenacetin	100	15	Acetaminophen
CYP2B6	Phenobarbital	Bupropion	500	20	Hydroxybupropion
CYP3A4	Rifampicin	Testosterone	200	15	6β-hydroxytestosterone

containing ITS, dexamethasone, and penicillin–streptomycin as media supplements (*see* **Note 8**).

3. Dissolve the test compound preferably in DMSO (v/v 0.1 %) whenever possible and incubate at three or more different concentrations (in triplicate), spanning anticipated or known therapeutic concentration range including a concentration at least an order of magnitude higher. Minimally three concentrations of the test compound should be assessed spanning the in vivo or predicted C_{max} exposure levels, such as 0.1×, 1×, and 10× C_{max}.

4. Use recommended positive controls such as 3-methylchloranthrene (2 µM), phenobarbital (1000 µM), and rifampicin (10 µM) for CYP1A2, 2B6, and 3A4, respectively, (dissolved in 20 mL of incubation medium) at concentrations known to elicit maximal induction response (Table 1).

5. Prepare working concentrations of the probe substrates as follows: (a) 100 mM phenacetin in DMSO, (b) 500 mM bupropion in methanol, (c) 200 mM testosterone in methanol.

6. Prepare a negative control by adding 20 µL DMSO into 20 mL warm incubation medium.

7. In separate conical tubes, dissolve the test compound in warm incubation medium to yield final inducing concentrations 1000-fold lower than the respective working concentrations.

8. Remove the hepatocyte plate(s) from the incubator and replace the medium in the appropriate wells with 0.5 mL of the negative control, inducers, or test compound solutions, preferably each in triplicate.

9. Return plate(s) to the incubator for 24 h.

10. After 24 h, replace the medium with freshly prepared inducing solutions, repeating **steps 4** through **8**.

11. Fresh solutions should be prepared daily and applied every 24 h during the duration of the experiment, which is typically 48–72 h prior to harvest.

12. On the day of harvest, warm the incubation medium to 37 °C, as usual and remove the induction plate(s) from the incubator.

13. Aspirate the medium from each well, wash and replace with warm incubation medium. Take precautions to avoid the hepatocytes drying out, as mentioned in the notes (*see* **Note 9**).

14. Prepare probe substrates in 20 mL warm incubation medium to yield final incubation concentrations 1000-fold lower than working concentrations.

15. Remove the wash medium from the hepatocytes and replace with 0.5 mL of the probe substrate solutions prepared in the previous step.

16. Incubations should be conducted in triplicate with the respective CYP450 treatment group.

17. Incubate on an orbital shaker in the incubator for the time specified in table. A speed of 120–150 rpm is recommended for a 24-well plate.

18. Stop the incubations at the appropriate time points by collecting and freezing the sample medium, or by removing the sample medium and replacing it with an organic stop solution. Samples can be stored at −80 °C before analysis.

19. Extract and analyze samples by method of choice, typically HPLC/LC-MS by monitoring the formation of metabolite.

20. If mRNA analyses are desired, upon removal of the probe substrates, wash hepatocytes twice with Hank's Balanced Salt Solution (HBSS).

21. Remove the HBSS from the wells and replace with an appropriate volume of lysis buffer. Collect the cell lysates in RNAse-free microcentrifuge tubes and extract RNA for subsequent PCR reactions.

22. CYP activity can be expressed as pmol/min/million cells where pmol is defined as the amount of metabolite formed during the reaction. The fold-induction enzyme activity is determined by the ratio: CYP activity (induced)/CYP activity (vehicle). The percent adjusted positive control is determined by: ([CYP activity (induced)-CYP activity (vehicle)]/[CYP activity (positive control)-CYP activity (vehicle)]) × 100. A drug that produces a change that is equal to or greater than 40 % of the positive control can be considered as an enzyme inducer in vitro and further in vivo evaluation is warranted (*see* **Note 10**).

3.6 In Vitro Drug Stability Assay

1. Incubate three concentrations of the test drug (spanning anticipated or known therapeutic concentration range including a concentration at least an order of magnitude higher) in purified mouse plasma.

2. Analyze triplicates of the samples that have been exposed to different conditions (time and temperature) (*see* **Note 5**).

3. Freeze-thaw stability: The samples are stored at –20 °C for 24 h. They are then removed from the freezer and allowed to thaw unassisted at room temperature. When completely thawed, the samples are refrozen for 24 h under the same conditions. This freeze-thaw cycle is repeated twice more, for a total of three cycles. Then the samples are extracted using a validated extraction procedure and subjected to HPLC/LC-MS analysis.

4. Short-term and long-term stability: the short-term stability in plasma is assessed by analyzing samples kept at 37 °C for 2, 4, 6, 8, 12, and 24 h and samples kept at 4 °C for 4, 8, 12, and 24 h. The long-term stability is determined by assaying samples after storage at –80 °C for 2, 4, 6 and 8 weeks. Then the samples are extracted using a validated extraction procedure and subjected to HPLC/LC-MS analysis.

5. Bench-top stability: to assess the bench-top stability of teat compound in a particular biological matrix at room temperature, samples are incubated in plasma or tissue homogenates and kept at ambient temperature (25 °C) for 4, 8, 12, and 24 h, which far exceeded the routine sample preparation time.

6. Homogenization stability: samples are injected into equal weights of blank control tissues (brain, heart, liver, kidneys, lungs, spleen, tumor, and muscle) in the collection tube. Care is taken to ensure the drug solution is contained inside the tissue/organs. The tissue samples are homogenized, and extracted and immediately analyzed.

7. Stock solution stability: fresh and stored at room temperature (25 °C) stock solutions of samples are tested at 6, 12, and 24 h after sample preparation. The extraction and analytical methods are identical to the above sample preparation.

3.7 Subcutaneous Xenograft Models

1. Culture the desired breast cancer cells in 10 or 15 cm dishes according to specified ATCC guidelines (*see* **Note 11**).

2. Harvest tumor cells in logarithmic growth phase by trypsinization and transfer to 50 mL centrifuge tubes. Centrifuge them at $500 \times g$ for 5 min (*see* **Note 12**).

3. Remove supernatant and wash cells twice with medium lacking serum and antibiotics.

4. Remove a small aliquot of cell suspension before the last centrifugation and count the number of cells using a hemocytometer. Calculate the total cell number by multiplying cell density (cells/mL) by the volume of cell suspension.

5. Centrifuge tubes again for 5 min at $500 \times g$ to collect cells.

6. Add necessary volume of culture medium lacking serum to the cells. Pipet thoroughly to obtain a homogeneous cell suspension (*see* **Note 13**).

7. Mix the cell suspension with Matrigel basement membrane matrix (2:1, v:v, final 2.5×10^7 cells/mL) (*see* **Note 14**).

8. Transfer the cell suspension to a 1 mL syringe with a 27 gauge needle. Inject 0.2 mL of the cell suspension (5×10^6 cells/per animal) subcutaneously into the inguinal area of the mouse (*see* **Notes 15–17**).

9. To establish MCF7 human breast cancer xenografts, each of the female nude mice is first subcutaneously implanted with slow release estrogen pellet (SE-121, 17-β estradiol/pellet; Innovative Research of America, Sarasota, FL).

10. Label each cage card with date and type of cell injected.

11. Observe animals for tumor growth and clinical signs every day. When tumors are easily visible, begin to measure tumor diameter with calipers and record tumor measurements (*see* **Note 18**).

12. Tumor mass and volume are calculated using the formula: Mass (mg) = 0.5 (short diameter) (long diameter)$^2 \times 1000$. When tumors reach ~50–150 mg, initiate the treatment with test drug.

13. Divide animals randomly into treatment groups and controls (10–15 animals/group). Calculate the mean, SD, and SE in mass and volume for each group.

14. Label the cage card with the date, experiment, and treatment group numbers, drug, dose, and frequency.

15. The test compound is administered with i.p., i.v., or p.o. The injection volume is based on body weight (5 μL/g of body weight) and the concentrations of the drug should be adjusted on the basis of the doses.

16. Tumors should be measured at least twice a week. For rapidly growing tumors, it may be necessary to measure tumors every other day.

17. Monitor animal closely for distress, lethargy, tumor morphology (presence of ulceration) and body weight. In general, when tumor reaches 10 % of the body weight or when severe toxicity due to tumor growth or treatment toxicity develops, animals should be sacrificed.

18. Final measurements and tumor size calculations should be performed prior to sacrificing animals and removing tumors.

19. At the end of the experiment, confirm that final measurements and calculations have been performed for each animal.

3.8 Orthotopic Xenograft Tumor Models

1. Establish orthotopic models using procedures described above. Only different is transfer the luciferase tagged cell suspension to a 1 mL syringe with a 30 gauge needle. Inject 30 μL of the cell suspension (5×10^6 cells/per animal) into mammary fat pad (*see* **Notes 19** and **20**).

2. Immediately after injection and every 3 days thereafter, monitor bioluminescence using a Caliper IVIS imaging instrument.

3. Inject Luciferin i.p. at 150 mg/kg. Wait 5–10 min after injection and then place each mouse into the anesthesia chamber, and anesthetize with isoflurane inhalant. Once the mouse is fully anesthetized (approximately 5–10 min), transfer it to the imaging chamber and keep under mild anesthesia with oxygen mixture through a nose cone until image is acquired, and then analyze optical images using the IVIS Living Image software package.

4. Begin treatment of mice with drug when all tumors have reached an average bioluminescence intensity of ~1×10^7 (*see* **Note 21**).

3.9 In Vivo PK Study

3.9.1 Animals

1. Establish xenograft or orthotopic models using procedures described earlier.

2. Use normal model, e.g., CD-1 mice (no tumors).

3. Animals with tumor size of ~500–1000 mg are randomly divided into treatment groups for various time points (3–5 animals/group). Calculate the mean, SD, and SE of tumor mass and body weight for each group.

4. Label the cage card with the date, experiment, and treatment group numbers, and the drug, dose, and time point. Typically, the following time points are included: 5, 15, 30, and 60 min, and 2, 4, 6, 8, 24, 48, and 72 h (*see* **Note 22**).

5. Each animal should be identified by an ear tag.

3.9.2 Dosing for Pharmacokinetic Study

1. Measure body weight immediately prior to dosing.

2. Drug maybe given ip, i.v (via a tail vein), or p.o. (oral gavage) at one or more dose levels, e.g., 5, 10, and 20 mg/kg of body weight (*see* **Notes 23** and **24**).

3. The injection volume should be based on body weight (usually 5 μL/g of body weight). The concentrations of the drug should be adjusted on the basis of the doses.

4. Each dosed animal should be placed in a glass metabolic cage and given commercial diet and water ad libitum.

3.9.3 Urine and Fecal Sample Collection

1. Urine samples should be collected using a metabolic cage at various intervals, e.g., 0–6, 6–12, 12–24, and 24–48 h. The urine volume should be measured and recorded.

2. Fecal samples should be collected using a metabolic cage at various intervals e.g., 0–12, 12–24, and 24–48 h. The weight of feces should be measured and recorded.

3. At the end of each interval, the inner wall of the metabolic cage should be rinsed twice with 50 mL of PBS. The cage wash should be collected, measured, and placed in properly labeled containers.

4. All samples and rinse water should be kept at –20 °C until analysis.

3.9.4 Blood and Tissue Sample Collection

1. For each animal, check the general conditions. Weigh animal and record body weight.

2. Record time of treatment. At the designated time interval post dosing, draw blood from retro-orbital plexus and collect on ice in a heparin tube. Centrifuge at 14,000 rpm for 10 min. Transfer supernatant to a clean Eppendorf tube.

3. Collect various tissues including tumor tissue at necropsy and immediately trim of extraneous fat and connective tissue.

4. Blot on filter paper.

5. Weigh and store in labeled plastic tubes in –80 °C freezer until homogenization.

3.10 PK Data Analysis

1. Plasma, urine, and tissue concentrations should be determined on the basis of quantitation of the drug via appropriate established and validated analytical method. The drug concentration is usually expressed as ng/mL for plasma and ng/g for various tissues.

2. Pharmacokinetic parameters should be estimated, including the maximal concentration (Cmax), the time at Cmax (Tmax), half-life ($t_{1/2}$), the area under the curve (AUC), volume of distribution (Vd), and mean residue time (MRT) using an appropriate modeling software such as Phoenix WinNonlin (Certara, NJ).

3.11 PD Data Analysis

1. Various assay methods such as real-time reverse transcription-PCR (RT-PCR, to measure changes in the target gene/protein expression at mRNA level), Western blotting (to measure activation or inhibition of target proteins, cellular markers of processes such as metastases, inflammation), flow cytometry (to measure changes in DNA content, stem cell markers, etc.) are used to measure these outcomes [26–31].

2. A chosen end point (i.e., a biomarker or the target gene/protein is measured at different time points). Time intervals are chosen according to the aim and objective of the experiment. Therapeutic efficacy is determined by the ratio of % T/C (mean tumor mass of treated group divided by that of control).

3. The biomarkers may involve markers obtained from easily sampled fluids such as blood/urine and include enzymes (e.g., SOD, catalase, LDH, ALT, AST), blood count variables (e.g., erythrocytes, leukocytes, platelet count), hormones (e.g., estrogen) that do not require sacrifice of the animal.

4. Alternatively, the animals can be sacrificed at desired time points, tumors excised and subjected to RT-PCR or Western blotting or immunohistochemical staining (IHC) to identify effect of the drug on the target gene/protein at transcriptional and translational levels.

5. The most common PD parameters are the maximum effect (E_{max}) that can be reached and the half maximal effective concentration (EC_{50}).

$$E = E_{max} \frac{C}{EC_{50} + C}$$

E = effort; C = concentration; E_{max} = max response; EC_{50} = concentration capable of producing 50 % of the maximum effect.

4   Notes

1. For drug uptake and cellular transport studies, ensure that drug concentrations used are lesser than those required for cytotoxic activities.

2. A reliable and sensitive analytical method must be established and validated as per regulatory guidelines—USFDA, EMEA, etc.

3. It is very important to establish a bioanalytical method where the limit of quantitation is lesser than the expected concentration of the test compound in the biological matrix. Otherwise, low concentrations in the biomatrix will not be detected and this will result in incorrect AUC determination.

4. Avoid using ultra low volumes of the biological matrix, wherever possible, (less than 5 μL) for sample preparation as there can be errors in volume due to personnel error.

5. Use freshly prepared samples for HPLC/LC-MS, especially in stability and metabolism determination experiments.

6. If a particular metabolite (endogenous or exogenous) is to be determined, ensure appropriate sampling procedure is carried out. Variations in collection and storage of samples can lead to variability in analyte quantitation.

7. The protein binding of the drug must be determined since extensive binding to plasma proteins can impede its in vivo distribution/disposition and affect its therapeutic performance.

8. If any test compound is dissolved in a solvent other than DMSO, a negative control with an equivalent organic solvent

percentage must be included in the study design for fair comparison.

9. In order to prevent hepatocytes from drying out in, each treatment group should be aspirated separately and replaced with the medium before dosing the next set.

10. If performing protein determinations to normalize data across wells, keep in mind that the protein is not solely from hepatocytes, but also includes protein in the extracellular matrix used to overlay the hepatocyte monolayers.

11. A selective subpopulation of a parental cancer cell line with migration/colonization affinity towards specific organs can be used to mimic far site metastases. For example, the MDA-MB-231BR (brain seeking) cell lines, upon injection into the bloodstream of immunocompromised mice (via intracardiac route), travel into and colonize brain tissue; thus mimicking brain metastases [13].

12. Ensure that the cells are passaged the day before the injection and are no more than 80 % confluent on the day of the experiment.

13. Ensure cells are viable once harvested (use a trypan blue test for counting viable cells).

14. Always gently but thoroughly mix the cells before loading the syringe. Ensure there are no cell clumps in the solution that is being injected.

15. The mice used for developing the xenograft models should not be older than 6–8 weeks.

16. The use of a needle usually results in a strong, negative pressure, which can cause cell damage and death. Therefore, while withdrawing cell solution to the syringe, the needle should be removed to prevent negative pressure stress on the cells.

17. An alternative area for subcutaneous injection of the cancer cells is the upper region of the back near the neck where there is a lot of fat. This is a vascularized area for good tumor growth and the mice cannot reach the tumor as it grows.

18. The rate of tumor growth is dependent on individual cell lines and the amount of cells injected into each mouse. The optimum number of cells to be injected is determined by three variables: (a) type of tumor cell (slow growing vs. metastatic, with or without additional cells such as stromal cells); (b) location of injection (s.c., vs. orthotopic); (c) type of mouse (athymic nude vs. NOD/SCID). It is recommended using over 10^6 cells at tumor site if not sure of tumor growth activity.

19. Recently patient derived (PDX) models have been introduced that direct implantation of breast tumor tissues derived from patients into mouse mammary fat pads of immunocompromised

mice and closely resemble the original tumors in the patient maintaining original histopathology as well as biomarkers, hormone dependence or independence, gene expression, etc. [32].

20. In order to achieve, successful injection of tumor cells orthotopically into organ site, ensure minimum leaking at the site after injection.

21. Bioluminescence intensity also varies from tumor to tumor. The mean photons emitted over time by a specific tumor are highly comparable with the corresponding tumor volumes. It is estimated that a bioluminescence signal at 1×10^7 is roughly equal to a tumor volume of 50–150 mm^3, which is the starting point to initiate drug therapy.

22. Ensure that appropriate sampling procedures and times are chosen. For example, in studies where in vivo plasma (or tissue) drug concentrations are to be measured, ensure that sampling time is chosen according to route of administration. Include earlier sampling points (e.g., 3, 5, and 15 min post iv dosing).

23. While doing PK/PD experiments, care should be taken to ensure uniform dispersion of drug in carrier vehicle. Improper drug distribution in vehicle will lead to improper dosing.

24. Ensure that the carrier vehicle used for drug administration is itself, nontoxic, and does not affect tumor growth. Use of a sufficiently large control group (negative control) is the solution.

Acknowledgements

The project was partially supported by the National Institutes of Health R01 CA186662 (to R.Z.). The content is solely the responsibility of the authors, and does not necessarily represent the official views of the National Institutes of Health. This work was also supported by the American Cancer Society (ACS) grant RSG-15-009-01-CDD (to W.W.).

References

1. Danhof M, de Lange EC, Della Pasqua OE, Ploeger BA, Voskuyl RA (2008) Mechanism-based pharmacokinetic-pharmacodynamic (PK-PD) modeling in translational drug research. Trends Pharmacol Sci 29:186–191

2. Siegel R, Ma J, Zou Z, Jemal A (2014) Cancer statistics. CA Cancer J Clin 64:9–29

3. Richie RC, Swanson JO (2003) Breast cancer: a review of the literature. J Insur Med 35:85–101

4. Perou CM, Sorlie T, Eisen MB, van de Rijn M, Jeffrey SS, Rees CA et al (2000) Molecular portraits of human breast tumours. Nature 406:747–752

5. Genetic A (2012) Comprehensive molecular portraits of human breast tumours. Nature 490:61–70

6. Bombonati A, Sgroi DC (2011) The molecular pathology of breast cancer progression. J Pathol 223:307–317

7. Murawa P, Murawa D, Adamczyk B, Polom K (2014) Breast cancer: actual methods of treatment and future trends. Rep Pract Oncol Radiother 19:165–172

8. Nagar S (2010) Pharmacokinetics of anti-cancer drugs used in breast cancer chemotherapy. Adv Exp Med Biol 678:124–132

9. Widmer N, Bardin C, Chatelut E, Paci A, Beijnen J, Leveque D et al (2014) Review of therapeutic drug monitoring of anticancer drugs part two--targeted therapies. Eur J Cancer 50:2020–2036

10. Dienstmann R, Brana I, Rodon J, Tabernero J (2011) Toxicity as a biomarker of efficacy of molecular targeted therapies: focus on EGFR and VEGF inhibiting anticancer drugs. Oncologist 16:1729–1740

11. Zhou Q, Gallo JM (2011) The pharmacokinetic/pharmacodynamic pipeline: translating anticancer drug pharmacology to the clinic. AAPS J 13:111–120

12. Wang S, Zhou Q, Gallo JM (2009) Demonstration of the equivalent pharmacokinetic/pharmacodynamic dosing strategy in a multiple-dose study of gefitinib. Mol Cancer Ther 8:1438–1447

13. Taskar KS, Rudraraju V, Mittapalli RK, Samala R, Thorsheim HR, Lockman J et al (2012) Lapatinib distribution in HER2 overexpressing experimental brain metastases of breast cancer. Pharm Res 29:770–781

14. Ruoslahti E, Bhatia SN, Sailor MJ (2010) Targeting of drugs and nanoparticles to tumors. J Cell Biol 188:759–768

15. Neidle S, Thurston DE (2005) Chemical approaches to the discovery and development of cancer therapies. Nat Rev Cancer 5:285–296

16. Di L, Kerns EH (2005) Application of pharmaceutical profiling assays for optimization of drug-like properties. Curr Opin Drug Discov Devel 8:495–504

17. Nag S, Qin JJ, Voruganti S, Wang MH, Sharma H, Patil S et al (2015) Development and validation of a rapid HPLC method for quantitation of SP-141, a novel pyrido[b]indole anticancer agent, and an initial pharmacokinetic study in mice. Biomed Chromatogr 29:654

18. Nag S, Qin JJ, Patil S, Deokar H, Buolamwini JK, Wang W et al (2014) A quantitative LC-MS/MS method for determination of SP-141, a novel pyrido[b]indole anticancer agent, and its application to a mouse PK study. J Chromatogr B Analyt Technol Biomed Life Sci 969:235–240

19. Biomarkers Definitions Working Group (2001) Biomarkers and surrogate endpoints: preferred definitions and conceptual framework. Clin Pharmacol Ther 69:89–95

20. Colburn WA (2003) Biomarkers in drug discovery and development: from target identification through drug marketing. J Clin Pharmacol 43:329–341

21. Yip CH, Rhodes A (2014) Estrogen and progesterone receptors in breast cancer. Future Oncol 10:2293–2301

22. Sueta A, Yamamoto Y, Hayashi M, Yamamoto S, Inao T, Ibusuki M et al (2014) Clinical significance of pretherapeutic Ki67 as a predictive parameter for response to neoadjuvant chemotherapy in breast cancer: is it equally useful across tumor subtypes? Surgery 155:927–935

23. Wang W, Qin JJ, Voruganti S, Srivenugopal KS, Nag S, Patil S et al (2014) The pyrido[b]indole MDM2 inhibitor SP-141 exerts potent therapeutic effects in breast cancer models. Nat Comm 5:5086

24. Wang W, Zhang X, Qin J, Nag S, Voruganti S, Wang M et al (2012) Natural product ginsenoside 25-OCH3-PPD inhibits breast cancer growth and metastasis through downregulating MDM2. PLoS One 7:e41586

25. Rayburn E, Wang W, Li M, Zhang X, Xu H, Li H et al (2012) Preclinical pharmacology of novel indolecarboxamide ML-970, an investigative anticancer agent. Cancer Chemother Pharmacol 69:1423–1431

26. Wang W, Rayburn ER, Velu SE, Nadkarni DH, Murugesan S, Zhang R (2009) In vitro and in vivo anticancer activity of novel synthetic makaluvamine analogues. Clin Cancer Res 15:3511–3518

27. Wang W, Rayburn ER, Velu SE, Chen D, Nadkarni DH, Murugesan S et al (2010) A novel synthetic iminoquinone, BA-TPQ, as an anti-breast cancer agent: in vitro and in vivo activity and mechanisms of action. Breast Cancer Res Treat 123:321–331

28. Wang W, Zhao Y, Rayburn ER, Hill DL, Wang H, Zhang R (2007) In vitro anti-cancer activity and structure-activity relationships of natural products isolated from fruits of Panax ginseng. Cancer Chemother Pharmacol 59:589–601

29. Lindhagen E, Nygren P, Larsson R (2008) The fluorometric microculture cytotoxicity assay. Nat Prot 3:1364–1369

30. Vichai V, Kirtikara K (2006) Sulforhodamine B colorimetric assay for cytotoxicity screening. Nat Prot 1:1112–1116

31. Alley MC, Scudiero DA, Monks A, Hursey ML, Czerwinski MJ, Fine DL et al (1988) Feasibility of drug screening with panels of human tumor cell lines using a microculture tetrazolium assay. Cancer Res 48:589–601

32. DeRose YS, Gligorich KM, Wang G, Georgelas A, Bowman P, Courdy SJ et al. (2013) Patient-derived models of human breast cancer: protocols for in vitro and in vivo applications in tumor biology and translational medicine. Current Protoc Pharmacol Chapter 14: Unit 14.23

Chapter 24

Intracellular Delivery of Fluorescently Labeled Polysaccharide Nanoparticles to Cultured Breast Cancer Cells

Derek Rammelkamp, Weiyi Li, and Yizhi Meng

Abstract

Nanoparticle delivery is becoming an increasingly more valuable technique in cancer drug treatments. The use of fluorescent probes, in particular, can provide noninvasive strategies to interrogate the internalization mechanisms of cancer cells and aid in drug design. Here we describe the delivery of fluorescently labeled polysaccharide-based nanoparticles to breast cancer cells in vitro and their subsequent immunofluorescence microscopy examination. The description of the synthesis, preparation, and delivery of the nanoparticles can be widely applicable to other in vitro drug delivery studies.

Key words Nanoparticles, Delivery, Breast cancer cells, Fluorescence microscopy

1 Introduction

Nanoparticle-mediated drug delivery has become an emerging platform in cancer nanomedicine with a high potential to overcome clinical barriers by improving cellular penetration, tissue selectivity and therapeutic efficacy [1, 2]. To achieve targeted delivery, various delivery systems have utilized the tumor-homing characteristic of the nanocarriers [3, 4] by exploiting the enhanced permeability and retention (EPR) of the tumor microenvironment, a result of rapid growth that prevents the formation of fully functional vasculature and proper lymphatic drainage [5–9]. Anticancer drugs such as doxorubicin, paclitaxel and camptothecin show poor water solubility and may cause systemic toxicity, an issue that can be resolved by encapsulation in amphiphilic polymeric nanoparticles containing both hydrophilic and lipophilic domains.

The real-time, noninvasive monitoring of internalized nanoparticles in live cells can provide valuable information regarding the internalization pathway as well as the intracellular fate of the nanoparticles, and can substantially contribute to the knowledge of

Jian Cao (ed.), *Breast Cancer: Methods and Protocols*, Methods in Molecular Biology, vol. 1406,
DOI 10.1007/978-1-4939-3444-7_24, © Springer Science+Business Media New York 2016

drug specificity and delivery efficiency [10, 11]. Here we describe the preparation and delivery of polymer-based nanoparticles with an average diameter of 200 nm to a culture of 4T1 murine breast carcinoma cells. The nanoparticles are labeled with a red fluorescent molecule, Cy3, to enable their visualization in the cytoplasm.

2   Materials

Cell culture materials in this protocol are specific for the 4T1 *Mus musculus* mammary breast cancer cell line. Use of other cell lines may require different culture medium and/or conditions. It is important to check with ATCC for specifications specific to the cell line of use (http://www.atcc.org).

2.1   Cell Culture Reagents

1. Dulbecco's Modification of Eagle's Medium (DMEM) containing 4.5 g/L glucose and 1 mM sodium pyruvate without L-glutamine and phenol red (*see* **Note 1**).

2. L-glutamine (200 mM).

3. Hyclone heat-inactivated fetal bovine serum (FBS).

4. Penicillin–streptomycin (Pen-Strep) 10,000 U/mL in 0.85 % saline.

5. Phosphate buffered saline (PBS) pH 7.4.

6. Trypsin–EDTA 0.05 %, phenol red.

7. 70 % ethanol.

2.2   Preparation of Complete Growth Medium

1. In a sterile biosafety cabinet, open a new 500 mL bottle of DMEM and remove a 60 mL aliquot using a sterile 25 mL disposable polystyrene serological pipet. This aliquot can be stored at 4 °C to be used later.

2. Add 5 mL of Pen-Strep solution (10,000 U/mL).

3. Add 5 mL of L-glutamine solution (200 mM).

4. Add 50 mL of sterile-filtered FBS.

5. The final concentrations of Pen-Strep, L-glutamine, and FBS are 100 U/mL, 2 mM and 10 % (v/v), respectively.

6. Store the complete growth medium at 4 °C until ready to use.

2.3   Cell Culturing Supplies

1. 100 mm × 20 mm tissue-culture-treated culture dishes.

2. 15 mL conical centrifuge tubes.

3. Safe-Lock 1.5 mL centrifuge tubes.

4. Disposable polystyrene serological pipets, standard tip, 5 mL capacity.

5. Disposable polystyrene serological pipets, standard tip, 10 mL capacity.

6. Fisherbrand™ motorized pipet filler/dispenser—Powerpette Plus Turbo: Blue.

7. Eppendorf Research® adjustable micropipette (100–1000 µL).

8. Micropipette tip 200–1000 µL.

9. Eppendorf Research® adjustable micropipette (20–200 µL).

10. Micropipette tip 2–200 µL.

2.4 Nanoparticle Synthesis Materials/ Reagents

1. Glycol chitosan (250 kDa molecular weight, degree of deacetylation >60 %).

2. 5β-cholanic acid.

3. *N*-hydroxy-succinimide (NHS).

4. *N*-(3-Dimethylaminopropyl)-*N'*-ethylcarbodiimide hydrochloride (EDC).

5. Monoreactive hydroxysuccinimide ester of Cyanine 3 (Cy3-NHS).

6. Dimethyl sulfoxide (DMSO)—anhydrous.

7. Methanol—analytical grade.

8. Slide-A-Lyzer™ Dialysis Cassette—10 kDa molecular weight cutoff.

9. Mortar and pestle.

2.5 Reagents for Immunofluorescence Microscopy

1. Fluoroshield™ with DAPI histology mounting medium.

2. Formaldehyde solution ACS reagent, 37 wt% in water, with 10–15 % methanol as stabilizer.

3. Triton™ X-100—laboratory grade—diluted to 0.4 % in PBS.

4. Alexa Fluor® 546 protein/antibody labeling kit—diluted to final working concentration of 6.67 µg/mL in PBS.

5. DAPI nuclear counterstain—diluted to a final working concentration of 2.5 µg/mL in PBS.

2.6 Other Miscellaneous Supplies

1. 24-well flat-bottom with lid, standard tissue-culture polystyrene microplate.

2. Round glass coverslips 12 mm diameter × 0.13–0.16 mm thick.

3. Acrodisc® syringe filter 0.8 µm Supor® membrane low protein binding, non-pyrogenic.

4. Filtropur S 0.2 µm syringe filter for sterile filtration.

5. 10 mL syringe Luer-Lok™ Tip.

6. Precision glide needles—sterile single use 20 gauge (20 G).

7. 50 mL conical centrifuge tubes.

8. Plain glass microscope slides—25 mm × 75 mm × 1.2 mm.

9. Autoclave tape.

2.7 Equipment	1. Labconco purifier logic class II Type A2 biosafety cabinet.
	2. Forma™ Steri-Cult™ CO_2 incubator (95 % relative humidity, 5 % CO_2, 37 °C).
	3. Medline Vac-Assist vacuum pump.
	4. Tuttnauer automatic autoclave.
	5. Olympus inverted light microscope with 10×, 20× and 40× objectives.
	6. CLAY ADAMS® brand DYNAC® Centrifuge—4-place 50 mL horizontal head.
	7. Lab-Line Aquabath—general purpose.
	8. Branson 2510 Ultrasonic Cleaner—40 kHz, ½ gallon capacity.
	9. Soniprep 150 probe-type sonifier.

3 Methods

Unless otherwise noted, carry out all procedures at room temperature and conduct all cell culture work in a sterile biosafety cabinet.

3.1 Sterilization of Glass Coverslips	1. Place 12 mm (diam.) circular glass coverslips in a 10 cm glass petri plate. Cover with lid and tape the lid to the dish using autoclave tape.
	2. Autoclave for 20 min at 134 °C and 15 psi.
	3. Let the coverslips cool in the glass petri plate on the bench, covered, until room temperature. The coverslips can be stored in a sterile biosafety cabinet prior to use (*see* **Note 2**).
3.2 Thawing of Frozen Cells	1. Warm up growth medium in 37 °C water bath for 15–20 min, then wipe the outside of the containers with 70 % ethanol and place in the biosafety cabinet.
	2. Working in a biosafety cabinet, use a 10 mL disposable sterile serological pipet to transfer 9 mL growth medium into a sterile 15 mL conical centrifuge tube.
	3. Remove cryotube containing frozen cell suspension from storage (*see* **Note 3**).
	4. Using the warmth of your hands, defrost the cell suspension until it is completely liquid.
	5. Immediately wipe the outside of the cryotube with 70 % ethanol and place in the biosafety cabinet.
	6. Transfer the thawed cell suspension from the cryotube into a 15 mL conical centrifuge tube already containing 9 mL of growth medium.
	7. Close the screw-top cap and invert the 15 mL conical centrifuge 2–3 times.

8. Centrifuge the cell suspension at 1000 RPM ($180 \times g$) for 5 min.

9. Using a sterile Pasteur pipet, carefully aspirate the supernatant and resuspend the pellet in 1 mL of growth medium.

10. Using a 10 mL disposable sterile serological pipet, transfer 9 mL of fresh growth medium into a 100 mm × 20 mm tissue culture treated dish.

11. Transfer the 1 mL cell suspension prepared in fresh growth medium into the tissue culture dish.

12. Gently swirl the tissue culture dish to mix the cells evenly with the growth medium.

13. Immediately place tissue culture dish in the incubator at 37 °C and 5 % CO_2 with 95 % relative humidity.

3.3 Passaging Cells

1. Cells should be passaged when they reach about 90 % confluency in the tissue culture dish (*see* **Note 4**).

2. Warm up growth medium, phosphate buffered saline solution (PBS), and trypsin–EDTA solution in a 37 °C water bath for 15–20 min. When all the reagents have reached 37 °C, wipe the outside of the containers with 70 % ethanol and place in the biosafety cabinet.

3. Place a 100 mm × 20 mm tissue culture dish containing fully attached cells in the biosafety cabinet (*see* **Note 5**).

4. Carefully aspirate most of the growth medium from the tissue culture treated dish.

5. Using a 5 mL disposable sterile serological pipet, transfer 5 mL of PBS into the culture dish, swirl gently, and aspirate. Repeat one more time.

6. Using a 5 mL disposable sterile serological pipet, transfer 2 mL of trypsin–EDTA into the culture dish and swirl gently.

7. Return the culture dish to the 37 °C incubator and leave for 5 min.

8. Remove the culture dish from incubator and check under the microscope to confirm cell detachment (*see* **Note 6**).

9. If all the cells appear to be freely floating, proceed by placing the culture dish to the biosafety cabinet.

10. Using a 5 mL disposable sterile serological pipet, add 3 mL of fresh growth medium to the dish (*see* **Note 7**). Swirl gently and transfer the entire cell suspension into a 15 mL conical centrifuge tube.

11. Close the screw-top cap and centrifuge the cell suspension at 1000 RPM ($180 \times g$) for 5 min.

12. Return the centrifuge tube to the biosafety cabinet and carefully aspirate the supernatant.

13. Resuspend the pellet in 3–4 mL growth medium (*see* **Note 8**) and homogenize the cell suspension by pipetting up and down 4–5 times with the serological pipet. Place the screw-top cap on the centrifuge tube and invert 2–3 times.

14. If passaging cells for maintenance only, follow **steps 16–19**.

15. If plating the cells for an experiment, go to the next section (Subheading 3.4).

16. Using a 10 mL disposable sterile serological pipet, transfer 9 mL of fresh growth medium each into a 100 mm × 20 mm tissue culture treated dish (*see* **Note 9**).

17. Into each tissue culture dish containing 9 mL of growth medium, add 1 mL of cell suspension.

18. Gently swirl each dish to evenly mix the cell suspension.

19. Incubate the cells at 37 °C, 95 % relative humidity, and 5 % CO_2.

3.4 Plating Cells

1. Follow **steps 1–13** in the previous section (Subheading 3.3).

2. Using a 100–1000 μL sterile micropipette, transfer a 0.5 mL aliquot of the homogenized cell suspension into a 1.5 mL microcentrifuge tube to use for calculating cell concentration (*see* **Note 10**).

3. Using the same 100–1000 μL micropipette, pipet up and down to homogenize the cell suspension one more time.

4. Place a hemocytometer grid on the bench next to an inverted phase-contrast microscope.

5. Place the quartz coverslip over the hemocytometer grid so that it encompasses both counting chambers (*see* Fig. 1).

6. Using a micropipette, place a single 10 μL drop of the cell suspension into each of the V-shaped grooves (*see* Fig. 1). Make sure the quartz coverslip stays in place.

7. Carefully place the hemocytometer grid containing the two drops of cell suspension onto the microscope stage.

8. Using a 10 × or 20 × microscope objective, look for the center grid lines of each counting chamber (*see* Fig. 1).

9. Count the total number of cells present in ten 1 mm × 1 mm grids (5 grids per counting chamber) (*see* Fig. 1) (*see* **Note 11**).

10. Calculate the total volume of cell suspension needed to reach the desired concentration (*see* **Note 12**).

11. To dilute the cell suspension to the final working concentration, first place a sterile 15 mL conical centrifuge tube inside the biosafety cabinet.

12. Add the proper volume of complete growth medium that is needed to reach the desired final cell concentration (*see* **Note 13**).

13. Add 0.5 mL cell suspension prepared at the appropriate cell density into each well of a 24-well microplate.

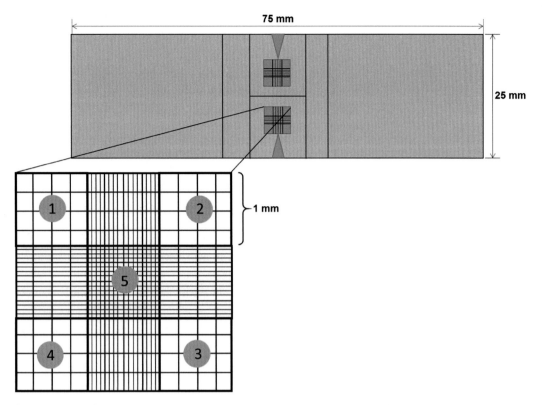

Fig. 1 Top view of a hemocytometer grid containing dual counting chambers. Next to each counting chamber is a V-shaped groove for inserting the cell suspension. Each counting chamber contains grid lined areas for counting the attached cells

14. If seeding cells for fluorescence microscopy experiments, a sterile glass coverslip should be placed inside each well prior to cell seeding so that the cells can be mounted to a microscope slide at each time point. The glass coverslips should be sterilized beforehand (by autoclaving) and handled using a pair of sterile tweezers.

15. Place the microplate in the 37 °C incubator and let the cells attach completely for 12–18 h before nanoparticle delivery.

3.5 Nanoparticle Synthesis

1. Dissolve 500 mg of glycol chitosan in a 250 mL glass beaker containing 60 mL of deionized water.

2. In a 100 mL glass beaker, add 150 mg of 5β-cholanic acid in 60 mL methanol. Allow to stir until 5β-cholanic acid is dissolved. Add 72 mg of NHS and 120 mg of EDC (*see* **Note 14**).

3. Slowly add the 5β-cholanic acid solution into the 250 mL beaker containing the glycol chitosan solution and stir for 18 h at room temperature. This step will create a hydrophobically modified glycol chitosan, or HGC [12].

4. Dialyze the mixture using four 10 kDa molecular weight cut-off dialysis cassettes for 24 h against a water–methanol mixture

(1:4 v/v). Remove the dialysis medium and add new water–methanol mixture (1:4 v/v) after 2 h, and then again after 4 h.

5. Remove the water–methanol mixture at 24 h and continue to dialyze the solution against deionized water for 24 h. Remove the dialysis medium and add new water after 2 h, and again after 4 h.

6. Discard the deionized water and lyophilize the purified retentate for 3 days.

7. Grind the lyophilized HGC into a fine powder using a mortar and pestle.

8. Dissolve 100 mg of HGC in a 100 mL glass beaker containing 40 mL of DMSO. Allow solution to stir overnight until HGC is dissolved.

9. Dissolve 1 mg of Cy3-NHS in 250 μL DMSO in a 1.5 mL Eppendorf tube. Pipet up and down 2–3 times with a 1 mL micropipette to evenly mix.

10. With the same micropipette, add the Cy3-NHS solution dropwise to the HGC solution.

11. Stir the mixture for 6 h at room temperature shielded from light.

12. Dialyze the mixture using two 10 kDa molecular weight cutoff dialysis cassette for 2 days against deionized water. Keep the container covered to prevent photobleaching. At the beginning of each day change the dialysis medium every 2 h during a 4-h interval (a total of three medium changes).

13. Discard the deionized water and lyophilize the purified retentate for 3 days.

14. Grind the lyophilized Cy3-labeled HGC into a fine powder using a mortar and pestle.

3.6 Preparation of the Nanoparticle Suspension

1. Suspend the lyophilized nanoparticles (synthesized in Subheading 3.5) in serum-free DMEM to yield a concentration between 0.1 and 1 mg/mL in either a 15 or 50 mL centrifuge tube (see **Note 15**).

2. Using both a probe-type sonifier and a sonicating water bath, homogenize the nanoparticles suspension for 15 min (see **Note 16**).

3. Place the centrifuge tube containing the homogenized nanoparticles inside the biosafety cabinet.

4. Bring three 20 G needles, three appropriately sized syringes (see **Note 17**), three 50 mL conical centrifuge tubes, one 0.80 μm syringe filter, one 0.45 μm syringe filter, and one 0.20 μm syringe filter into the biosafety cabinet.

5. Attach a 20 G needle onto a syringe and place the needle inside the centrifuge tube containing the nanoparticle suspension, so that the tip of the needle is immersed in the liquid (see Fig. 2).

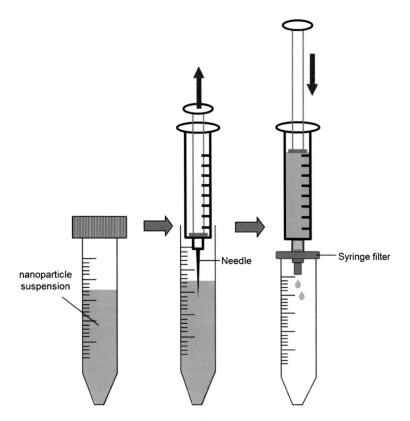

nanoparticle
suspension

Needle

Syringe filter

Fig. 2 Schematic depicting the process of sterile filtering the nanoparticle suspension. First, the nanoparticles are suspended in growth medium in a centrifuge tube. Next a syringe and needle are used to draw the suspension into the barrel of the syringe by pulling the plunger upwards. The needle is then discarded and a filter is attached. The syringe is placed directly above a new centrifuge tube and the plunger is pushed downward to pass the suspension through the filter. The sterile nanoparticle suspension is collected at the bottom of the centrifuge tube

6. Slowly pull the plunger upwards until all the nanoparticle suspension is drawn up inside the syringe. Be careful not to draw too much, or an air bubble might form (*see* **Note 18**).

7. While holding the syringe horizontally, remove the needle and discard it in an appropriate sharps container.

8. Attach a 0.80 μm syringe filter to the syringe.

9. Place the syringe directly above a new 50 mL conical centrifuge tube so that the syringe filter just rests on the edge of the tube (*see* Fig. 2).

10. Slowly pass the nanoparticle suspension through the 0.80 μm syringe filter by applying a gentle pressure on the plunger. The syringe and filter can now be discarded.

11. Assemble a new needle and a new syringe and repeat **steps 5–10** for the 0.45 and 0.20 μm syringe filters.

12. Once the nanoparticle suspension passes through the 0.20 μm filter, the nanoparticle suspension is now ready to be delivered to the breast cancer cells.

3.7 Delivery of the Nanoparticles

1. Place the 24-well microplate containing the breast cancer cells inside the biosafety cabinet.

2. Using a Pasteur pipet, aspirate the culture medium from each well in the 24-well plate.

3. Add a 0.5 mL aliquot of the nanoparticle suspension to each well (*see* **Note 19**).

4. Place the plate in the incubator and allow cells to be incubated with the nanoparticles for up to 24 h (*see* **Note 20**).

3.8 Immuno-fluorescence Microscopy

1. Remove the 24-well microplate containing the cells from the incubator.

2. Carefully aspirate the culture medium in each well using a 100–1000 μL micropipette.

3. Add 0.5 mL PBS to each well and aspirate. Repeat.

4. Add 0.5 mL 3.7 % formaldehyde (in PBS) to each well and incubate for 15 min to fix the cells.

5. Aspirate the formaldehyde solution and dispose in a proper hazardous waste container.

6. Rinse each well twice with fresh PBS.

7. If performing a stain for F-actin fibers, continue on to **step 9** (*see* **Note 21**). If not, directly proceed to **step 14**.

8. Remove the PBS from each well and permeabilize the cells by adding 0.5 mL Triton X-100 working solution. Incubate for 7 min.

9. Aspirate the Triton X-100 solution and rinse each well twice with PBS.

10. Remove the PBS from the well and add 0.5 mL of Alexa Fluor® phalloidin working solution into each well. Incubate for 20 min with the plate covered and protected from heat and light.

11. Remove the Alexa Fluor® phalloidin solution and dispose of it in a proper hazardous waste container.

12. Rinse twice with PBS and leave the last rinse of PBS inside the well.

13. If staining for nuclei, continue on to **step 15** (*see* **Note 22**). If not, directly proceed to **step 18**.

14. Remove the PBS from each well and add 0.5 mL DAPI working solution (*see* **Note 23**). Incubate for 5 min with the plate covered and protected from heat and light.

15. Remove the DAPI solution from each well and dispose of it in a proper hazardous waste container.

16. Rinse twice with PBS and leave the last rinse of PBS inside the well.

17. To mount the glass coverslips, first place a drop of mounting medium (with or without DAPI) on a clean glass microscope slide.

18. Using a pair of tweezers, remove each glass coverslip from the 24-well microplate and transfer to the microscope slide, cell-side down, directly on top of the drop of mounting medium (*see* Fig. 3).

19. Place the microscope slide and coverslip inside a dark container and allow the mounting medium to dry completely (about 24 h) before imaging.

20. Follow the manufacturer's instructions when operating the fluorescence microscope.

21. The fluorescently labeled nanoparticles should appear as small punctate clusters within the cytoplasm, and the DAPI-stained nuclei should appear blue (*see* Fig. 4).

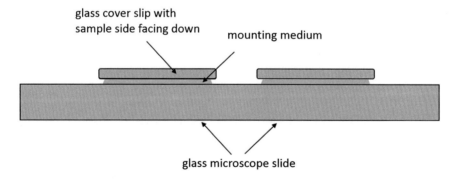

Fig. 3 Side view of two glass coverslips mounted cell-side down onto a microscope slide

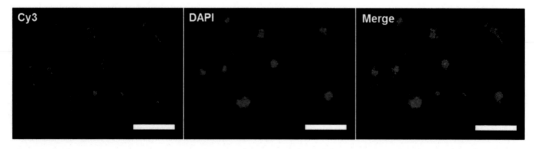

Fig. 4 Fluorescence microscopy images of Cy3-labeled glycol chitosan nanoparticles (visualized in the *red* channel) internalized by 4T1 murine breast carcinoma cells (nuclei visualized in the *blue* channel) after 2 h of delivery. Scale, 50 µm

4 Notes

1. Typical growth medium consists of Dulbecco's Modified Eagles Medium (DMEM) supplemented with 10 % fetal bovine serum (FBS) and 1 % Penicillin Streptomycin solution (10,000 U/mL). Check ATCC for growth medium requirements if using a different cell line.

2. It is recommended to store the sterile coverslips inside the sterile glass petri plates at all times in the biosafety cabinet.

3. ATCC typically ships cells in a 2.0 mL cryotube, which should be immediately stored in either a liquid nitrogen dewar or −80 °C freezer.

4. Frequency of passages depends on the growth rate of the cells, which can vary greatly between cancer cell lines. For 4T1 cell cultures grown on 100 mm × 20 mm tissue culture dishes, it is recommended to passage them approximately every 2 days to avoid over-confluency.

5. If the cells are not fully attached or appear slightly rounded, return them to the incubator to allow more time for attachment.

6. If cells are not detached, additional time or gentle agitation will help in cell detachment. However, it is not recommended to expose the cells to trypsin for more than 15 min.

7. The purpose of this step is to dilute the trypsin, which could become toxic to the cells after long-term exposure.

8. The volume of growth medium used for resuspension depends on size of pellet, which depends on the number of cells originally in the culture dish.

9. The suspension should be evenly divided between the desired number of plates (dilution factor), which depends on how quickly the cells grow and varies between cell lines. For 4T1 cells, it is recommended to passage at 1:6–1:8.

10. This aliquot will be used outside of the biosafety cabinet and will be discarded upon completion of cell counting since it will no longer remain sterile.

11. Each 1 mm × 1 mm grid contains 10^{-4} mL of liquid. Therefore the total number of cells counted in 10 grids is the concentration of cells per 10^{-6} mL or μL. For simplicity, convert cell concentration to cells/mL for calculations. The surface area of a well in a 24-well microplate is 200 mm². Total number of cells needed is seeding density multiplied by total surface area. Volume of cell suspension needed is total number of cells divided by concentration of cell suspension counted with hemocytometer.

12. For 4T1 cells, it is recommended that to seed the cells at a density of 7500 cells/cm² approximately 24 h prior to nanoparticle delivery.

13. For example, seeding a single well of a 24-well microplate at 7500 cells/cm² typically requires 0.5 mL of liquid volume per well. Since the growth area of a single well is 1.9 cm², this is equivalent to 7500 cells/cm² × 1.9 cm² = 14,250 cells total per well. The volume density of the suspension should then be 14,250 cells per 0.5 mL = 28,500 cells/mL.

14. If scaling up or down the nanoparticle batch size, it is recommended to use the same molar ratios of 5β-cholanic acid, NHS, and EDC. For example, 150 mg of 5β-cholanic acid (MW = 360.57) is equivalent to 0.000416 mol, whereas 72 mg of NHS (MW = 115.09) and 120 mg of EDC (MW = 191.70) are both equivalent to 0.0006256 mol. Therefore the optimal 5β-cholanic acid–NHS–EDC molar ratio is 1:1.5:1.5.

15. Optimal concentration should be determined experimentally through trial and error. We found that the fluorescence signal of the nanoparticles prepared in this range of concentration was most easily observed in the cytoplasm after internalization.

16. The sonication process can be broken down into 5-min intervals to avoid overheating of the samples. Depending on the nanoparticle type, the total duration that is needed to break apart any clusters and homogenize the solution will differ.

17. For example, if preparing 8 mL of nanoparticle suspension, a 10 mL syringe would be optimal.

18. If an air bubble should appear inside the syringe, hold the syringe plunger-side down so that the air bubble appears near the top. Slowly push the plunger upward to drive out the air bubble.

19. Some wells should be assigned as a control group receiving only serum-free DMEM without any nanoparticles. The purpose of this is to ascertain that the cells are in a viable state with the expected phenotype.

20. The exposure time will depend on factors such as the length of experiment and the concentration of nanoparticles. For 4T1 cells, nanoparticles have been internalized as soon as 15 min.

21. A common fluorescent dye for F-actin is Alexa Fluor® phalloidin, which works best if the cell membranes are permeabilized first using diluted Triton X-100, a mild detergent.

22. If using mounting medium that does not contain DAPI, a separate DAPI staining step may be performed right before the cells are mounted.

23. Different working concentrations for DAPI can be found in the literature. We found 2.5 μg/mL to work well for 4T1 breast cancer cells.

Acknowledgement

This work was supported by the Research Foundation for the State University of New York.

References

1. Cabral H, Matsumoto Y, Mizuno K et al (2011) Accumulation of sub-100 nm polymeric micelles in poorly permeable tumours depends on size. Nat Nanotechnol 6: 815–823

2. Peer D, Karp JM, Hong S et al (2007) Nanocarriers as an emerging platform for cancer therapy. Nat Nanotechnol 2:751–760

3. Kim K, Kim JH, Park H et al (2010) Tumor-homing multifunctional nanoparticles for cancer theragnosis: simultaneous diagnosis, drug delivery, and therapeutic monitoring. J Control Release 146:219–227

4. Morille M, Passirani C, Vonarbourg A et al (2008) Progress in developing cationic vectors for non-viral systemic gene therapy against cancer. Biomaterials 29:3477–3496

5. Cho K, Wang X, Nie S et al (2008) Therapeutic nanoparticles for drug delivery in cancer. Clin Cancer Res 14:1310–1316

6. Greish K (2007) Enhanced permeability and retention of macromolecular drugs in solid tumors: a royal gate for targeted anticancer nanomedicines. J Drug Target 15:457–464

7. Matsumura Y, Maeda H (1986) A new concept for macromolecular therapeutics in cancer-chemotherapy - mechanisms of tumoritropic accumulation of proteins and the antitumor agent SMANCS. Cancer Res 46: 6387–6392

8. Maeda H (2010) Tumor-selective delivery of macromolecular drugs via the EPR effect: background and future prospects. Bioconjug Chem 21:797–802

9. Taurin S, Nehoff H, Greish K (2012) Anticancer nanomedicine and tumor vascular permeability; Where is the missing link? J Control Release 164:265–275

10. Khalil IA, Kogure K, Akita H et al (2006) Uptake pathways and subsequent intracellular trafficking in nonviral gene delivery. Pharmacol Rev 58:32–45

11. Zeng CB, Vangveravong S, Xu JB et al (2007) Subcellular localization of sigma-2 receptors in breast cancer cells using two-photon and confocal microscopy. Cancer Res 67:6708–6716

12. Chin A, Suarato G, Meng Y (2014) Evaluation of physicochemical characteristics of hydrophobically modified glycol chitosan nanoparticles and their biocompatibility in murine osteosarcoma and osteoblast-like cells. J Nanotech Smart Mater 1:1–7

Chapter 25

Imaging Matrix Metalloproteinase Activity Implicated in Breast Cancer Progression

Gregg B. Fields and Maciej J. Stawikowski

Abstract

Proteolysis has been cited as an important contributor to cancer initiation and progression. One can take advantage of tumor-associated proteases to selectively deliver imaging agents. Protease-activated imaging systems have been developed using substrates designed for hydrolysis by members of the matrix metalloproteinase (MMP) family. We presently describe approaches by which one can optically image matrix metalloproteinase activity implicated in breast cancer progression, with consideration of selective versus broad protease probes.

Key words Collagen, Imaging agent, Matrix metalloproteinase, Near infrared, Triple-helix

1 Introduction

1.1 Matrix Metalloproteinases and Breast Cancer

The matrix metalloproteinases (MMPs) are a family of 23 human zinc endopeptidases that have been implicated in numerous diseases, including multiple aspects of cancer initiation and progression [1–3]. More specifically, MMP-2, MMP-9, and MMP-14/membrane type 1 matrix metalloproteinase (MT1-MMP) have been recognized for their role in tumor angiogenesis [1, 2]. MMP-9 regulates the bioavailability of vascular endothelial growth factor (VEGF), an inducer of angiogenesis [1, 2]. Localized MT1-MMP on the cell surface plays a critical role in tumor cell invasion of collagenous matrices [1].

Numerous studies have correlated MMP expression and/or production with breast cancer progression. The increased production of proMMP-9 was found in the early phases of breast cancer, while later stages had increased production and activation of MMP-9 [4]. Increased MMP-2 levels are associated with reduced survival of patients with node-negative breast cancer and adverse prognosis in patients with node-positive primary breast cancer [5]. Measurement of MMP-2 and MMP-9 concentration and activity in the sera of 345 patients found these two enzymes to serve as

Jian Cao (ed.), *Breast Cancer: Methods and Protocols*, Methods in Molecular Biology, vol. 1406,
DOI 10.1007/978-1-4939-3444-7_25, © Springer Science+Business Media New York 2016

biomarkers for breast disease classification, as higher levels of each correlated to disease stage (none, benign, and malignant) [6]. Examination of a database of 295 breast cancer patients indicated that high mRNA expression of MMP-1, MMP-9, MMP-12, MT1-MMP, and MMP-15 was associated with poor overall survival [7]. Analysis of gene expression and protein production for five normal breast tissues, 10 grade two breast cancer tissues, and 10 grade three breast cancer tissues found (a) higher MMP-1, MMP-9, MMP-11, MMP-13, and MMP-28 mRNA expression in breast cancer versus normal breast tissue, (b) lower MMP-7 mRNA expression in breast cancer versus normal breast tissue, and (c) higher protein production of MMP-1, MMP-8, MMP-10, MMP-11, MMP-12, and MMP-15 in breast cancer versus normal breast tissue [8]. Examination of the enzyme/inhibitor balance in breast cancer tumors for MMP-2, MMP-9, and MT1-MMP and their inhibitors tissue inhibitor of metalloproteinase-1 (TIMP-1), TIMP-2, and RECK indicated that mRNA levels were higher for the MMPs and lower for the inhibitors in primary breast tumors compared with adjacent non-tumor tissue [9].

The roles of specific MMPs in breast cancer have also been examined. Increased MMP-1 mRNA expression was correlated with MDA MB-231 breast cancer cell angiogenesis and subsequent bone osteolysis and tumor invasion of bone [10]. Short hairpin RNA-mediated stable knockdown of MMP-1 inhibited MDA MB-231 breast cancer cell invasion and tumor growth and metastasis to the brain [11, 12]. MDA MB-231 breast cancer metastasis to the bone and subsequent degradation was correlated to mRNA expression of MMP-13 and MT1-MMP [13].

Protective roles of MMPs have also been described. In vivo studies using two human cell lines derived from the parental MDA MB-435 breast cancer cell line found a 20-fold increase in MMP-8 expression in the non-metastatic cell line compared with the metastatic one [14, 15]. Plasma levels of MMP-8 showed a negative correlation with the risk of distant breast cancer metastasis [14]. With the recognition that MMPs can have protective roles in cancer, the concept of elevated expression of an MMP always being associated with worsened disease outcome is not always valid [16]. For example, primary breast cancer tumors in MT1-MMP-deficient mice developed faster than in their wild type counterparts, but showed a 50 % reduction in metastasis [17]. Nonetheless, the correlation of MT1-MMP upregulation with patients with metastatic breast cancer, and the inverse correlation of high MT1-MMP expression with patient survival time [18], strongly indicates an important role for MT1-MMP in breast cancer.

Numerous MMP probes have been developed for in vitro and in vivo analyses of MMP activities in disease states, and they have been comprehensively reviewed [19, 20]. The present discussion will focus on probes applied for MMPs implicated in breast cancer progression, i.e., general MMP probes and those selective for MMP-2, MMP-9, MMP-13, and MT1-MMP.

1.2 Near-infrared Probes for Matrix Metalloproteinases

Molecular optical imaging has become of great interest for its use in noninvasive diagnostics and intraoperative fluorescence-guided surgery [21–24]. Near-infrared (NIR) encompasses light in the $\lambda = 650$–900 nm range. The advantage of NIR compared with visible light for imaging is in the potential for tissue penetration. It has been estimated that, in the NIR window, 7–14 cm of tissue penetration can be achieved [25]. An additional advantage is that autofluorescence artifacts in the NIR range are minimal, providing a good signal to background ratio. A great variety of fluorochromes and quenchers have been developed for NIR optical imaging of protease activities [25]. Agents developed primarily for optical-based molecular imaging have potential medical applications at the forefront [25–27].

MMPSense™ 680 is a commercially available pan-matrix metalloproteinase imaging agent that has been utilized in a variety of animal models to investigate MMP activity as a function of disease. MMPSense™ 680 has molecular weight of 450 kDa and consists of relatively short peptides bearing fluorophores conjugated to a polymer backbone. Within the peptide is an MMP sensitive sequence. More specifically, Cy5.5-Gly-Pro-Leu-Gly ~ Val-Arg-Gly-Lys(FITC)-Cys-NH$_2$ is attached to a synthetic graft copolymer composed of poly-L-Lys (PL) and methoxyPEG (MPEG) (PL-MPEG$_{92}$) [28, 29]. The close proximity of the fluorophores results in quenching of the fluorescence. Fluorescence dequenching is observed after enzyme-mediated hydrolysis as the peptide fragments bearing the fluorophores are liberated from the polymer. A control probe was generated by attaching Cy5.5-Gly-Val-Arg-Leu-Gly-Pro-Gly-Lys(FITC)-Cys-NH$_2$ [29]. The MMP-targeted probe was able to image HT-1080 human fibrosarcoma tumors in mice, while much lower signal was obtained from low MMP-2-producing BT20 breast adenocarcinoma tumors [28, 29]. Inhibition of MMP activity by intraperitoneal administration of prinomastat could also be imaged in mice, and correlated with MMP activity in tumors [29].

The MMPSense™ 680 probe has subsequently been utilized for imaging of MMP hyperactivity in adenomas, considered an early tumor biomarker [30]. Overexpression of MMP-7 have been found in 85 % of colorectal adenocarcinomas and are associated with poor prognosis, while elevated levels of MMP-9 have been associated with severely dysplastic polyps, colorectal adenomas, and colorectal cancer metastases. Adenomatous polyposis coli (Apc)$^{+/Min-FCCC}$ mice that develop spontaneous colorectal adenomas were injected with MMPSense™ 680, and colons were imaged ex vivo [30]. A strong correlation was observed between a positive signal and the presence of pathologically confirmed colonic adenomas. 92.9 % of the 350 areas of interest examined were classified correctly, based on the comparison of MMPSense™ 680 results to immunohistochemical staining of MMP-7 and MMP-9.

Imaging of MMP activity in a mouse breast cancer model using 4T1-luc2 cells was pursued using MMPSense™ 680 [31]. 4T1-luc2 cells were implanted into the lateral thoracic mammary fat pad of nude mice. Tumors were detected on day 4 following cell implantation, and signal intensity increased up to day 18. The tumor to background ratio on day 18 was found to be 2.62. Immunohistological analysis indicated that fluorescence originating from MMPSense™ 680 correlated to the presence of MMP-9 [31].

Early MMP activity can be imaged using MMPSense™ 750, which has a lower MW than MMPSense™ 680 (43 kDa versus 450 kDa). More specifically, MMPSense™ 750 is recommended for imaging in the range of 6 h after injection, as opposed to 24 h or greater for MMPSense™ 680. However, it has been observed that MMPSense™ 750 can be activated by proteases other than MMPs [32].

Another NIR-based activatable probe was designed for MMP-13 for use in in vivo imaging. The probe was based on a previously determined MMP-13 substrate, Gly-Pro-Leu-Gly ~ Met-Arg-Gly-Leu-Gly-Lys, linked to Cy5.5 and BHQ3 (Probe 1) [33]. BHQ3 has no native fluorescence and minimal absorption, and can efficiently quench fluorophores that emit between $\lambda = 620–730$ nm. Other dyes typically have their own fluorescent properties and do not provide complete quenching, contributing to limited image resolution attributable to low signal to background ratio. Probe 1 was tested in vitro against MMP-13 and MMP-7 [33]. Although Probe 1 displayed a 32-fold increase of fluorescent signal when treated with MMP-13, it was also significantly cleaved by MMP-7. Three other probes with modified sequences were designed, and all probes were tested against MMP-13, MMP-2, MMP-9, and MMP-7 [34]. Only Probe 2, with the sequence Gly-Val-Pro-Lys-Ser ~ Leu-Thr-Met-Gly-Lys-Gly-Gly, displayed improved signal to background ratio in vitro compared with Probe 1 (36-fold) and was not cleaved by other enzymes tested. Probe 2 was then tested in vivo, with a 3.8-fold increase in fluorescence emission where MMP-13 activity was present [34].

MMP-13 activity has also been imaged using MMP13ap (QSY-21-Gly-Gly-Pro-Ala-Gly ~ Leu-Tyr-Glu-Lys(Cy5.5)-Gly-OH) [35]. The substrate was developed based on reverse design from a phosphinate inhibitor library [36]. MMP13ap was cleaved two times more rapidly by MMP-13 than by MMP-12, and 10–20 times more rapidly by MMP-13 than by MMP-1, MMP-2, or MMP-3 [36]. When applied to arthritis models, the correlation between MMP13ap activation and histological damage was variable [35, 36]. It was suggested that probe improvements might include increased specificity, increased rate of hydrolysis by MMP-13, and increased retention in tissues of interest [35].

1.3 Collagen-Based Triple-Helical Substrates

Collagen is the most abundant protein in animals, and is the major structural protein found in the connective tissues such as basement membranes, tendons, ligaments, cartilage, bone, and skin. The most defining feature of collagen is the supersecondary structure, composed of three parallel extended left-handed polyproline type II alpha chains of primarily repeating Gly-Xxx-Yyy triplets. Three left-handed strands intertwine in a right-handed fashioned around a common axis to form a triple-helix. In the collagen Gly-Xxx-Yyy triplet, the residue in the Xxx position is often L-proline (Pro) and the residue in the Yyy position is often 4(R)-hydroxy-L-proline (Hyp), accounting for 20 % of the total amino acid composition in collagen [37]. The other commonly found amino acids are Ala, Lys, Arg, Leu, Val, Ser, and Thr [37]. The packing of the triple-helical coiled-coil structure requires Gly in every third position.

For several decades triple-helical peptides (THPs) consisting of collagen-model sequences and their three-dimensional folds have been constructed and studied to fully investigate the structural and biological roles of collagenous proteins [38–45]. Our laboratory previously described a number of Förster resonance energy transfer (FRET) triple-helical peptide (fTHP) substrates that are either suitable for most collagenolytic MMPs or selective for different collagenolytic MMPs [46–52]. Amongst these was a selective MMP-2/MMP-9/MMP-12 FRET THP substrate, $\alpha 1(V)436\text{-}447$ fTHP [(Gly-Pro-Hyp)$_5$-Gly-Pro-Lys(Mca)-Gly-Pro-Pro-Gly ~ Val-Val-Gly-Glu-Lys(Dnp)-Gly-Glu-Gln-(Gly-Pro-Hyp)$_5$-NH$_2$] [48, 53]. The type V collagen sequence Gly-Pro-Pro-Gly ~ Val-Val-Gly-Glu-Lys-Gly-Glu-Gln, as a single-stranded peptide, was hydrolyzed extremely slowly by either MMP-2 or MMP-9 [48]. Therefore, it is the triple-helical structure, along with the sequence, that imparted the substrate specificity amongst the various MMPs [48].

1.4 Triple-Helical MMP Probes

$\alpha 1(V)436\text{-}447$ fTHP has been utilized to develop a potentially selective NIR imaging agent by replacing the Mca and Dnp groups with either LS276 or cypate. LS276 is a highly fluorescent, monofunctional, water-soluble heptamethine cyanine dye [54]. The single-stranded peptides were assembled by solid-phase methodologies and LS276 was incorporated while the peptide was resin-bound (Fig. 1) [55]. Upon self-assembly of the triple-helical structure, the 3 peptide chains intertwined, bringing the fluorophores into close proximity and reducing fluorescence via self-quenching [55]. Upon enzymatic cleavage of the THP, six labels were released, resulting in an amplified fluorescent signal. The fluorescence yield of the probe increased 3.8-fold upon activation. Kinetic analysis showed a rate of LS276-THP hydrolysis by MMP-2

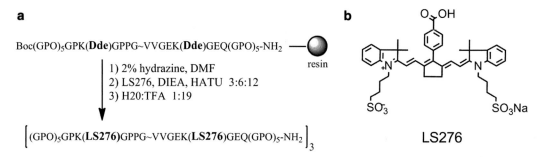

Fig. 1 Assembly of LS276-THP. (**a**) Solid-phase synthesis of LS276-THP (O = Hyp). (**b**) Chemical structure of LS276. Reprinted with permission from *Bioconjugate Chemistry*, copyright 2012, American Chemical Society

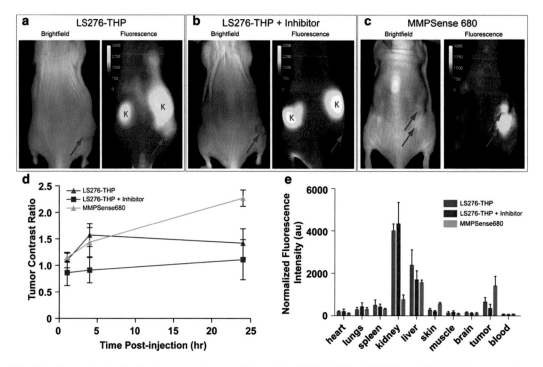

Fig. 2 In vivo analysis of mice tumor protease activity using LS276-THP or MMPSense™ 680. Representative in vivo whole-body images of mice bearing HT-1080 tumor xenografts 24 h after injection of (**a**) LS276-THP; $n=4$, (**b**) LS276-THP and inhibitor; $n=3$, or (**c**) MMPSense™ 680; $n=3$. Tumors (*arrows*) and kidney (K) regions are marked. (**d**) The ratio of tumor and contralateral thigh ROI fluorescence with respect to time show the time dependent activation of the molecular probes. (**e**) Ex vivo fluorescence biodistribution confirmed the high fluorescence in the non-inhibited tumors and the high retention of LS276-THP in the mouse kidneys relative to the larger MMPSense™ 680. Error bars represent standard deviation; au = arbitrary units. Reprinted with permission from *Bioconjugate Chemistry*, copyright 2012, American Chemical Society

($k_{cat}/K_M = 30,000$ s^{-1}M^{-1}) similar to that of MMP-2 catalysis of α1(V)436-447 fTHP. Administration of LS276-THP to mice bearing a human fibrosarcoma xenografted tumor resulted in a tumor fluorescence signal more than fivefold greater than muscle (Fig. 2) [55]. These results were the first to demonstrate that

THPs were suitable for highly specific in vivo detection of tumor-related MMP-2 and MMP-9 activity.

Subsequently, cypate$_3$-THP and the polyethylene glycol (PEG) modified variant, cypate$_3$-(PEG)$_2$-THP, were developed to create two activatable reporter probes for the selective imaging of MMP-2/MMP-9 activity in HT-1080 fibrosarcoma xenografted tumors [56]. Cypate, a cyanine-based NIR dye, was chosen for its facile synthesis, high extinction coefficient, high absorption peak in the NIR region, where tissue absorption and light scattering are at a minimum, and overall biocompatibility. Further, cypate is compatible with the acidic conditions of solid-phase peptide synthesis and the photo-physical properties of cypate conjugated THPs can be easily tuned by structure modifications and/or bio-conjugation [57]. Cypate also incorporates carboxylic acid groups for conjugation to amino groups of resin-bound peptides and is also reported to have good binding affinity to albumin [58].

To assemble cypate$_3$-THP, the carboxylic acid group on cypate was conjugated to the ε-amino groups of Lys in the THP backbone. Self-assembly of three cypate-conjugated peptides into a THP led to the quenching of the fluorescence from the cypate molecules until hydrolysis by MMP-2 and/or MMP-9. The ratio of cypate to THP was determined to be 3:1. Cypate$_3$-(PEG)$_2$-THP was synthesized and characterized in the same manner.

To determine whether cypate$_3$-THP could visualize MMP-2/MMP-9 activity in vivo, the quenched probe was administered intravenous to mice bearing HT-1080 xenografts, which express high levels of these MMPs, with and without the pan-MMP inhibitor Ilomastat [56]. Since no improvements to the quenching or kinetic efficiency were observed with the PEG-ylated probe, only cypate$_3$-THP was used in imaging studies. Ilomastat was administered at both 24 and 1 h pre-administration of cypate$_3$-THP and again 4 h post cypate$_3$-THP administration to account for the putative widely different pharmacokinetics of a ~15 kDa THP and a 0.3 kDa inhibitor. Normalized fluorescence intensity values from MMP-2 and MMP-9 mediated hydrolysis of cypate$_3$-THP (pmol cypate/mm^3 tumor volume) were relatively low initially at 1 h post-cypate$_3$-THP injection, which peaked at 4 h post-cypate$_3$-THP injection, and slowly cleared from most tissues after 24 h (Fig. 3 Top, Bottom). The fluorescence intensity was significantly reduced when Ilomastat was co-administered ($p < 0.05$; Fig. 3 Middle, Bottom). The average tumor to background ratio at 4 h post-injection was 10:2. Moreover, the most intense fluorescence signal was shown at the edge of the tumors, where MMP-2/MMP-9 activity is generally in the greatest abundance [59, 60]. Some of the fluorescent signal observed in mice treated with Ilomastat was attributed to non-quenched cypate$_3$-THP probe. However, the Ilomastat treated group served as a good control, allowing for visualization of the fluorescence signal from cypate$_3$-THP

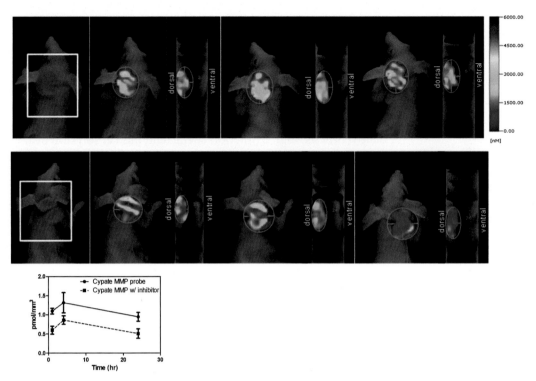

Fig. 3 In vivo analysis of mice tumor protease activity using cypate$_3$-THP. (*Top*) FMT images of a representative mouse injected with 2 nmol cypate$_3$-THP. From *left* to *right*: bright-field image with the scan area highlighted, fluorescence images at 1, 4, and 24 h post-injection of cypate$_3$-THP. (*Middle*) FMT images of a representative mouse injected with 2 nmol cypate$_3$-THP with llomastat administered 24 and 1 h pre-cypate$_3$-THP injection and again at 4 h post-cypate$_3$-THP injection. From *left* to *right*: bright-field image with the scan area highlighted, and fluorescence images at 1, 4, and 24 h post-injection of cypate$_3$-THP. (*Bottom*) ROI analysis in pmol cypate$_3$-THP/mm^3 of tumors from mice injected with cypate$_3$-THP ± llomastat. Reprinted with permission from *Bioorganic & Medicinal Chemistry Letters*, copyright 2014, Elsevier

in the absence of MMP activity. The significantly different results between the cypate$_3$-THP group and the cypate$_3$-THP plus MMP inhibitor group demonstrated the efficient and selective hydrolysis of cypate$_3$-THP by the MMPs.

A third study was undertaken to determine whether single-stranded peptides bearing 5-carboxyfluorescein (5FAM) dyes would assemble into a triple-helix and function as a substrate for MMP-2/MMP-9 with kinetic parameters suitable for the in vivo detection of MMP activity [61]. 5FAM was chosen due to its commercial availability, amenability to solid-phase peptide synthesis, high quantum yield, and the ubiquity of fluorescein filter sets in a variety of fluorescence imaging equipment [62]. The on-resin conjugation procedure of 5FAM to single-stranded peptides was modified to increase conjugation efficiency (Fig. 4). The addition of 4-dimethylaminopyridine (DMAP) was necessary to result in complete coupling of 5FAM to the Lys ε-amino groups. Previous

a

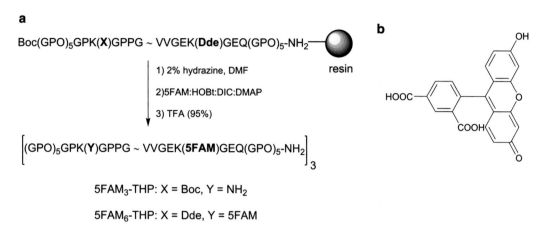

b

Boc(GPO)$_5$GPK(**X**)GPPG ~ VVGEK(**Dde**)GEQ(GPO)$_5$-NH$_2$——⬤

resin

1) 2% hydrazine, DMF

2)5FAM:HOBt:DIC:DMAP

3) TFA (95%)

$$\left[(\text{GPO})_5\text{GPK}(\textbf{Y})\text{GPPG} \sim \text{VVGEK}(\textbf{5FAM})\text{GEQ}(\text{GPO})_5\text{-NH}_2\right]_3$$

5FAM$_3$-THP: X = Boc, Y = NH$_2$

5FAM$_6$-THP: X = Dde, Y = 5FAM

Fig. 4 Assembly of 5FAM$_3$-THP. (**a**) Solid phase synthesis of 5FAM$_3$-THP and 5FAM$_6$-THP (O = Hyp). The tilde represents the site of hydrolysis by MMP-2 and MMP-9. (**b**) 5-carboxyfluorescein (5FAM). Reprinted with permission from *Molecules*, copyright 2014, MDPI

work had shown that during coupling of activated 5FAM to resin-bound peptides, more than one 5FAM molecule was conjugated via ester formation between the carboxylic acid and the phenolic oxygen of 5FAM [62]. Treatment with piperidine was required to remove these adducts. Here, DMAP may function as a nucleophile to remove the undesired esters during coupling of the 5FAMs. Alternatively, DMAP may act as an acyl transfer-reagent and facilitate the reaction between the Lys ε-amino groups and 5FAM carboxy groups, making the unwanted attack of the 5FAM phenolic oxygen that resulted in ester bond formation less favored [63, 64]. When DMAP was not present in the coupling cocktail, the amide bond formation was too slow and the side-reaction (esterification) occurred. The kinetic parameters and structural properties of 5FAM-THPs were evaluated, as well as whether 5FAM-THPs bearing homodimeric dyes would visualize MMP-2/MMP-9 activity secreted by human tumor cells with confocal fluorescence microscopy (Fig. 5).

1.5 Other MMP Probes

Genetically encoded fluorescence indicators have been designed for MMPs. One encoded a yellow fluorescence protein (YFP, the quencher) and a cyan fluorescence protein (CFP, the fluorophore) linked by the peptide Leu-Glu-Gly-Gly-Ile-Pro-Val-Ser-Leu-Arg-Pro-Val, which contains the MMP Substrate Site (mss). This sensor, designated YFP–mss–CFP[display], was used to detect secreted MMP-2 activity of MCF-7 breast cancer cells [65]. When tested against MMP-3, MMP-8, and MMP-9, it was cleaved by all three but to a lesser extent than MMP-2.

The same research group later designed a probe designated DMC (DsRed2-mss-CFP expressed from pDisplay vector) and measured the decrease of DsRed2/CFP ratio in MDA-MB435

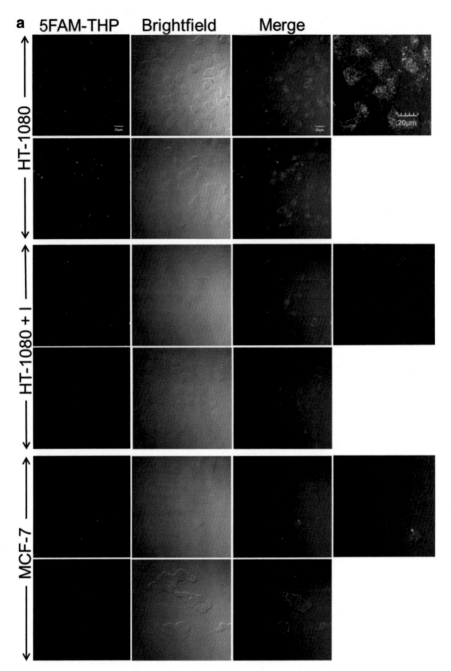

Fig. 5 In vitro analysis of tumor cell protease activity using 5FAM$_3$-THP. (**a**) Confocal fluorescence microscopy showing the hydrolysis of 5FAM$_6$-THP by HT-1080 human fibrosarcoma cells. HT-1080 or MCF-7 human breast cancer cells were treated with 5FAM$_6$-THP (1 μM) for 1 h. For inhibition studies, cells were pre-incubated with SB-3CT (1 μM) for 30 min prior to 5FAM$_6$-THP incubation. Two image fields are shown per condition. The first column represents the fluorescence from 5FAM-THP. The second column represents the corresponding bright-field image. Overlays of the fluorescence and bright-field images are shown in the third column (taken using the 10× objective) with higher magnification images in the fourth column (taken using the 60× objective) to show cell internalization of 5FAM-THP. Scale bars are indicated in the first image of each column. (**b**) Mean cell associated fluorescence quantified with FV1000 software ($n=10$ cells, "I" denotes SB-3CT, "1080" denotes HT-1080 cells).

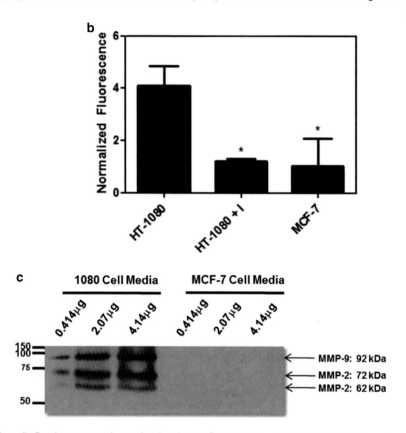

Fig. 5 (continued) *P*-values were determined using a Student's two-tailed *t*-test; "*" denotes $p < 0.05$. (**c**) Zymogram depicting MMP-2 and MMP-9 of cell-associated media with varying total protein loading. Varying levels of total protein from the conditioned media (0.41, 2.07, and 4.14 μg) were loaded into each well of a 10 % Zymogram Gel with gelatin. Proteins were electrophoresed at 100 V for 90 min and after electrophoresis, the proteins were renatured in 2.5 % Triton X-100. The gel was then incubated overnight at 37 °C in development solution and stained with Coomassie Blue. Reprinted with permission from *Molecules*, copyright 2014, MDPI

breast carcinoma cells as an indicator of MMP activity [66]. The red-shifted spectrum of the DsRed fluorophore provided a brighter signal and less fluorescent cross-talk with the CFP molecule than YFP–mss–CFP^{display}. This probe was also tested in shell-less culture of chick embryo chorioallantoic membrane (CAM) as an in vivo tumor model for MMP detection and proved to be readily detectable by fluorescence stereomicroscopy.

An MT1-MMP NIR probe was designed based on a peptide sequence identified in a phage display substrate library. A non-substrate peptide His-Trp-Lys-His-Leu-His-Asn-Thr-Lys-Thr-Phe-Leu (denoted as MT1-AF7p) displayed high binding affinity to the MT-loop region of MT1-MMP and was shown to interact with the enzyme through hydrogen bonding and hydrophobic interactions [67]. The loop is located within the catalytic domain and contains an eight amino acid insertion unique to MT-MMPs

(MT1-, 2-, 3-, and 5-MMP). This insert is absent from all other MMPs. The assay design was distinct, as the phage display library was screened against a unique sequence from MT1-MMP ([160]Arg-Glu-Val-Pro-Tyr-Ala-Tyr-Ile-Arg-Glu-Gly-His-Glu-Lys-Gln[174], designated MT1-160p) and not the entire enzyme or catalytic domain. Only the peptides that bound specifically to this sequence (which is located within the loop structure as mentioned above) were evaluated. MT1-AF7 displayed the highest affinity towards MT1-160p ($K_d = 0.075$ nM), and was labeled with Cy5.5 (Cy5.5-MT1-AF7p) and chosen for further validation in vivo [67]. The evaluation was performed in mice carrying MDA-MB-435 breast cancer xenografts (expressing high levels of MT1-MMP) and A549 xenografts (low MT1-MMP levels). MDA-MB-435 xenografts had significantly higher signal accumulation and better tumor contrast than the A549 xenografts. However, more precise quantitative data on tumor uptake and pharmacokinetics will be needed to determine the further utility of this probe.

Another MT1-MMP fluorogenic probe was designed based on the substrate sequence Gly-Arg-Ile-Gly-Phe ~ Leu-Arg-Thr-Ala-Lys-Gly-Gly and labeled with Cy5.5 and BHQ3 quencher (MT-P) [68] in an effort to target the membrane-bound form of MT1-MMP. MT1-MMP is found in both bound and soluble forms, and targeting the membrane-bound form could prove useful in localization of MT1-MMP-overexpressing tumor environments. MT-P was tested against MMP-2 and MMP-9 as well as against MT1-, MT2-, and MT3-MMPs. The probe displayed moderate selectivity. In vivo evaluation of probe activity in mice bearing MDA-MB-435 xenografts indicated strong NIR activation in the MT1-MMP-positive tumor region. Although MT-P displayed good specificity in the tumors, nonspecific activation and accumulation was also observed in the liver. This result suggests further optimization of the probe is needed.

A radiotracer was designed to detect the activity of MMP-2 and MMP-9 using single-photon emission computed tomography (SPECT). The cyclic decapeptide cyclo[Cys-Thr-Thr-His-Trp-Gly-Phe-Thr-Leu-Cys] (CTT) was designed based on the considerations that (1) the His-Trp-Gly-Phe motif of the peptide displays inhibition of MMP-2 and MMP-9 activity, (2) hydrophobic radiolabeled peptides have shown low-level accumulation in liver, and (3) negatively charged radiolabeled peptides previously exhibited good renal clearance [69]. A hydrophilic and negatively charged radiolabel, indium-111-diethylenetriaminepentaacetic acid (^{111}In-DTPA), was attached N-terminal to the His-Trp-Gly-Phe motif (^{111}In-DTPA-CTT). This probe exhibited significant inhibition of MMP-2 activity in vitro, and when injected into normal mice, displayed fast clearance from liver and kidneys. In mice bearing MDA-MB-231 breast tumors, the probe displayed increased accumulation within the tumor as compared to MDA-MB-435S tumor-bearing mice. However, the relatively low tumor contrast found in this study warrants further exploration of the molecular interactions between the inhibitor and the enzymes. A similar conclusion was found in a study of ^{64}Cu-DOTA-CTT with MDA MB-435 tumor bearing mice, where the low affinity of the ligand for the enzymes and the in vivo instability discouraged its use for in vivo tumor evaluation [70].

In order to evaluate MT1-MMP activity in MDA-MB-231 cells using SPECT, a tripartite probe was designed containing the following components: (1) a positively charged D-Arg octamer (r8) cell penetrating peptide (CPP) attached with single amino acid chelate (SAAC) for technetium-99m; (2) a MT1-MMP specific substrate (Ser-Gly-Arg-Ile-Gly-Phe-Leu-Arg-Thr-Ala); and (3) a negatively charged D-Glu attenuation sequence [71]. Several attenuation sequences were evaluated in order to achieve linear conformation of the cleavable sequence that would be available for MT1-MMP docking. The sequence comprising of four D-Glu-Gly-Gly repeats (4egg) was chosen for cell-based studies. Probe activation was determined by treatment of transfected MDA-MB-231 cells with and without a broad spectrum MMP inhibitor (GM1489). The average uptake of the 4egg probe was two times greater in untreated cells indicating successful cleavage and increased uptake of the activated probe into MT1-MMP expressing cells. Negative results from cells treated with free ^{99m}Tc-tricarbonyl complex ($[^{99m}Tc(CO)_3]^+$) suggested that there was no nonspecific uptake of the radiolabel and that the ^{99m}Tc accumulation in cells treated with the intact probe was the result of cleavage rather than leakage of free radiolabel into these cells. Further in vivo imaging analysis will be needed to determine clinical potential of this probe.

Antibody radiolabeling has also been explored as a potential novel probe for MT1-MMP activity [72]. ^{99m}Tc-anti-MT1-MMP monoclonal antibody was evaluated in breast tumor-bearing rodents. MT1-MMP was highly expressed in all malignant cells tested. Tumor radioactivity increased over time and displayed three to fivefold increase at 24 h as compared to 1 h after antibody injection. The radio-antibody cleared other organs and blood rapidly suggesting a promising in vivo application of this probe.

2 Materials

All commercially available materials and reagents are used as received.

2.1 Manual Peptide Synthesis

For manual peptide synthesis standard peptide reaction vessels are used. They can be purchased from Chemglass Life Sciences (Vineland, USA), Peptides International (Louisville, USA), or other commercial sources.

2.2 Automated Peptide Synthesis	For automated peptide synthesis a conventional peptide synthesizer such as the PS3 (Protein Technologies Inc., Tucson, USA) or a microwave assisted peptide synthesizer such as the Liberty Blue (CEM Corp., Matthews, USA) can be used.
2.3 Solid phase for Peptide Synthesis	Resins for solid-phase synthesis, including Fmoc-Rink amide 4-methylbenzhydrylamine (MBHA), can be purchased from Rapp-Polymere (Germany), EMD Millipore (Darmstadt, Germany), Advanced ChemTech (Louisville, USA), or other suppliers.
2.4 Amino Acids and Other Peptide Synthesis Reagents	Fmoc-protected amino acids are commercially available from Sigma-Aldrich (St. Louis, USA), EMD Millipore (Darmstadt, Germany), Protein Technologies Inc., (Tucson, USA), Peptides International (Louisville, USA), or other commercial sources. Other reagents such as O-(6-chlorobenzotriazol-1-yl)-N,N,N',N-tetramethyluronium hexafluorophosphate (HCTU), 1-hydroxybenzotriazole (HOBt), N,N'-diisopropylcarbodiimide (DIC), N,N-diisopropylethylamine (DIEA), 1,8-diazabicyclo[5.4.0]undec-7-ene (DBU), hydrazine hydrate, [dimethylamino(triazolo[4,5-b]pyridin-3-yloxy) methylidene]-dimethylazanium hexafluorophosphate (HATU), 4-dimethylaminopyridine (DMAP), trifluoroacetic acid (TFA, peptide synthesis grade), piperidine, ammonia solution, and acetic acid can be purchased from the same sources. The ninhydrin test kit can be purchased from Anaspec (USA).
2.5 Organic Solvents and Solutions	Dimethylformamide (DMF), N-methylpyrrolidone (NMP), dimethylsulfoxide (DMSO), dichloromethane (DCM), methanol (MeOH), and acetonitrile can be purchased from Sigma-Aldrich (St. Louis, USA), EMD Millipore (Darmstadt, Germany), or Fisher Scientific (Atlanta, USA).
2.5.1 Fmoc Deprotection Solution	This solution is prepared by mixing DBU, piperidine, and NMP in a ratio of 1:5:44 (v/v). 1 L of such solution is composed of 20 mL DBU, 100 mL of piperidine, and 880 mL of NMP.
2.5.2 Dde Deprotection Solution	The 1-(4,4-dimethyl-2,6-dioxocyclohexylidene)ethyl (Dde) protecting groups are removed using 2 % hydrazine in DMF. This solution is prepared by mixing 2 mL hydrazine hydrate and 98 mL DMF.
2.5.3 Peptide Cleavage Cocktail	Under the fume hood prepare the peptide cleavage solution containing 5 % H_2O in TFA. To prepare 5 mL of such a solution mix 250 μL of H_2O and 4.75 mL of TFA. For every 100 mg of resin use 1.5 mL cleavage cocktail (*see* **Note 1**).
2.6 HPLC Characterization and Purification	The analytical and preparative reversed-phase HPLC (RP-HPLC) runs are carried out on an Agilent 1200 Infinity series liquid chromatograph from Agilent Technologies (Germany) or equivalent. All analytical runs are performed using a Vydac C18 column

(5 μm, 300 Å, 150 mm×4.6 mm) (Grace Davison, Columbia, USA) or similar. Preparative runs are carried out using a Vydac C18 column (15–20 μm, 300 Å, 250 mm×22 mm) (Grace Davison, Columbia, USA) or similar. Water is of HPLC or Milli-Q (Millipore) quality. The following solvent system is used: solvent A contains 0.1 % TFA in H_2O while solvent B contains 0.1 % TFA in acetonitrile. The analytical gradient ranging from 2 to 100 % B over 20 min using 1 mL/min flow rate is used. The preparative HPLC gradient of 5–50 % B in 60 min with flow rate of 10 mL/min is used. In both cases peptide detection is carried out at $\lambda = 220$ and 280 nm.

2.7 Mass Spectrometry

The molecular weight of peptides and peptide conjugates are determined using matrix-assisted laser desorption/ionization time-of-flight mass spectrometry (MALDI-TOF MS). MALDI-TOF MS analysis is performed using an Applied Biosystems (Carlsbad, USA) Voyager DE-PRO Biospectrometry workstation or Bruker Microflex LF mass spectrometer (Bruker Daltonics, Inc., Billerica, USA) using α-cyano-4-hydroxycinnamic acid (HCCA) or 2,5-dihydroxybenzoic acid (DHB) as matrix. Matrices can be purchased from Sigma-Aldrich (St. Louis, USA) or Fisher Scientific (Atlanta, USA).

2.8 Circular Dichroism (CD) Spectroscopy

Triple-helical peptide structure is evaluated by near-UV CD spectroscopy using a Jasco J-810 spectropolarimeter (Easton, USA) with a path length of 1 mm. Typically a CD spectra of collagen-like peptides are recorded over the range $\lambda = 190–250$ nm and sample concentration of 0.1 mg/mL in 0.5 % acetic acid solution (v/v). Thermal transition curves are obtained by recording the molar ellipticity ($[\theta]$) at $\lambda = 225$ nm with an increase in temperature of 20 °C/h over the range of 5–80 °C. The temperature is controlled by a JASCO PTC-348WI temperature control unit.

2.9 Peptide Lyophilization

Synthetic peptides and conjugates are lyophilized using a Labconco FreeZone lyophilizer (freeze dryer) system (Labconco Corp., Kansas City, USA) or similar.

2.10 Absorption Spectrophotometry

Peptide concentration and fluorescent probe substitution is measured by UV–Vis absorption spectrophotometry using a NanoDrop 2000c UV–Vis spectrophotometer (Thermo Fisher Scientific, Wilmington, USA). Fluorescent dye-conjugated peptide is measured using specific wavelength (λ) and molar absorption coefficient (ε) characteristic for the fluorophore. Triple-helical peptide-conjugate concentration is calculated according to the Beer-Lambert law, using the following formula:

$$C = A / (l \times 3 \times e)$$

where C = peptide concentration (M), A = measured absorbance, and l = cell pathlength (cm). Please note that in case of triple-helical peptides the molar absorption coefficient ε has to be multiplied by a number of fluorophore groups present in a molecule.

2.11 Confocal Microscopy

Confocal microscopy is performed using the Olympus FV1000 microscope (Center Valley, USA) equipped with a 20×/0.95 W water immersion objective. Fluorescence emission of $5FAM_6$-THP is detected from λ = 490–550 nm using a 488 nm laser at 5 % power.

2.12 Fluorophore Dyes

1. MMPSense™ 680 is purchased from PerkinElmer, Inc. (Waltham, USA).

2. 5FAM (5-carboxyfluorescein, single isomer) is commercially available from Invitrogen (Carlsbad, USA) or Sigma-Aldrich (St. Louis, USA).

3. LS276 is prepared by multistep organic synthesis as described [54].

4. Cypate is prepared by multistep organic synthesis as described [73].

2.13 Animal Imaging Systems

1. LS276-THP probe is imaged using the Kodak Image Station 4000MM multimodal imaging system (Carestream Health, New Heaven, USA). Fluorescence images are acquired using $\lambda_{excitation}$ = 755 ± 35 nm and $\lambda_{emission}$ = 830 ± 75 nm.

2. MMPSense™ 680 probe is imaged with the Pearl NIR fluorescence imaging system (LiCor Biosciences, Lincoln, USA) with $\lambda_{excitation}$ = 685 nm and $\lambda_{emission}$ = 720 nm.

3. $Cypate_3$-THP probe is imaged using FMT2500 fluorescence molecular tomography system (PerkinElmer, USA) using $\lambda_{excitation}$ = 790 nm and $\lambda_{emission}$ = 805 nm.

2.14 Cell Culture Media

Cell culture media is commercially available and used as received. DMEM medium is available from American Type Culture Collection (ATCC, cat. no. 30-2002). Complete DMEM medium is prepared by addition of 10 % fetal bovine serum (FBS) and 100 units/mL of penicillin G.

2.15 Cells

Cancer cell lines are cultured accordingly. A variety of cancer cell lines including HT-1080 one can be obtained from American Type Culture Collection (ATCC). General cell culture protocols are available in *Methods in Molecular Biology* volumes 731 and 806.

2.16 Animals

Animal models including NCR nude and athymic nude mice are available from Taconic Farms (Hudson, USA). For mice anesthesia a 2 % isoflurane/oxygen mixture is used. The isoflurane mixture must be administered using an anesthesia apparatus such as SomnoSuite low-flow animal anesthesia system available from Kent Scientific (Torrington, USA).

2.17 MMP Inhibitors

1. Ilomastat (GM6001) is commercially available from Sigma-Aldrich (St. Louis, USA).

2. MMP-2 and MMP-9 inhibitor SB-3CT is purchased from Enzo Life Sciences (Farmingdale, USA).

2.18 Buffers and Stock Solutions

1. Phosphate buffered saline (PBS) solution is commercially available from Fisher Scientific (Atlanta, USA).

2. LS276-THP stock solution for in vivo imaging is prepared by dissolving 20 nmol of peptide in 250 μL PBS buffer.

3. Ilomastat stock solution is prepared by dissolving 100 mg of ilomastat in 1 mL of DMSO.

4. For the cell viability assay ethidium homodimer-1 can be obtained from Invitrogen (USA) as a 2 mM 25 % DMSO solution. Use this stock to obtain the final 1 μM solution during the viability assay.

5. SB-3CT stock solution can be prepared by dissolving 10 mg of SB-3CT in 1 mL of DMSO.

3 Methods

3.1 Peptide Synthesis Protocols

For the incorporation of individual amino acids by Fmoc-solid-phase methodology a HCTU/HOBt activation is recommended [74]. Reagent amounts including Fmoc amino acids (Fmoc-AA), HCTU, HOBt, and DIEA are calculated in relation to peptide synthesis scale (typically expressed in mmol).

3.1.1 General Amino Acid Coupling Cycles

For Fmoc-amino acid couplings the following molar excess is recommended: Fmoc-AA = 3 equiv., HCTU = 2.5 equiv., and DIEA = 5.5 equiv. Recommended coupling time using conventional automatic peptide synthesizer is 1 h.

3.1.2 General Fmoc Deprotection Cycle

The Fmoc protecting group is removed by gentle treatment of peptidyl-resin using standard Fmoc deprotection solution. This step can be performed on an automated peptide synthesizer using appropriate deprotection program or manually as follows:

1. Place the resin in a peptide synthesis reaction vessel.

2. Add the appropriate amount of Fmoc deprotection solution (3–5 times the resin volume).

3. Agitate the resin for 5 min. Drain the solution.

4. Repeat **step 2**.

5. Agitate the resin for another 15 min and drain the solution.

6. Wash resin (three times) using 5 mL of DMF.

3.1.3 Removal of the Dde Protecting Group

To remove the Dde protecting group from lysine residues the following steps are required:

1. Place the resin in a peptide synthesis reaction vessel.

2. Add the appropriate amount of Dde deprotection solution (four times the resin volume).

3. Agitate the resin for 5 min and then drain the solution.

4. Repeat the treatment with Dde deprotection solution two more times. Drain the solution.

5. Wash resin (three times) using 5 mL of DMF.

3.2 Synthesis of Triple-Helical Peptides Containing LS276

The synthetic scheme is depicted in Fig. 1. This protocol is adapted from [55].

1. Synthesize peptide using an automated peptide synthesizer. For the incorporation of individual amino acids use general coupling conditions (see above). For the Fmoc removal, use Fmoc removal solution and Fmoc removal protocol (*see* above). *See* **Note 2**.

2. Remove the side chain Dde protecting groups by using the Dde deprotection solution and conditions (see above).

3. After Dde deprotection, wash the resin in the peptide synthesis reaction vessel sequentially using 5 mL DMF, 10 mL of 10 % aqueous DMF, 10 mL DCM, and 15 mL MeOH.

4. Place peptide-resin containing reaction vessel (no cap) in a desiccator connected to a vacuum pump. Dry peptide-resin *in vacuo* for 4–6 h.

5. Conjugate LS276 fluorescent dye to the ε-amino groups of Lys using the following conditions:

   (a) Place 25 mg of resin in a dry reaction vessel

   (b) Prepare a mixture of 150 mg LS276 dye, 800 μL of DIEA, and 300 mg of HATU in 3 mL DMF.

   (c) Add this mixture to the resin and agitate for 3.5 h. After that filter the resin and wash 3×5 mL DMF, DCM, and MeOH.

   (d) Perform a ninhydrin test. A negative result indicates complete coupling of LS276.

   (e) Cleave peptide from the resin using 5 % water in TFA (mix 50 μL of H_2O with 950 μL of TFA). Cleavage and side-chain deprotection of the peptide is accomplished by treating the resin for 3 h. Perform this step under the fume hood.

   (f) Collect the filtrate and dilute it with 10–15 mL of water. Perform this step under the fume hood.

   (g) Lyophilize the peptide.

(h) Purify the peptide by preparative RP-HPLC. Characterize by analytical HPLC and MALDI-TOF mass spectrometry.

(i) LS276-THP peptide concentration can be quantified using absorption spectroscopy ($\lambda = 780$ nm, $\varepsilon = 220{,}000$ cm^{-1}M^{-1}) on a NanoDrop spectrophotometer.

3.3 Synthesis of Triple-Helical Peptides Containing Cypate

This protocol is adapted from [56].

1. Synthesize cypate-THP and/or cypate$_3$-(PEG)$_2$-THP peptide using an automated peptide synthesizer. For the incorporation of individual amino acids use general coupling conditions (*see above*). For the Fmoc removal, use Fmoc removal solution and Fmoc removal protocol (see above). *See* **Note 2**.

2. Remove the side chain Dde protecting groups by using the Dde deprotection solution and conditions (see above).

3. After Dde deprotection, wash the resin in peptide synthesis reaction vessel sequentially using 5 mL DMF, 10 mL of 10 % aqueous DMF, 10 mL DCM, and 15 mL MeOH.

4. Place peptide-resin containing reaction vessel (no cap) in a desiccator connected to a vacuum pump. Dry peptide-resin *in vacuo* for 4–6 h.

5. Conjugate cypate dye to the ε-amino groups of Lys using the following conditions:

   (a) Place 25 mg of resin in a dry reaction vessel.

   (b) Prepare a mixture of 125 mg cypate dye, 150 μL of DIC, and 125 mg of HOBt in 3 mL of DMF.

   (c) Add this mixture to the resin and agitate overnight. After that filter the resin and wash 3×5 mL DMF, DCM, and MeOH.

   (d) Perform a ninhydrin test. A negative result indicates complete coupling of cypate dye.

   (e) Cleave peptide from the resin using 5 % water in TFA (mix 50 μL of H$_2$O with 950 μL of TFA). Cleavage and side-chain deprotection of the peptide is accomplished by treating the resin for 3 h. Perform this step under the fume hood.

   (f) Collect the filtrate and dilute it with 10–15 mL of water. Perform this step under the fume hood.

   (g) Lyophilize the peptide.

   (h) Purify the peptide by preparative RP-HPLC. Characterize by analytical HPLC and MALDI-TOF mass spectrometry.

   (i) Cypate$_3$-THP peptide concentration can be determined using absorption spectroscopy ($\lambda = 780$ nm, $\varepsilon = 200{,}000$ cm^{-1}M^{-1}) on a NanoDrop spectrophotometer.

3.4 Synthesis of Triple-Helical Peptides Containing 5FAM

The synthetic scheme is depicted in Fig. 4. This protocol is adapted from [61].

1. Synthesize collagen-like peptide using an automated peptide synthesizer. For the incorporation of individual amino acids use general coupling conditions (see above). For the Fmoc removal, use Fmoc removal solution and Fmoc removal protocol (see above). *See* **Note 2**.

2. Remove the side chain Dde protecting groups by using the Dde deprotection solution and conditions (see above).

3. After the Dde deprotection, wash the resin in peptide synthesis reaction vessel sequentially using 5 mL DMF, 10 mL of 10 % aqueous DMF, 10 mL DCM, and 15 mL MeOH.

4. Place peptide-resin containing reaction vessel (no cap) in a desiccator connected to a vacuum pump. Dry peptide-resin *in vacuo* for 4–6 h.

5. Conjugate 5FAM fluorescent dye to the ε-amino groups of Lys using the following conditions:

   (a) Place 30 mg of resin in a dry reaction vessel.

   (b) Prepare a mixture containing 90 mg of 5FAM dye, 90 mg HOBt, 220 μL of DIC, and 90 mg of DMAP in 3 mL of DMF.

   (c) Add this mixture to the resin and agitate for 16 h. After that filter the resin and wash 3×5 mL DMF, DCM, and MeOH.

   (d) Perform a ninhydrin test. A negative result indicates complete coupling of 5FAM.

   (e) Cleave peptide from the resin using 5 % water in TFA (mix 50 μL of H_2O with 950 μL of TFA). Cleavage and side-chain deprotection of the peptide is accomplished by treating the resin for 3 h. Perform this step under the fume hood.

   (f) Collect the filtrate and dilute it with 10–15 mL of water. Perform this step under the fume hood.

   (g) Lyophilize the peptide.

   (h) Purify the peptide by preparative RP-HPLC. Characterize by analytical HPLC and MALDI-TOF mass spectrometry.

   (i) 5FAM-THP peptide concentration can be quantified using absorption spectroscopy ($\lambda = 475$ nm, $\varepsilon = 29,000$ cm^{-1}M^{-1}) on a NanoDrop spectrophotometer.

3.5 Characterization of Triple-Helical Peptides

1. The peptide molecular mass is confirmed by MALDI-TOF mass spectrometry using HCCA or SA MALDI matrices.

2. Triple-helicity is monitored by CD spectroscopy in the far UV wavelengths [46, 74–81]. The general protocol is as follows.

   (a) Dissolve the peptide in an appropriate buffer to final concentration of 0.1 mg/mL. Leave the peptide in solution for 24–48 h at 4 °C to allow for triple-helix formation. *See* **Note 3**.

   (b) Record a series of spectra ($n = 5$–10) over the range of $\lambda = 190$–250 nm. A typical spectrum for a triple-helix shows a positive molar ellipticity at λ ~225 nm and a negative molar ellipticity at λ ~205 nm.

   (c) Determine the melting temperature by monitoring the change in molar ellipticity at $\lambda = 225$ nm with a constant change in temperature (20 °C/h) from 5 to 80 °C. For samples exhibiting sigmoidal melting curves, the inflection point in the transition region (first derivative) is defined as T_m. The first derivative from the transition curve can be obtained using JASCO or GraphPad Prism software (GraphPad, USA). Alternatively, T_m is evaluated from the midpoint of the transition.

3.6 In Vivo Imaging Using LS276-THP and MMPSense™ 680

This protocol is adapted from [55, 56].

1. Human fibrosarcoma xenografts are grown by subcutaneous injection of 200,000 HT-1080 cells (ATCC) in the flanks of 6-week old male NCR nude mice (Taconic Farms, Hudson, NY) (*see* **Note 4**).

2. Tumor-bearing mice receive 1 mg/kg LS276-THP probe *i.p.* ($n = 4$).

3. A second group ($n = 3$) is treated with Ilomastat (1 mg/kg in DMSO, *i.p*) 2 h before LS276-THP injection and again 4 and 20 h after injection.

4. A third group ($n = 3$) receive 2 nmol MMPSense™ 680 *i.v.* (*see* **Note 5**).

5. Mice in the LS276-THP groups are imaged using the Kodak IS4000MM multimodal imaging system (Carestream Health, New Haven, CT) immediately and at 1, 4, and 24 h after LS276-THP injection, followed by ex vivo fluorescence biodistribution imaging of organ tissues. Fluorescence images are acquired using $\lambda_{excitation} = 755 \pm 35$ nm and $\lambda_{emission} = 830 \pm 75$ nm detection, 60 s exposure with 2×2 binning.

6. Mice in the MMPSense™ 680 group are imaged with the Pearl NIR fluorescence imaging system (LiCor Biosciences, Lincoln, NE) with $\lambda_{excitation} = 685$ nm and $\lambda_{emission} = 720$ nm collection.

7. Region of interest (ROI) analysis is performed using ImageJ software [82] (LS276-THP groups) or Pearl Cam Software (MMPSense™ 680). Mean fluorescence intensity values for

tumor and contralateral flank ROIs are plotted versus time to analyze biodistribution and activation kinetics.

8. Fluorescence values for ex vivo tissues are normalized to equalize blood fluorescence levels due to differences in absolute values between imaging systems and detection wavelengths (*see* **Note 6**).

9. Statistical significance is calculated with a one-tailed, unpaired *t*-test (GraphPad Prism software, GraphPad, San Diego, USA). Outlying data is analyzed with the Grubb's test.

3.7 In Vivo Imaging Using Cypate$_3$-THP

The protocol is adopted from [56].

1. Human fibrosarcoma xenografts are grown by subcutaneous injection of 2×10^6 HT-1080 cells (ATCC) in the neck region of 6-week old female athymic nude mice (Taconic). Two to three weeks after implantation, tumor-bearing mice are divided into two groups.

2. One group receives cypate$_3$-THP i.v. ($n = 3$) and a second group ($n = 4$) is treated with Ilomastat (1 mg/kg in DMSO, i.p.) 24 and 1 h prior to cypate$_3$-THP injection and again 4 h after cypate$_3$-THP injection.

3. Mice are anesthetized with 2 % isoflurane and oxygen mixture and placed in an imaging cassette (VisEN Medical) for image acquisition.

4. Mice are imaged using the FMT2500 fluorescence molecular tomography system (PerkinElmer) at 1, 4, and 24 h after cypate$_3$-THP (or cypate$_3$-THP + Ilomastat) injection. Fluorescence images were acquired using $\lambda_{excitation} = 790$ nm and $\lambda_{emission} = 805$ nm.

5. Prior to imaging of tumor-bearing mice, the Fluorescence Molecular Tomography (FMT) system is calibrated with the cypate dye. Region of interest (ROI) analyses are conducted using the FMT system software TrueQuant. ROIs are selected as the tumor location according to its location in the bright-field image.

6. The picomoles of cypate are normalized to tumor volume (pmol/mm^3; calculated as $1/2 L \times W^2$, length (L) and the width (W) measured with calipers).

7. Data is analyzed with *p*-values determined from two-way ANOVA analysis to determine statistical differences between experimental groups.

3.8 Confocal Microscopy

The protocol is adopted from [61].

1. Cells are grown on Lab-Tek slides in complete DMEM.

2. The medium is replaced and cells are incubated with 5FAM$_6$-THP (1 μM, 1 h). For the inhibitor studies, cells are pretreated

with a MMP-2 and MMP-9 inhibitor (SB-3CT, 1 µM, 30 min) after which $5FAM_6$-THP (1 µM, 1 h) is added to the wells so that the inhibitor is present for the entire incubation. The concentration of $5FAM_6$-THP is chosen to ensure significant saturation of MMP-2/MMP-9 to produce a detectable signal that would rise above level of autofluorescence in a reasonable amount of time before cell viability was lost.

3. After rinsing, a cell viability assay is performed with ethidium homodimer-1 (1 µM, 1 h). The slides are rinsed with PBS, mounted, coverslipped, and visualized using an Olympus FV1000 microscope (Center Valley, PA, USA) equipped with a 20×/0.95 W water immersion objective. Fluorescence emission of $5FAM_6$-THP is detected from $\lambda = 490$–550 nm using a 488 nm laser at 5 % power. Images of each group of slides are acquired with the same microscope settings during a single imaging session, allowing for quantitative analysis of cellular fluorescence.

4. Relative fluorescence of the cells is quantified with FV1000 software in terms of fluorescence per unit area. Relative fluorescence of ten cells of each image is determined and the results averaged (with standard deviation; $n = 10$). The results are analyzed with an unpaired, two-tailed t-test (GraphPad Prism software, GraphPad, San Diego, USA).

4   Notes

1. When crude peptide shows unsatisfactory yield and/or purity a different cleavage cocktail could be used. Please refer to [83].

2. During peptide synthesis the last, N-terminal amino acid to be incorporated must be coupled as Boc and not Fmoc-derivative in order to be compatible with Dde group removal.

3. Due to aggregation of triple-helical peptides a concentration of 2–500 µM can be used.

4. Although the specific examples given here are for HT-1080 cells, the same imaging approach can be utilized following implantation of breast cancer cells [31].

5. The molecular weight of MMPSense™ 680 is relatively large; therefore it was administered intravenously rather than impose the additional barrier of peritoneal absorption with $i.p.$ administration.

6. The tumor-specific fluorescence contrast with MMPSense™ 680 increased over time for 24 h, indicating that the higher molecular weight resulted in greater residence time in the tumor tissue and therefore greater activation. By 24 h post-injection, the tumor contrast for MMPSense™ 680 was about

twofold higher than the contralateral thigh while the contrast ratio for LS276-THP remained at about 1.5, unchanged from the 4 h time point. Another factor that could have contributed to the higher tumoral activation of MMPSense™ 680 was the lack of MMP selectivity. Although a great many probes exist, their actual selectivity is often problematic [19, 20].

Acknowledgments

The methods described in this chapter reflect the pioneering work of the laboratories of Drs. Ralph Weissleder and W. Barry Edwards. We gratefully acknowledge the National Institutes of Health (EB000289 and CA098799) and the Texas Higher Education STAR Award Program for support of our laboratory's research on matrix metalloproteinases.

References

1. Kessenbrock K, Plaks V, Werb Z (2010) Matrix metalloproteinases: regulators of the tumor microenvironment. Cell 141:52–67

2. Deryugina EI, Quigley JP (2010) Pleiotropic roles of matrix metalloproteinases in tumor angiogenesis: contrasting, overlapping and compensatory functions. Biochim Biophys Acta 1803:103–120

3. Gialeli C, Theocharis AD, Karamanos NK (2011) Roles of matrix metalloproteinases in cancer progression and their pharmacological targeting. FEBS J 278:16–27

4. Rha SY, Kim JH, Roh JK, Lee KS, Min JS, Kim BS, Chung HC (1997) Sequential production and activation of matrix metalloproteinase-9 (MMP-9) with breast cancer progression. Breast Cancer Res Treat 43:175–181

5. Vihinen P, Ala-aho R, Kahari V-M (2005) Matrix metalloproteinases as therapeutic targets in cancer. Curr Cancer Drug Targets 5:203–220

6. Somiari SB, Somiari RI, Heckman CM, Olsen CH, Jordan RM, Russell SJ, Shriver CD (2006) Circulating MMP2 and MMP9 in breast cancer - potential role in classification of patients into low risk, high risk, benign disease and breast cancer categories. Int J Cancer 119:1403–1411

7. McGowan PM, Duffy MJ (2008) Matrix metalloproteinase expression and outcome in patients with breast cancer: analysis of a published database. Ann Oncol 19:1566–1572

8. Köhrmann A, Kammerer U, Kapp M, Dietl J, Anacker J (2009) Expression of matrix metalloproteinases (MMPs) in primary human breast cancer and breast cancer cell lines: new findings and review of the literature. BMC Cancer 9:188

9. Figueira RCS, Gomes LR, Neto JS, Silva FC, Silva IDCG, Sodayar MC (2009) Correlation between MMPs and their inhibitors in breast cancer tumor tissue specimens and in cell lines with different metastatic potential. BMC Cancer 9:20

10. Eck SM, Hoopes PJ, Petrella BL, Coon CI, Brinckerhoff CE (2009) Matrix metalloproteinase-1 promotes breast cancer angiogenesis and osteolysis in a novel in vivo model. Breast Cancer Res Treat 116:79

11. Wyatt CA, Geoghegan JC, Brinckerhoff CE (2005) Short hairpin RNA-mediated inhibition of matrix metalloproteinase-1 in MDA-231 cells: effects on matrix destruction and tumor growth. Cancer Res 65:11101–11108

12. Liu H, Kato Y, Erzinger SA, Kiriakova GM, Qian Y, Palmieri D, Steeg PS, Price JE (2012) The role of MMP-1 in breast cancer growth and metastasis to the brain in a xenograft model. BMC Cancer 12:583

13. Ohshiba T, Miyaura C, Inada M, Ito A (2003) Role of RANKL-induced osteoclast formation and MMP-dependent matrix degradation in bone destruction by breast cancer metastasis. Br J Cancer 88:1318–1326

14. Decock J, Thirkettle S, Wagstaff L, Edwards DR (2011) Matrix metalloproteinases: protective roles in cancer. J Cell Mol Med 15:1254–1265

15. Martin MD, Matrisian LM (2007) The other side of MMPs: protective roles in tumor progression. Cancer Metastasis Rev 26:717–724

16. Dufour A, Overall CM (2013) Missing the target: matrix metalloproteinase antitargets in inflammation and cancer. Trends Pharm Sci 34:233–242

17. Morrison C, Mancini S, Cipollone J, Kappelhoff R, Roskelley C, Overall C (2011) Microarray and proteomic analysis of breast cancer cell and osteoblast co-cultures: role of osteoblast matrix metalloproteinase (MMP)-13 in bone metastasis. J Biol Chem 286:34271–34285

18. Zarrabi K, Dufour A, Li J, Kuscu C, Pulkoski-Gross A, Zhi J, Hu Y, Sampson NS, Zucker S, Cao J (2011) Inhibition of matrix metalloproteinase-14 (MMP-14)-mediated cancer cell migration. J Biol Chem 286:33167–33177

19. Fields GB (2008) Protease-activated delivery and imaging systems. In: Edwards D, Hoyer-Hansen G, Blasi F, Sloane B (eds) The cancer degradome – proteases in cancer biology. Springer, New York, NY, pp 827–851

20. Knapinska A, Fields GB (2012) Chemical biology for understanding matrix metalloproteinase function. ChemBioChem 13:2002–2020

21. Li C, Wang W, Wu Q, Ke S, Houston J, Sevick-Muraca E, Dong L, Chow D, Charnsangavej C, Gelovani JG (2006) Dual optical and nuclear imaging in human melanoma xenografts using a single targeted imaging probe. Nucl Med Biol 33:349–358

22. Piao D, Xie H, Zhang W, Krasinski JS, Zhang G, Dehghani H, Pogue BW (2006) Endoscopic, rapid near-infrared optical tomography. Opt Lett 31:2876–2878

23. Rudin M, Weissleder R (2003) Molecular imaging in drug discovery and development. Nat Rev Drug Discov 2:123–131

24. Weissleder R (2002) Scaling down imaging: molecular mapping of cancer in mice. Nat Rev Cancer 2:11–18

25. Tung C-H (2004) Fluorescent peptide probes for in vivo diagnostic imaging. Biopolymers 76:391–403

26. Weissleder R, Mahmood U (2001) Molecular imaging. Radiology 219:316–333

27. Bremer C, Ntzachristos V, Weisslender R (2003) Optical-based molecular imaging: contrast agents and potential medical applications. Eur Radiol 13:231–243

28. Bremer C, Bredow S, Mahmood U, Weissleder R, Tung CH (2001) Optical imaging of matrix metalloproteinase-2 activity in tumors: feasibility study in a mouse model. Radiology 221:523–529

29. Bremer C, Tung C-H, Weissleder R (2001) In vivo molecular target assessment of matrix metalloproteinase activity. Nat Med 7:743–748

30. Clapper ML, Hensley HH, Chang WC, Devarajan K, Nguyen MT, Cooper HS (2011) Detection of colorectal adenomas using a bio-activatable probe specific for matrix metalloproteinase activity. Neoplasia 13:685–691

31. Xie BW, Mol IM, Keereweer S, van Beek ER, Que I, Snoeks TJ, Chan A, Kaijzel EL, Löwik CW (2012) Dual-wavelength imaging of tumor progression by activatable and targeting near-infrared fluorescent probes in a bioluminescent breast cancer model. PLoS One 7:e31875

32. Barber PA, Rushforth D, Agrawal S, Tuor UI (2012) Infrared optical imaging of matrix metalloproteinases (MMPs) up regulation following ischemia reperfusion is ameliorated by hypothermia. BMC Neurosci 13:76

33. Lee S, Park K, Lee S-Y, Ryu JH, Park JW, Ahn HJ, Kwon IC, Youn I-C, Kim K, Choi K (2008) Dark quenched matrix metalloproteinase fluorogenic probe for imaging osteoarthritis development in vivo. Bioconjug Chem 19:1743–1747

34. Ryu JH, Lee A, Na JH, Lee S, Ahn HJ, Park JW, Ahn CH, Kim BS, Kwon IC, Choi K, Youn I, Kim K (2011) Optimization of matrix metalloproteinase fluorogenic probes for osteoarthritis imaging. Amino Acids 41: 1113–1122

35. Lim NH, Meinjohanns E, Meldal M, Bou-Gharios G, Nagase H (2014) In vivo imaging of MMP-13 activity in the murine destabilised medial meniscus surgical model of osteoarthritis. Osteoarthritis Cartilage 22:862–868

36. Lim NH, Meinjohanns E, Bou-Gharios G, Gompels LL, Nuti E, Rossello A, Devel L, Dive V, Meldal M, Nagase H (2014) In vivo imaging of matrix metalloproteinase 12 and matrix metalloproteinase 13 activities in the mouse model of collagen-induced arthritis. Arthritis Rheum 66:589–598

37. Woodhead-Galloway J (1980) Collagen: the anatomy of a protein. Edward Arnold Limited, London, pp 10–19

38. Shoulders MD, Raines RT (2009) Collagen structure and stability. Annu Rev Biochem 78:929–958

39. Fields GB, Prockop DJ (1996) Perspectives on the synthesis and application of triple-helical, collagen-model peptides. Biopolymers 40:345–357

40. Fields GB (2010) Synthesis and biological applications of collagen-model triple-helical peptides. Org Biomol Chem 8:1237–1258

41. Jenkins CL, Raines RT (2002) Insights on the conformational stability of collagen. Nat Prod Rep 19:49–59

42. Brodsky B, Shah NK (1995) The triple-helix motif in proteins. FASEB J 9:1537–1546

43. Koide T (2005) Triple helical collagen-like peptides: engineering and applications in matrix biology. Connect Tissue Res 46:131–141

44. Koide T (2007) Designed triple-helical peptides as tools for collagen biochemistry and matrix engineering. Phil Trans R Soc B 362:1281–1291

45. Brodsky B, Thiagarajan G, Madhan B, Kar K (2008) Triple-helical peptides: an approach to collagen conformation, stability, and self-association. Biopolymers 89:345–353

46. Lauer-Fields JL, Broder T, Sritharan T, Nagase H, Fields GB (2001) Kinetic analysis of matrix metalloproteinase triple-helicase activity using fluorogenic substrates. Biochemistry 40:5795–5803

47. Lauer-Fields JL, Kele P, Sui G, Nagase H, Leblanc RM, Fields GB (2003) Analysis of matrix metalloproteinase activity using triple-helical substrates incorporating fluorogenic L- or D-amino acids. Anal Biochem 321:105–115

48. Lauer-Fields JL, Sritharan T, Stack MS, Nagase H, Fields GB (2003) Selective hydrolysis of triple-helical substrates by matrix metalloproteinase-2 and -9. J Biol Chem 278:18140–18145

49. Minond D, Lauer-Fields JL, Nagase H, Fields GB (2004) Matrix metalloproteinase triple-helical peptidase activities are differentially regulated by substrate stability. Biochemistry 43:11474–11481

50. Minond D, Lauer-Fields JL, Cudic M, Overall CM, Pei D, Brew K, Visse R, Nagase H, Fields GB (2006) The roles of substrate thermal stability and P2 and P1′ subsite identity on matrix metalloproteinase triple-helical peptidase activity and collagen specificity. J Biol Chem 281:38302–38313

51. Minond D, Lauer-Fields JL, Cudic M, Overall CM, Pei D, Brew K, Moss ML, Fields GB (2007) Differentiation of secreted and membrane-type matrix metalloproteinase activities based on substitutions and interruptions of triple-helical sequences. Biochemistry 46:3724–3733

52. Lauer-Fields JL, Chalmers MJ, Busby SA, Minond D, Griffin PR, Fields GB (2009) Identification of specific hemopexin-like domain residues that facilitate matrix metalloproteinase collagenolytic activity. J Biol Chem 284:24017–24024

53. Bhaskaran R, Palmier MO, Lauer-Fields JL, Fields GB, Van Doren SR (2008) MMP-12 catalytic domain recognizes triple-helical peptide models of collagen V with exosites and high activity. J Biol Chem 283:21779–21788

54. Lee H, Mason JC, Achilefu S (2006) Heptamethine cyanine dyes with a robust C-C bond at the central position of the chromophore. J Org Chem 71:7862–7865

55. Akers WJ, Xu B, Lee H, Sudlow GP, Fields GB, Achilefu S, Edwards WB (2012) Detection of MMP-2 and MMP-9 activity in vivo with a triple-helical peptide optical probe. Bioconjug Chem 23:656–663

56. Zhang X, Bresee J, Fields GB, Edwards WB (2014) Near-infrared triple-helical peptide with quenched fluorophores for optical imaging of MMP-2 and MMP-9 proteolytic activity in vivo. Bioorg Med Chem Lett 24:3786–3790

57. Achilefu S, Jimenez HN, Dorshow RB, Bugaj JE, Webb EG, Wilhelm RR, Rajagopalan R, Johler J, Erion JL (2002) Synthesis, in vitro receptor binding, and in vivo evaluation of fluorescein and carbocyanine peptide-based optical contrast agents. J Med Chem 45:2003–2015

58. Berezin MY, Guo K, Akers W, Livingston J, Solomon M, Lee H, Liang K, Agee A, Achilefu S (2011) Rational approach to select small peptide molecular probes labeled with fluorescent cyanine dyes for in vivo optical imaging. Biochemistry 50:2691–2700

59. Zhang Z, Fan J, Cheney PP, Berezin MY, Edwards WB, Akers WJ, Shen D, Liang K, Culver JP, Achilefu S (2009) Activatable molecular systems using homologous near-infrared fluorescent probes for monitoring enzyme activities in vitro, in cellulo, and in vivo. Mol Pharm 6:416–427

60. Gonzalez LO, Pidal I, Junquera S, Corte MD, Vazquez J, Rodriguez JC, Lamelas ML, Merino AM, Garcia-Muniz JL, Vizoso FJ (2007) Overexpression of matrix metalloproteinases and their inhibitors in mononuclear inflammatory cells in breast cancer correlates with metastasis-relapse. Br J Cancer 97:957–963

61. Zhang X, Bresee J, Cheney PP, Xu B, Bhowmick M, Cudic M, Fields GB, Edwards WB (2014) Evaluation of a triple-helical peptide

with quenched fluorophores for optical imaging of MMP-2 and MMP-9 proteolytic activity. Molecules 19:8571–8588

62. Fischer R, Mader O, Jung G, Brock R (2003) Extending the applicability of carboxyfluorescein in solid-phase synthesis. Bioconjug Chem 14:653–660

63. Höfle G, Steglich W, Vorbrüggen H (1978) 4-Dialkylaminopyridines as highly active acylation catalysts. Angew Chem Int Ed Engl 17:569–583

64. Xu S, Held I, Kempf B, Mayr H, Steglich W, Zipse H (2005) The DMAP-catalyzed acetylation of alcohols—a mechanistic study. Chemistry 11:4751–4757

65. Yang J, Zhang Z, Lin J, Lu J, Liu BF, Zeng S, Luo Q (2007) Detection of MMP activity in living cells by a genetically encoded surface-displayed FRET sensor. Biochim Biophys Acta 1773:400–407

66. Zhang Z, Yang J, Lu J, Lin J, Zeng S, Luo Q (2008) Fluorescence imaging to assess the matrix metalloproteinase activity and its inhibitor in vivo. J Biomed Opt 13:011006

67. Zhu L, Wang H, Wang L, Wang Y, Jiang K, Li C, Ma Q, Gao S, Wang L, Li W, Cai M, Wang H, Niu G, Lee S, Yang W, Fang X, Chen X (2011) High-affinity peptide against MT1-MMP for in vivo tumor imaging. J Control Release 150:248–255

68. Zhu L, Zhang F, Ma Y, Liu G, Kim K, Fang X, Lee S, Chen X (2011) In vivo optical imaging of membrane-type matrix metalloproteinase (MT-MMP) activity. Mol Pharmaceut 8:2331–2338

69. Hanaoka H, Mukai T, Habashita S, Asano D, Ogawa K, Kuroda Y, Akizawa H, Iida Y, Endo K, Saga T, Saji H (2007) Chemical design of a radiolabeled gelatinase inhibitor peptide for the imaging of gelatinase activity in tumors. Nucl Med Biol 34:503–510

70. Sprague JE, Li WP, Liang K, Achilefu S, Anderson CJ (2006) In vitro and in vivo investigation of matrix metalloproteinase expression in metastatic tumor models. Nucl Med Biol 33:227–237

71. Watkins GA, Jones EF, Shell MS, VanBrocklin HF, Pan M-H, Hanrahan SM, Feng JJ, He J, Sounni NE, Dill KA, Contag CH, Coussens LM, Franc BL (2009) Development of an optimized activatable MMP-14 targeted SPECT imaging probe. Bioorg Med Chem 17:653–659

72. Temma T, Sano K, Kuge Y, Kamihashi J, Takai N, Ogawa Y, Saji H (2009) Development of a radiolabeled probe for detecting membrane type-1 matrix metalloproteinase on malignant tumors. Biol Pharm Bull 32:1272–1277

73. Ye Y, Bloch S, Achilefu S (2004) Polyvalent carbocyanine molecular beacons for molecular recognitions. J Am Chem Soc 126:7740–7741

74. Fields CG, Lovdahl CM, Miles AJ, Matthias-Hagen VL, Fields GB (1993) Solid-phase synthesis and stability of triple-helical peptides incorporating native collagen sequences. Biopolymers 33:1695–1707

75. Fields CG, Mickelson DJ, Drake SL, McCarthy JB, Fields GB (1993) Melanoma cell adhesion and spreading activities of a synthetic 124-residue triple-helical "mini-collagen". J Biol Chem 268:14153–14160

76. Grab B, Miles AJ, Furcht LT, Fields GB (1996) Promotion of fibroblast adhesion by triple-helical peptide models of type I collagen-derived sequences. J Biol Chem 271:12234–12240

77. Lauer-Fields JL, Tuzinski KA, Shimokawa K, Nagase H, Fields GB (2000) Hydrolysis of triple-helical collagen peptide models by matrix metalloproteinases. J Biol Chem 275:13282–13290

78. Lauer-Fields JL, Nagase H, Fields GB (2000) Use of Edman degradation sequence analysis and matrix-assisted laser desorption/ionization mass spectrometry in designing substrates for matrix metalloproteinases. J Chromatogr A 890:117–125

79. Malkar NB, Lauer-Fields JL, Borgia JA, Fields GB (2002) Modulation of triple-helical stability and subsequent melanoma cellular responses by single-site substitution of fluoroproline derivatives. Biochemistry 41:6054–6064

80. Yu Y-C, Tirrell M, Fields GB (1998) Minimal lipidation stabilizes protein-like molecular architecture. J Am Chem Soc 120:9979–9987

81. Yu Y-C, Berndt P, Tirrell M, Fields GB (1996) Self-assembling amphiphiles for construction of protein molecular architecture. J Am Chem Soc 118:12515–12520

82. Schneider CA, Rasband WS, Eliceiri KW (2012) NIH image to ImageJ: 25 years of image analysis. Nat Methods 9(7):671–675

83. Guy CA, Fields GB (1997) Trifluoroacetic acid cleavage and deprotection of resin-bound peptides following synthesis by Fmoc chemistry. Methods Enzymol 289:67–83

INDEX

Jian Cao (ed.), *Breast Cancer: Methods and Protocols*, Methods in Molecular Biology, vol. 1406,
DOI 10.1007/978-1-4939-3444-7, © Springer Science+Business Media New York 2016